D1165358
WITHDRAWN
UTSA LIBRARIES

Modelling and Simulation

Louis G. Birta and Gilbert Arbez

Modelling and Simulation

Exploring Dynamic System Behaviour

Louis G. Birta
School of Information Technology
and Engineering
University of Ottawa
Ottawa, Ontario K1N 6N5
Canada
e-mail: lbirta@site.ottawa.in

Gilbert Arbez
School of Information Technology
and Engineering
University of Ottawa
Ottawa, Ontario K1N 6N5
Canada
e-mail: garbez@site.uottawa.in

British Library Cataloguing in Publication Data
A catalogue record for this book is available from the British Library

Library of Congress Control Number: 2007922719

ISBN-10: 1-84628-621-2 e-ISBN-10: 1-84628-622-0
ISBN-13: 978-1-84628-621-6 e-ISBN-13: 978-1-84628-621-2

Printed on acid-free paper

9 8 7 6 5 4 3 2 1

Springer Science+Business Media
springer.com

To the next and future generations:

Christine, Jennifer, Alison

Amanda, Julia, Jamie

and

Mika

PREFACE

Overview

Modelling and simulation is a tool that provides support for the planning, design, and evaluation of systems as well as the evaluation of strategies for system transformation and change. Its importance continues to grow at a remarkable rate, in part because its application is not constrained by discipline boundaries. This growth is also the consequence of the opportunities provided by the ever-widening availability of significant computing resources and the expanding pool of human skill that can effectively harness this computational power. However, the effective use of any tool and especially a multifaceted tool such as modelling and simulation involves a learning curve. This book addresses some of the challenges that lie on the path that ascends that curve.

Consistent with good design practice, the development of this textbook began with several clearly defined objectives. Perhaps the most fundamental was the intent that the final product provide a practical (i.e., useful) introduction to the main facets of a typical modelling and simulation project. This objective was, furthermore, to include projects emerging from both the discrete-event and the continuous-time domains. In addition, the work was not to evolve into a treatise on any particular software tool, nor was it to be overly biased towards the statistical notions that play a key role in handling projects from the discrete-event domain. To a large extent, these objectives were the product of insights acquired by the first author over the course of several decades of teaching a wide range of modelling and simulation topics. Our view is that we have been successful in achieving these objectives.

Features

We have taken a project-oriented perspective of the modelling and simulation enterprise. The implication here is that modelling and simulation is, in fact, a collection of activities that are all focused on one particular objective, namely, providing a credible resolution to a clearly stated goal, a goal that is formulated within a specific system context. There can be no project unless there is a goal. All the constituent subactivities work in concert to achieve the goal. Furthermore the 'big picture' must always be clearly in focus when dealing with any of the subactivities. We have striven to reflect this perspective throughout our presentation.

The notion of a conceptual model plays a central role in our presentation. This is not especially significant for projects within the continuous-time domain inasmuch as the differential equations that define the system dynamics can be correctly regarded as the conceptual model. On the other hand, however, this is very significant in the case of projects from the discrete-event domain because there is no generally accepted view of what actually constitutes a conceptual model in that context. This invariably poses a significant hurdle from a pedagogical point of view because there is no abstract framework in which to discuss the structural and behavioural features of the system under investigation. The inevitable (and unfortunate) result is a migration to the semantic and syntactic formalisms of some programming environment.

We have addressed this problem by presenting a conceptual modelling environment which we call the ABCmod framework. Its basis is the identification of 'units of behaviour' within the system under investigation and their subsequent synthesis into individual activities. The approach is a version of the activity-based world view that is often mentioned in the modelling and simulation literature as one of three standard approaches for organising a computer program that captures the time evolution of a discrete-event dynamic system. In our ABCmod (Activity-Based Conceptual modelling) framework the underlying notions are elevated from the programming level to a more abstract and hence more conceptual level. The inherent implementation difficulty with the notions does not arise because there is no execution requirement at the conceptual level. The examples that are presented in the text illustrate conceptual model development using the ABCmod framework. Furthermore we demonstrate the utility of the ABCmod framework by showing how its constructs conveniently map onto those that are required in both the event-scheduling and the process-oriented programming environments.

Audience

This textbook is intended for students (and indeed, anyone else) interested in learning about the problem-solving methodology called modelling and simulation. The book makes no pretence at being a research monograph, (although our ABCmod conceptual modelling framework is novel and previously unpublished). A meaningful presentation of the topics involved necessarily requires a certain level of technical maturity and our reference in this regard is a science or engineering background at the senior undergraduate or the junior graduate level.

More specifically our readers are assumed to have a reasonable comfort level with standard mathematical notation which we frequently use to concisely express relationships. There are no particular topics from mathematics that are essential to the discussion but some familiarity with the basic notions of probability and statistics plays a key role in the material in Chapters 3 and 6. (In this regard, a probability primer is provided in Annex 1). A reasonable level of computer programming skills is assumed in the discussions of Chapter 5 and 8. We use Java as our environment of choice in developing event-scheduling simulation models. The GPSS programming environment is used to illustrate the process-oriented approach to developing simulation models and we provide a GPSS primer in Annex 2. Our discussion of the modelling and simulation enterprise in the continuous-time domain is illustrated using the Open Desire programming environment and we provide an Open Desire primer in Annex 3.

Organisation

This book is organised into three parts. The first part has two chapters and serves to provide an overview of the modelling and simulation discipline. It provides a context for the subsequent discussions and, as well, the process that is involved in carrying out a modelling and simulation study is presented. Important notions such as quality assurance are also discussed.

The four chapters of Part 2 explore the various facets of a modelling and simulation project within the realm of discrete-event dynamic systems (DEDS). We begin by pointing out the key role of random (stochastic) phenomena in modelling and simulation studies in the DEDS realm. This, in particular, introduces the need to deal with data models as an integral part of the modelling phase. Furthermore there are significant issues that must be recognised when handling the output data resulting from experiments with DEDS models. These topics are explored in some detail in the discussions of Part 2.

As noted earlier, we introduce in this book an activity-based conceptual modelling framework that provides a means for formulating a description of the structure and behaviour of a model that evolves from the DEDS domain. An outline of this framework is provided in Part 2. A conceptual model is intended to provide a stepping stone for the development of a computer program that will serve as the 'solution engine' for the project. We show how this can be accomplished for both the event-scheduling and the process-oriented program development perspectives (i.e., world views).

There are three chapters in Part 3 of the book and these are devoted to an examination of various important aspects of the modelling and simulation activity within the continuous-time dynamic system (CTDS) domain. We begin by showing how conceptual models for a variety of relatively simple systems can be formulated. Most of these originate in the physical world that is governed by familiar laws of physics. However, we also show how intuitive arguments can be used to formulate credible models of systems that fall outside the realm of classical physics.

Inasmuch as a conceptual model in the CTDS realm is predominantly a set of differential equations, the 'solution engine' is a numerical procedure. We explore several options that exist in this regard and provide some insight into important features of the alternatives. Several properties of CTDS models that can cause numerical difficulty are also identified.

Determining optimal values for some set of parameters within a CTDS model is a common project goal in a modelling and simulation study. The last chapter in Part 3 explores this task in some detail. We outline two particular numerical procedures that can be applied to optimisation problems that arise in this context.

Web Resources

A Web site has been established at the URL address http: //modsimbook. site.uottawa.ca to provide access to a variety of supplementary material that accompanies this textbook. Included are the following.

1. A set of PowerPoint slides from which presentation material can be developed
2. An ABCmod tool that supports the development of discrete-event conceptual models based on the framework
3. A methodology for organising student projects
4. A set of links to other Web sites that provide modelling and simulation tools and information

This site is dynamic and it is anticipated that material will be updated on a regular basis.

Acknowledgements

We would, first of all, like to acknowledge the privilege we have enjoyed over the years in having had the opportunity to introduce so many students to the fascinating world of modelling and simulation. The publication of this book is, furthermore, evidence that the impact of this experience was not strictly unidirectional.

We would also like to acknowledge the contribution made by the student project group called Luminosoft whose members (Mathieu Jacques Bertrand, Benoit Lajeunesse, Amélie Lamothe, and Marc-André Lavigne) worked diligently and capably in developing a software tool that supports the ABCmod conceptual modelling framework that is presented in this textbook. Our many discussions with the members of this group fostered numerous enhancements and refinements that would otherwise not have taken place. Their work provides the basis for the support tool called ABCMtool that can be found at the Web site that has been established for material relating to this textbook.

The development of this book has consumed substantial amounts of time. To a large extent this has been at the expense of time we would otherwise have shared with our families. We would like to express our gratitude to our families for their patience and their accommodation of the disruptions that our preoccupation with this book project has caused on so many occasions. Thank you all!

Finally we would like to express our appreciation for the help and encouragement provided by Mr Wayne Wheeler, senior editor (computer science) for Springer-Verlag. His enthusiasm for this project fuelled our determination to meet the various deadlines that were established for the timely completion of our work. Thanks also to Ms Catherine Brett (editorial assistant, computing) who always provided quick and comprehensive responses to our concerns and queries. The guidance and support of Mr Herman Makler through the often exasperating copyediting stage, has been most helpful and much appreciated.

Table of Contents

PART 1
FUNDAMENTALS

Part 1 of this book establishes the foundations for our subsequent discussions about our topic of interest; namely, modelling and simulation. It consists of two chapters; that is, Chapter 1 and Chapter 2.

In Chapter 1 we briefly consider a variety of topics that can be reasonably regarded as background material. A natural beginning is a brief look at a spectrum of reasons why a modelling and simulation study might be undertaken. Inasmuch as the notion of a model is fundamental to our discussions, some preliminary ideas that relate to this notion are presented. A generic 'full-service' gas station is used to illustrate some of the key ideas. We then acknowledge that modelling and simulation projects can fail and suggest a number of reasons why this might occur.

Monte Carlo simulation and simulators are two topics which fall within a broadly interpreted perspective of modelling and simulation. In the interests of completeness, both of these topics are briefly reviewed. We conclude Chapter 1 with a brief look at the historical roots of the modelling and simulation discipline.

Modelling and simulation is a multifaceted, goal-oriented activity and each of the steps involved must be duly recognized and carefully carried out. Chapter 2 is concerned with outlining these steps and providing an appreciation for the modelling and simulation process. The discussion begins with an examination of the essential features of a dynamic model and with the abstraction process that is inherent in its construction. The basic element in this abstraction process is the introduction of variables. These provide the means for carrying out a meaningful dialogue about the model and its behaviour properties which must necessarily be consistent with the goals of the study. Variables fall into three categories: namely, input variables, output variables, and state variables. The distinctive features of each of these categories are outlined.

The modelling and simulation process gives rise to a number of artefacts and these emerge in a natural way as the underlying process evolves. These various artefacts are outlined together with their interrelationships. The credibility of the results flowing from a modelling and simulation project is clearly of fundamental importance. This gives rise to the topic of quality assurance and we conclude Part 1 by exploring various facets of this important topic. In particular, we examine the central role of verification and validation as it relates to the phases of the modelling and simulation activity.

Chapter 1 Introduction

1.1 Opening Perspectives

This book explores the use of modelling and simulation as a problem-solving tool. We undertake this discussion within the framework of a modelling and simulation project. This project framework embraces two key notions; first there is the notion of a 'system context'; that is, there is a system that has been identified for investigation, and second, there is a problem relating to the identified system that needs to be solved. Obtaining an acceptable solution to this problem is the goal of the modelling and simulation project. We use the term 'system' in its broadest possible sense; it could, for example, include the notions of a process or a phenomenon. Furthermore, physical existence of the system is not a prerequisite; the system in question may simply be a concept, idea, or proposal. What is a prerequisite, however, is the requirement that the system in question exhibit 'behaviour over time,' in other words, that it be a dynamic system.

Systems, or more specifically dynamic systems, are one of the most pervasive notions of our contemporary world. Broadly speaking, a dynamic system is a collection of interacting entities that produces some form of behaviour that can be observed over an interval of time. There are, for example, physical systems such as transportation systems, power generating systems, or manufacturing systems. On the other hand, in less tangible form, we have healthcare systems, social systems, and economic systems. Systems are inherently complex and tools such as modelling and simulation are needed to provide the means for gaining insight into features of their behaviour. Such insight may simply serve to provide the intellectual satisfaction of deeper understanding or, on the other hand, may be motivated by a variety of more practical and specific reasons such as providing a basis for decisions relating to the control, management, acquisition, or transformation of the system under investigation (the SUI).

The defining feature of the modelling and simulation approach is that it is founded on a computer-based experimental investigation that utilises an appropriate model for the SUI. The model is a representation or abstraction of the system. The use of models (in particular, mathematical models) as a basis for analysis and reasoning is well established in such disciplines as engineering and science. It is the emergence and widespread availability of computing power that has made possible the new dimension of experimentation with complex models and hence, the emergence of the modelling and simulation discipline.

It must be emphasised, furthermore, that there is an intimate connection between the model that is 'appropriate' for the study and the nature of the problem that is to be solved. The important corollary here is that there rarely exists a 'universal' model that will support all modelling and simulation projects that have a common system context. This is especially true when the system has some reasonable level of complexity. Consider, for example, the difference in the nature of a model for an airliner, first in the case where the model is intended for use in evaluating aerodynamic properties versus the case where it is simply a revenue-generating object within a business model. Identification of the most appropriate model for the project is possibly the most challenging aspect of the modelling and simulation approach to problem solving.

Although the word 'modelling' has a meaning that is reasonably confined in its general usage, the same cannot be said for the word 'simulation'. Nevertheless, the phrase 'modelling and simulation' does have a generally accepted meaning and implies two distinct activities. The modelling activity creates an object (i.e., a model) that is subsequently used as a vehicle for experimentation. This experimentation with the model is the simulation activity.

The word 'simulation' is frequently used alone in a variety of contexts. For example, it is sometimes used as a noun to imply a specialised computer program (as in, 'A simulation has been developed for the proposed system.'). It is also used frequently as an adjective (as in, 'The simulation results indicate that the risk of failure is minimal,' or 'Several extensions have been introduced into the language to increase its effectiveness for simulation programming'). These wide-ranging and essentially inconsistent usages of the word 'simulation' can cause regrettable confusion for neophytes to the discipline. As a rule, we avoid such multiplicity of uses of this word but, as will become apparent, we do use the word as an adjective in two specific contexts where the implication is particularly suggestive and natural.

1.2 Role of Modelling and Simulation

There is a wide range of possible reasons for undertaking a modelling and simulation study. Some of the most common are listed below (the order is alphabetical and hence should not be interpreted as a reflection of importance):

1. Education and training
2. Engineering design
3. Evaluation of decision or action alternatives
4. Evaluation strategies for transformation or change
5. Forecasting
6. Performance evaluation
7. Prototyping and concept evaluation
8. Risk/safety assessment
9. Sensitivity analysis
10. Support for acquisition/procurement decisions

It was noted earlier that the goals of a simulation project have a major impact on the nature of the model that evolves. However it's also important to observe that the goals themselves may be bounded by constraints. These typically are a consequence of limitations on the level of knowledge that is available about the SUI. The unavoidable reality is that the available knowledge about the underlying dynamics of systems varies considerably from system to system. There are systems whose dynamics can be confidently characterised in considerable detail and, in contrast, there are systems whose dynamics are known only in an extremely tentative fashion. An integrated circuit developed for some telecommunications application would fall into the first category whereas the operation of the stock market would reasonably fall into the second. This inherent range of knowledge level is sometimes reflected in the terminology used to describe the associated models. For example, an integrated circuit model might be referred to as a 'deep' model and a model of the stock market would be a 'shallow' model. The goals of a modelling and simulation project are necessarily restricted to being relatively qualitative when only a shallow model is feasible. Quantitative goals are feasible only for those situations where deep knowledge is available. In other words, the available knowledge level significantly influences the nature of the goals that can be realistically formulated for a modelling and simulation study.

The centrality of the goals of a modelling and simulation project has been recognised in terms of a notion called the 'experimental frame' (see References [1.7] and [1.13]). This notion is rather broadly based and implies a mechanism to ensure an appropriate compatibility among the SUI, the model, and the project goals. This usually includes such fundamental issues as the proper identification of the data the model must deliver, identification and representation of pertinent features of the environment in which the model functions, its parameters, and its granularity.

A model plays the role of a surrogate for the system it represents and its purpose (at least from the perspective of this textbook) is to replace the system in experimental studies. When the underlying system (i.e., SUI) does not exist (e.g., it may merely be an idea, concept, or proposal) then the model is the only option for experimentation. But even when the SUI does exist there is a variety of reasons why experimentation directly with it could be inappropriate. For example, such experimentation might be:

- Too costly (determining the performance benefit likely to be achieved by upgrading the hardware at all the switch nodes of a large data communications network)
- Too dangerous (exploring alternate strategies for controlling a nuclear reactor)
- Too time consuming (determining the ecological impact of an extended deer hunting season, implemented over several consecutive years, on the excessive deer population in a particular geographical region)
- Too disruptive (evaluating the effectiveness of a proposed grid of one-way streets within the downtown core of an urban area)
- Morally/ethically unacceptable (assessing the extent of radiation dispersion following a particular catastrophic failure at some nuclear generation facility)
- Irreversible (investigating the impact of a fiscal policy change on the economy of a country)

Behavioural data is almost always easier to acquire from a model than from the system itself and this is another important reason for favouring experimentation with a model. Consider, for example, the challenges inherent in monitoring the force exerted on the blades of a rotating turbine by escaping combustion gases. Furthermore, the fact that the platform for the experimentation is a computer (or more correctly, a computer program) ensures reproducibility of results which is an essential requirement for establishing credibility of experimental investigations. Such reproducibility can generally only be approximated when experiments are carried out directly with an existing system.

1.3 The Nature of a Model

A model is a specification for behaviour generation and the modelling process is concerned with the development of this specification. It is often suggested that the task is to ensure that the behaviour of the model is as indistinguishable as possible from the behaviour of the SUI. This assertion is only partially correct. A more appropriate statement of the task at hand is to develop the specification so that it captures the behaviour properties at a level of granularity that is appropriate to the goals of the study. The challenge is to capture all relevant detail and to avoid superfluous features. (One might recall here the quotation from Albert Einstein, 'Everything should be made as simple as possible, but not simpler.') For example, consider a project concerned with evaluating strategies for improving the operating efficiency of a fast-food restaurant. Within this context it would likely be meaningless (and indeed, nonsensical) to incorporate into the model information about the sequence in which a server prepares the hot and cold drinks when both are included in a customer's order.

The notion of 'behaviour' is clearly one that is fundamental to these discussions and in particular, we have suggested that there is usually a need to evaluate behaviour. But what does this mean and how is it done? At this point we have to defer addressing these important questions until a more detailed exploration of the features of models has been completed.

Modelling is a constructive activity and this raises the natural question of whether the product (i.e., the model) is 'good enough.' This question can be answered only if there is an identified context and as we show in the discussions to follow, there are many facets to this key issue. One facet that is most certainly fundamental is the goal(s) of the project. In other words, a key question is always whether the model is good enough from the point of view of the project goals. The corollary of this assertion is that it is not meaningful to undertake any modelling study without a clear understanding of the purpose for which the model will be used. Perhaps the most fundamental implication of the above discussion is that it is never meaningful to undertake a study whose goal is simply 'to develop a model of . . . .'

There is a variety of ways in which the specification of behaviour can be formulated. Included here are: natural language, mathematical formalisms, rule-based formalisms, symbolic/graphical descriptions, and combinations of these. It is typical for several distinct formulations of the model (or perhaps only portions of it) to evolve over the course of the study. These alternatives are generally created in formats that are best suited to capturing subtleties or providing clarification.

A particular format that plays a very special role is a specification formulated as a computer program. The importance of such a specification arises because that computer program provides the means for actually carrying out the experiments that are central to the modelling and simulation approach. This illustrates, furthermore, the important fact that some realisations of the specification (which, after all, is the model) are actually executable and produce the behaviour we seek to observe. This legitimises the implications in our frequent use of the phrase 'the model's behaviour.'

1.4 An Example (Full-Service Gas Station)

To illustrate some facets of the discussion above, we consider a modelling and simulation project whose system context (SUI) is a ‘full-service’ gas station with two islands and four service lanes (see Figure 1.1). A significant portion of the customers at this station drive small trucks and vans which typically have gas tank capacities that are larger than those of most passenger vehicles. Often the drivers of passenger cars find themselves queued behind these large-tank vehicles which introduce substantially longer wait times when they arrive at the gas pumps. This can cause aggravation and complaints. The station management is considering restricting these large-tank vehicles to two designated lanes. The goal of the modelling and simulation project could be to obtain performance data that would assist in determining whether such a policy would improve the flow of vehicles through the station.

Vehicles are obliged (via appropriate signage) to access the pumps from the east side and after their respective gas purchases they exit on the west side. Upon arrival, drivers always choose the shortest queue. In the case where two or more queues have the same shortest length, a random choice is made. An exception is when it is observed that a customer in one of the ‘shortest queues’ is in the payment phase of the transaction in which case that queue is selected by the arriving driver.

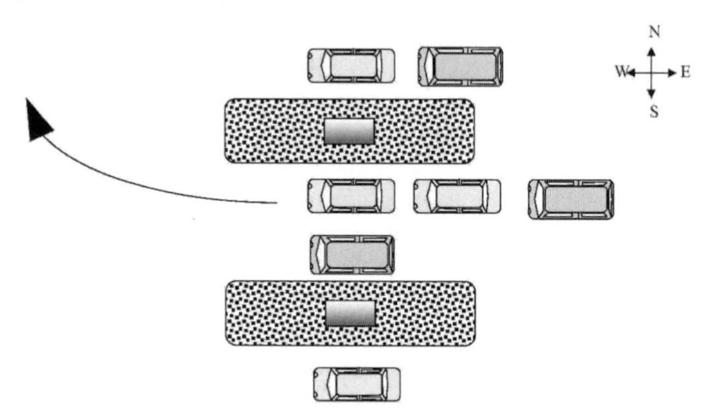

FIGURE 1.1. Gas station project.

Depending on the time of day, one or two attendants are available to serve the customers. The service activity has three phases. During the first, the attendant determines the customer’s requirement and begins the pumping of gas (the pumps have a preset delivery amount and automatically shut off when the preset amount has been delivered). In addition, any peripheral service such as cleaning of windshields and checking oil levels are carried out during this first phase. Phase two is the delivery phase during which the gas is pumped into the customer's gas tank. Phase three is the payment phase; the attendant accepts payment either in the form of cash or credit card. The duration of phase two is reasonably long and an attendant typically has sufficient time either to begin serving a newly arrived customer or to return to handle the phase three (payment) activity for a customer whose gas delivery is complete. The protocol is to give priority to a payment function before serving a newly arrived customer. It is standard practice for the payment function to be carried out by the same attendant who initiated the transaction.

The above text can be regarded as an initial phase in the model building process for this particular modelling and simulation project. It corresponds to the notion of a project description which we examine more carefully in Chapter 2. Notice, however, that much detail remains to be added; for example, the specification of the arrival rate of vehicles, the proportion of vehicles that fall into the small truck/van category, service times for each of the three service phases, and so on (these correspond to data requirements). Nor should it be assumed that it is entirely complete and adequately comprehensive.

Refinements to this description are almost certain to be necessary; these may simply provide clarification (what are the conditions that govern the attendant's options during phase two) or may introduce additional detail; such as what happens when a pump becomes defective or, under what conditions does an arriving customer 'balk,' that is, decide the queues are too long and leave. Or, in fact, is balking even a relevant occurrence? What about accommodating the possibility that drivers (or passengers) may need to use the washroom facilities and thereby 'hold' the pump position longer than is otherwise necessary? The merits of introducing such refinements must always be weighed against their relevance in terms of achieving the goals of the modelling and simulation project. (It may be useful for the reader to dwell on other possible refinements.) In fact, some refinement of the goals is most certainly necessary. (What exactly are the performance data that would enable a meaningful decision to be made?)

It is also important to observe that the model's features as outlined above have an orientation that is specific to the stated goal of the project. There is very little in the presentation that would allow a model formulated from the given description to be useful in, for example, an environmental assessment of the gas station's operation or indeed in an analysis of its financial viability.

1.5 Is There a Downside to the Modelling and Simulation Paradigm?

The implicit thrust of the presentation in this textbook is that of promoting the strengths of the modelling and simulation paradigm as a problem-solving methodology. However, one might reasonably wonder whether there exist inherent dangers or pitfalls. And the simple answer is that these do indeed exist! As with most tools (both technological and otherwise) modelling and simulation must be used with a good measure of care and wisdom. An appreciation for the limitations and dangers of any tool is a fundamental prerequisite for its proper use. We examine this issue within the modelling and simulation context somewhat indirectly by examining some reasons why modelling and simulation projects can fail.

(a) Inappropriate Statement of Goals
No project can ever be successful unless its objectives are clearly articulated and fully understood by all the stakeholders. This most certainly applies to any modelling and simulation project. The goals effectively drive all stages of the development process. Ambiguity in the statement of goals can lead to much wasted effort or yield conclusions that are unrelated to the expectations of the 'client' responsible for the initiation of the project.

A second, but no less important goal-related issue relates to the feasibility of achieving the stated goals. As suggested earlier, the project goals have to be consistent with the realities of the depth of knowledge that characterises the SUI. Any attempt to extract precise knowledge from a shallow model will most certainly fail. There are other feasibility issues as well. For example, the available level of resources may simply not be adequate to achieve the goals. Here resources include time (to complete the project), talent (skill set; see (d) below), and funding.

(b) Inappropriate Granularity of the Model
The granularity of the model refers to the level of detail with which it attempts to replicate the SUI. The level of granularity is necessarily bounded by the goals of the project and care must always be taken to ensure that the correct level has been achieved. Excessive detail increases complexity and this can lead to cost overruns and/or completion delays that usually translate into project failure. Too little detail, on the other hand, can mask the very effects that have substantial relevance to the behaviour that is of critical interest. This is particularly serious because the failure of the project only becomes apparent when undesired consequences begin to flow from the implementation of incorrect decisions based on the study.

(c) Ignoring Unexpected Behaviour
Although a validation process is recognised to be an essential stage in any modelling and simulation project, its main thrust generally is to confirm that expected behaviour does occur. On the other hand, testing for unexpected behaviour is never possible. Nevertheless such behaviour can occur and when it is observed there often is a tendency to dismiss it, particularly when validation tests have provided satisfactory results. Ignoring such counterintuitive, or unexpected observations can lay the foundation for failure.

(d) Inappropriate Mix of Essential Skills
A modelling and simulation project of even modest size can have substantial requirements in terms of both the range of skills and the effort needed for its completion. A team environment is therefore common; team members contribute complementary expertise to the intrinsically multi-faceted requirements of the project. The range of skills that needs to be represented among the team members can include: project management, documentation, transforming domain knowledge into the format of a credible dynamic model, development of data modules as identified in the data requirements, experiment design, software development, and analysis of results. The intensity of coverage of these various areas is very much dependent on the specific nature of the project. Nevertheless, an inappropriate mix of skills can seriously impede progress and can ultimately result in project failure.

(e) Inadequate Flow of Information to the Client
The team that carries out a modelling and simulation project often does so on behalf of a 'client' who is not a member of the team. In such cases, care must be taken to ensure that the client is fully aware of how the project is unfolding in order to avoid the occurrence of a 'disconnect' that results in the delivery of a product that falls short of expectations. For example, a minor misinterpretation of requirements, if left uncorrected, can have consequences that escalate to the point of jeopardising the project's success.

1.6 Monte Carlo Simulation

References to Monte Carlo simulation are often encountered in the modelling and simulation literature. This somewhat fanciful label refers to a problem-solving methodology that is loosely related to, but is very different from, the topic that we explore in this textbook. The term refers to a family of techniques that are used to find solutions to numerical problems. The distinctive feature of these techniques is that they proceed by constructing an artificial stochastic (probabilistic) system whose properties contain the solution of the underlying problem. The origins of the approach can be traced back to Lord Rayleigh who used it to develop approximate solutions to simple partial differential equations. The power of the methodology was exploited by von Neumann and colleagues in solving complex problems relating to their work in developing a nuclear arsenal in the latter years of the Second World War. The Monte Carlo label for the methodology is, in fact, attributed to this group.

Perhaps the simplest example of the method is its application to the evaluation of the definite integral:

$$I = \int_a^b f(x)\,dx \tag{1.1}$$

for the special case where $f(x) \geq 0$. The value of I is the area under $f(x)$ between $x = a$ and $x = b$. Consider now a horizontal line drawn at $y = K$ such that $f(x) \leq K$ for $a \leq x \leq b$ (see Figure 1.2). The rectangle R, enclosed by $x = a$, $x = b$, $y = 0$, and $y = K$ has the area $K(b - a)$ and furthermore $I \leq K(b - a)$. Suppose a sequence of points (x_i, y_i) is chosen at random within the rectangle R such that all points within R are equally likely to be chosen (e.g., by choosing from two uniform distributions oriented along the length and width of R). It can then be easily appreciated that the ratio of the number of points that fall either on the curve or under it (say, n) to the total number of points chosen (say, N) is an approximation of the ratio of I to the area of the rectangle R. In other words,

$$n/N \approx I/[K(b-a)] \tag{1.2}$$

or

$$I \approx nK(b-a)/N \tag{1.3}$$

In the procedure, a point (x_i, y_i) is included in the count n, if $y_i \leq f(x_i)$. The accuracy of the approximation improves as N increases.

The interesting feature in this example is that the original problem is entirely deterministic and yet the introduction of probabilistic notions can yield an approximation to its solution.

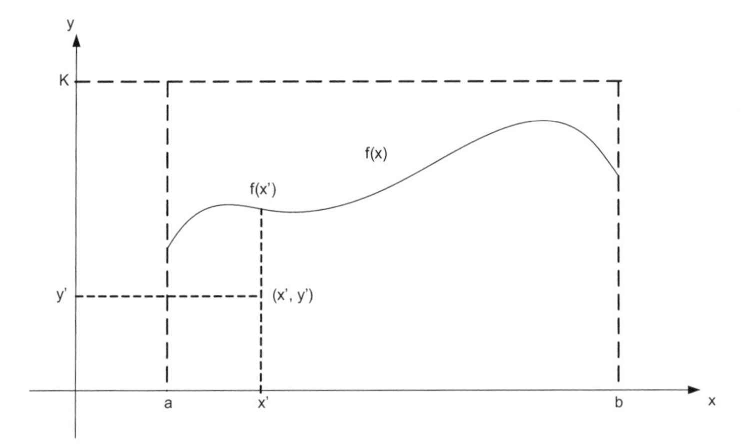

FIGURE 1.2. Example of Monte Carlo simulation.

The class of problems that can be effectively investigated by Monte Carlo simulation generally falls within the domain of numerical analysis. The approach provides an alternate, and often very effective, solution option for these problems. However, these problems do not fall within the scope of the modelling and simulation methodology because they lack the prerequisite of 'behaviour,' that is, an evolution over time. The reference to 'simulation' in the label for the approach could be regarded as a reflection of the dissimilarity between the solution mechanism and the inherent nature of the problem.

1.7 Simulators

There is frequent reference in the modelling and simulation literature to the notion of simulators. Most commonly the notion is a reference to a training device or platform and it is from that perspective that we explore the topic in the discussion that follows. Within the training context a simulator can be viewed as a device that replicates those operational features of some particular system that are deemed to be important for the training of operators of that system. A characteristic feature of any simulator is the incorporation of some physical parts of the system itself as a means of enhancing the realism of the training environment, for example, an actual control panel layout. Beginning with the development of flight simulators for training pilots (see Historical Overview which follows), the use of simulators has spread into a wide range of domains; for example, there exist power plant simulators (both nuclear and fossil), battlefield simulators, air traffic control simulators, and (human) patient simulators. An interesting presentation of contemporary applications of simulators in the training of health science professionals can be found in the various papers of the special journal issue of Reference [1.2].

The fundamental requirement of any simulator is the replication of system behaviour within a physical environment that is as realistic as possible from the perspective of an operator. Although the simulator incorporates some physical features of the system, substantial components of the system necessarily exist only in the form of models. In early simulators these models were themselves physical in nature but with the emergence of computing technology the modelled portions of the system have increasingly exploited the modelling power of this technology.

The development of a simulator can be viewed as a modelling and simulation project whose goal is to achieve an effective training environment. This, in particular, implies that the device must operate in realtime; that is, behaviour, as experienced by the operator, must evolve at a rate that corresponds exactly to that of the real system. This introduces additional important constraints on the models for those portions of the system that are being emulated, such as synchronisation of 'virtual' time within the model with 'real' (i.e., clock) time.

Simulators can also contribute in a variety of ways to enhancing the educational experience of students especially in circumstances where alternatives are precluded (e.g., by budgetary constraints) or alternately when the devices being examined are either hypothetical or are no longer available in the marketplace. The areas of computer organisation and operating systems are particularly well suited to this application and relevant discussions can be found in References [1.11] and [1.12].

Apart from their importance as training platforms and educational tools, it is interesting to observe that simulators also have a lighter side in their role within the entertainment industry. This is clearly apparent in the various versions of vehicular-oriented games that populate the entertainment arcades that have become commonplace in the shopping malls throughout North America. Less comprehensive versions of these games are available as well for home computers.

Simulators represent an application area of the modelling and simulation paradigm. Inasmuch as our intent in this textbook is to avoid a focus on any particular application area, the topic of simulators is not explicitly addressed in the sequel apart from one exception. This is a brief examination of their important role in the history of modelling and simulation as discussed in the following section.

1.8 Historical Overview

The birth of the modelling and simulation discipline can reasonably be associated with the development of the Link Trainer by Edward Link in the late 1920s. The Link Trainer is generally regarded as the first successful device designed specifically for the training of pilots and represents the beginning of an extensive array of training tools called flight simulators. The initial Link Trainer clearly predates the arrival of the modern computer and its behaviour-generating features were produced instead using pneumatic/hydraulic technology. As might be expected, flight simulators quickly embraced computer technology as it developed in the 1950s. The sophistication of flight simulators has continuously expanded and they have become indispensable platforms for training not only aircraft pilots (both commercial and military) but also the members of the Apollo Missions and, as well, the various space shuttle teams. In fact, the technology and methodology demands made by the developers of flight simulators have had a great impact upon the evolution of the modelling and simulation discipline itself.

Although the development and evolution of simulators in general represent the initial (and probably pivotal) application area for modelling and simulation, it is the analogue computer that represents the initial computing platform for the discipline. The commercial availability of these computers began in the early 1950s. The principles upon which this computing machine was based were originally formulated by Lord Kelvin in the latter part of the nineteenth century. The electronic realisation of the concept was developed by Vannevar Bush in the 1930s.

The analogue computer was primarily a tool for the solution of differential equations. The solution of such equations was obtained by direct manipulation of voltage signals using active elements called operational amplifiers. The computing environment was highly interactive and provided convenient graphical output. Although primarily electronic in nature, the early machines nevertheless relied on electromechanical devices to carry out basic nonlinear operations such as multiplication and division. This often introduced serious constraints in terms of both solution speed and accuracy.

Programming the analogue computer was a relatively complex and highly error-prone process inasmuch as it involved physically interconnecting the collection of processing elements that were required for solving the problem at hand. The range of processing element types was relatively narrow but did include one that directly carried out an integration operation. It was this device that provided the basis for the straightforward solution of differential equations. As a result the problems that were best suited for solution emerged from the realm of engineering (e.g., aerospace flight dynamics and control system design).

By the mid-1960s the speed of general-purpose digital computers and the software support for their use had improved to a point where it was apparent that they were going to have important consequences upon the modelling and simulation discipline. For example, their capabilities showed promise in providing an alternate solution tool for the same problem class that had previously fallen exclusively into the domain of the analogue computer (thereby setting the stage for the demise of the 'worthy predecessor'). But perhaps even more significantly it was clear that computing power was now becoming available to support a class of modelling and simulation projects that had been beyond the capabilities of the analogue computer, namely, the class of discrete event problems that incorporate stochastic phenomena.

Over the next two decades of the 1970s and the 1980s a wide variety of software support for modelling and simulation applications was developed. This made possible the initiation of increasingly more ambitious projects which, by and large, fall into two distinct realms: namely, the continuous (typically engineering design problems formulated around differential equation models) and the discrete event (typically process design problems incorporating stochastic phenomenon and queueing models).

Some interesting perspectives on the development of the modelling and simulation paradigm can be found in Nance and Sargent [1.4]. The evolution of specialised programming languages for modelling and simulation is an integral part of the history of the discipline and a comprehensive examination of such developments within the discrete event context (up to the year 1986) can be found in Nance [1.5, 1.6]. The overview given by Bowden [1.1] and the survey results presented by Swain [1.10] provide some insight into the wide spectrum of commercially available software products in this domain. A comprehensive collection of contemporary simulation software is listed and summarised in the document [1.3] prepared by the Society for Modeling and Simulation International. Making correct choices from among the available alternatives can be a challenging task and the work of Nikoukaran et al. [1.9] can provide useful guidance in this regard.

In spite of the relatively short time span of its history, the modelling and simulation discipline has given birth to a remarkably large number of professional organisations and associations that are committed to its advancement (see: www.site.uottawa.ca/~oren/links-MS-AG.htm where a comprehensive listing is maintained). These span a broad range of specific areas of application which is not unexpected because the concepts involved are not constrained by discipline boundaries.

The impact of decisions that are made on the basis of results that flow from a modelling and simulation study are often very significant and far reaching. In such circumstances it is critical that a high degree of confidence can be placed on the credibility of the results and this, in turn, depends on the expertise of the team that carries out the study. This has given rise to an accreditation process for those individuals who wish to participate in the discipline at a professional level. The accreditation option has obvious benefits for both the providers and the consumers of professional modelling and simulation services which are now widely available in the marketplace. This accreditation process has been developed under the auspices of the Modeling and Simulation Professional Certification Commission (see: www.simprofessional.org).

An integral part of professionalism is ethical behaviour and this has been addressed by Oren et al. [1.8] who have proposed a code of ethics for modelling and simulation professionals (see: www.scs.org/ethics/). This proposed code of ethics has already been adopted by numerous modelling and simulation professional organisations.

1.9 Exercises and Projects

1.1. A new apartment building is being designed. It will have ten floors and will have six apartments on each floor. There will, in addition be two underground parking levels. The developer needs to make a decision about the elevator system that is to be installed in the building. The choice has been reduced to two alternatives: either two elevators each with a capacity of 15 or three smaller elevators each with a capacity of 10. A modelling and simulation study is to be undertaken to provide a basis for making the choice between the two alternatives.

   a) Develop a list of possible performance criteria that could be used to evaluate the relative merits of the two alternative designs.
   b) Develop a list of behavioural rules that would be incorporated in the formulation of the model of the elevator system (e.g., when does an elevator change its direction of motion, which of the elevators responds to a particular call for service, where does an idle elevator 'park,' etc.?)
   c) A model's behaviour can 'unfold' only as a consequence of input data. Develop a list of input data requirements that would necessarily become an integral part of a study of the elevator system, for example, arrival rates of tenants at each of the floors, velocity of the elevators, and so on.

1.2. Consider the intersection of two roads in a rapidly expanding suburban area. Both of these roads have two traffic lanes. Because of the development of the area, the volume of traffic flow in each direction at this intersection has dramatically increased and so has the number of accidents. A large proportion of the accidents involve vehicles making a left turn which suggests that the 'simple' traffic lights at the intersection are no longer adequate because they do not provide a left-turn priority interval. The city's traffic department is evaluating alternatives for upgrading these traffic signals so that such priority intervals are provided to assist left-turning drivers. The need is especially urgent for traffic that is flowing in the north and south directions. If the upgrade is implemented then a particular parameter value that will need to be determined is the proportion of time allocated to the priority interval. The planning department has decided to explore solution alternatives for the problem by undertaking a modelling and simulation study.

a) Develop a list of possible performance criteria that would be used for evaluating the traffic flow consequences of the upgraded traffic signals.
b) Formulate the behaviour rules that would be needed in the construction of the dynamic model (e.g., how do the queues of cars begin to empty when the light turns from red of green, how will right-turning cars be handled, and so on).
c) Develop a list of input data requirements of the behaviour-generating process for the model.

1.10 References

1.1. Bowden, R., (1998), The spectrum of simulation software, *IEEE Solutions*, **30**(May): 44–54.

1.2. Endoscopy simulators for training and assessment skills, (2006), J. Cohen (Ed.), *Gastrointestinal Endoscopy Clinics of North America*, **16**(3): 389–610 (see also: www.theclinics.com).

1.3. Modeling and Simulation Resource Directory, (2005), *Modeling and Simulation*, **4**(1) (see also www.scs.org).

1.4. Nance, R.E. and Sargent, R.G., (2002), Perspectives on the evolution of simulation, *Operations Research*, **50**: 161–172.

1.5. Nance, R.E., (1993), A history of discrete event simulation programming languages, *ACM SIGPLAN Notices*, **28**(3):149–175.

1.6. Nance, R.E., (1995), Simulation programming languages: An abridged history, in *Proceedings of the 1995 Winter Simulation Conference*, Arlington, VA, pp. 1307–1313.

1.7. Ören, T.I. and Zeigler, B.P., (1979), Concepts for advanced simulation methodologies, *Simulation*, **32**: 69–82.
1.8. Ören, T.I, Elzas, M.S., Smit, I. and Birta, L.G., (2002), A code of professional ethics for simulationists, in *Proceedings of the 2002 Summer Computer Simulation Conference*, San Diego, pp. 434–435.
1.9. Nikoukaran, J., Hlupic, V., and Paul, R.J., (1999), A hierarchical framework for evaluating simulation software, *Simulation Practice and Theory*, **7**: 219–231.
1.10. Swain, J.J., (2005), Seventh biennial survey of discrete event simulation software tools, *OR/MS Today*, **32.**
1.11. Wolffe, G., Yurcik, W., Osborne, H., and Holloday, M., (2002), Teaching computer organization/architecture with limited resources using simulators, in *Proceedings of the 33rd Technical Symposium on Computer Science Education*, ACM Press, New York.
1.12. Yurcik, W., (2002), Computer architecture simulators are not as good as the real thing — They are better!, *ACM Journal of Educational Resources in Computing*, **1**(4) (guest editorial, Special Issue on General Computer Architecture Simulators).
1.13. Zeigler, B.P., Praehofer, H., and Kim, T.G., (2000), *Theory of Modeling and Simulation: Integrating Discrete Event and Continuous Complex Dynamic Systems* (2nd edn.), Academic Press, San Diego.

Chapter 2 The Modelling and Simulation Process

2.1 Some Reflections on Models

The use of models as a means of obtaining insight or understanding is by no means novel. One could reasonably claim, for example, that the pivotal studies in geometry carried out by Euclid were motivated by the desire to construct models that would assist in better understanding important aspects of his physical environment. It could also be observed that it is rare indeed for the construction of even the most modest of structures to be undertaken without some documented perspective (i.e., an architectural plan or drawing) of the intended form. Such a document represents a legitimate model for the structure and serves the important purpose of providing guidance for its construction. Many definitions of a model can be found in the literature. One that we feel is especially noteworthy was suggested by Shannon [2.16]: 'A model is a representation of an object, system or idea in some form other than itself.'

Although outside the scope of our considerations, it is important to recognise a particular and distinctive class of models called physical models. These provide the basis for experimentation activity within an environment that mimics the physical environment in which the problem originates. An example here is the use of scale models of aircraft or ships within a wind tunnel to evaluate aerodynamic properties; another is the use of 'crash-test dummies' in the evaluation of automobile safety characteristics. A noteworthy feature of physical models is that they can, at least in principle, provide the means for direct acquisition of relevant experimental data. However, the necessary instrumentation may be exceedingly difficult to implement.

A fundamental dichotomy among models can be formulated on the basis of the role of time; more specifically, we note that some models are dynamic whereas others are static. A linear programming model for establishing the best operating point for some enterprise under a prescribed set of conditions is a static model because there is no notion of time dependence embedded in such a model formulation. Likewise, the use of tax software to establish the amount of income tax payable by an individual to the government can be regarded as the process of developing a (static) model of one aspect of that individual's financial affairs for the particular taxation year in question. The essential extension in the case of a dynamic model is the fact that it incorporates the notion of 'evolution over time'. The difference between static and dynamic models can be likened to the difference between viewing a photograph and viewing a video clip. Our considerations throughout this book are concerned exclusively with dynamic models.

Another important attribute of any model is the collection of assumptions that are incorporated into its formulation. These assumptions usually relate to simplifications and their purpose is to provide a means for managing the complexity of the model. Assumptions are invariably present but often they are not explicitly acknowledged and this can have very serious consequences. The assumptions embedded in a model place boundaries around its domain of applicability and hence upon its relevance not only to the project for which it is being developed but also to any other project for which its reuse is being considered.

Making the most appropriate choices from among possible assumptions can be one of the most difficult aspects of model development. The underlying issue here is identifying the correct balance between complexity and credibility where credibility must always be interpreted in terms of the goals of the project. It's worth observing that an extreme, but not unreasonable, view in this regard is that the development of any model is simply a matter of making the correct selection of assumptions from among the available options (often a collection of substantial size).

The assumptions embedded in a model are rarely transparent. It is therefore of paramount importance to ensure, via the documentation for the project, that all users of the model are cognisant of its limitations as reflected in the assumptions that underlie its development.

As might be expected, the inherent restricted applicability of any particular model as suggested above has direct and significant consequences upon the simulation activity. The implication is simply that restrictions necessarily emerge upon the scope of the experiments that can be meaningfully carried out with the model. This is not to suggest that certain experiments are impossible to carry out but rather that the value of the results that are generated is questionable. The phenomenon at play here parallels the extrapolation of a linear approximation of a complex function beyond its region of validity. The need to incorporate in simulation software environments a means for ensuring that experimentation remains within the model's range of credibility has been observed. Realisation of this desirable objective, however, has proved to be elusive.

2.2 Exploring the Foundations

The discussion in this section provides some essential background for the development of the modelling and simulation process that is explored in Section 2.3.

2.2.1 The Observation Interval

In Chapter 1 we indicated that a modelling and simulation project has two main constituents. The most fundamental is the underlying 'system context', namely, the dynamic system whose behaviour is to be investigated (i.e., the SUI or system under investigation). The second essential constituent is the goals for the project which generally correspond to means for obtaining the solution to a problem that has been formulated around the SUI. A subordinate, but nonetheless important, third constituent is the *observation interval* which is the interval of time over which the behaviour of the SUI is of interest. Often the specification of this interval, which we denote I_0, is clearly apparent in the statement of the project goals. There are, however, many important circumstances where this does not occur simply because of the nature of the output data requirements. Nevertheless it is essential that information about the observation interval be properly documented in the project description.

The starting point of this interval (its left boundary) almost always has an explicitly specified value. The endpoint (the right boundary) may likewise be explicitly specified, but it is not uncommon for the right boundary to be only implicitly specified. The case where a service facility (e.g., a grocery store) closes at a prescribed time (say 9:00 PM) provides an example of an explicitly specified right boundary. Similarly a study of the morning peak-period traffic in an urban area may be required, by definition, to terminate at 10:00 AM. Consider, on the other hand a study of a manufacturing facility that ends when 5000 widgets have been produced. Here the right-hand boundary of the observation interval is known only implicitly. Likewise consider a study of the performance of a dragster that necessarily ends when the vehicle crosses the finish line of the racing track. In these examples the right boundary is implicitly determined by conditions defined on the state variables of the model; that is, it is not known when the experiment begins.

Another case of implicit determination occurs when the right-hand boundary is dependent on some integrity condition on the data that is being generated by the model's execution. The most typical such situation occurs when there is a need to wait for the dissipation of undesired transient effects. Data collection cannot begin until this occurs. As a result, what might be called the 'data collection interval' has an uncertain beginning. The situation is further complicated by the fact that the duration of the data collection interval (once it begins) is likewise uncertain because of the difficulty in predicting when sufficient data of suitable 'quality' have been accumulated. Both these effects contribute to an uncertain right boundary for I_0. For example, consider a requirement for the steady-state average throughput for a communications network model following the upgrade of several key nodes with faster technology. The initial transient period following the upgrade needs to be excluded from the acquired data because of the distortion which the transient data would introduce. These are important issues that relate to the design of simulation experiments and they are examined in detail in Chapter 6.

Nevertheless a basic point here is simply that only portions of the data that are available over the course of the observation interval may have relevance to the project goals. Consider, for example, a requirement for the final velocity achieved by a dragster when it crosses the finish line. The velocity values prior to the final value are not of any particular significance. The point illustrated here is that the observation interval and data collection interval are not necessarily the same. It is not uncommon for I_0 to be substantially larger than the data collection interval.

Figure 2.1 illustrates some of the various possibilities relating to the observation interval that have been discussed above.

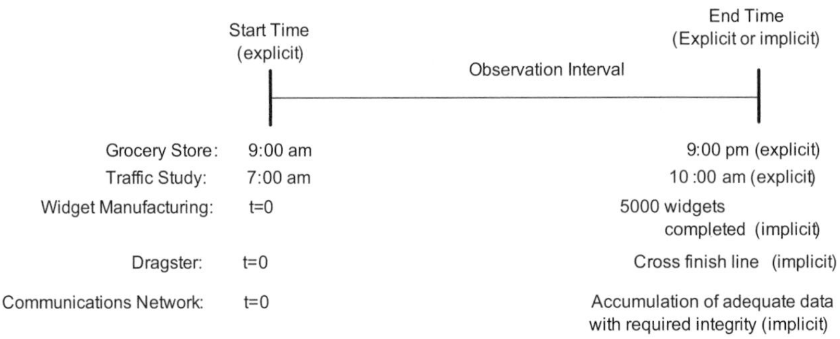

FIGURE 2.1. The observation interval.

2.2.2 Entities and Their Interactions

We have emphasised that a model is a specification of dynamic behaviour. This is a very general assertion and certainly provides no particular insight into the model building process itself. A useful beginning is an examination of some components that can be used as building blocks for the specification that we seek to develop. It is a dialogue about these components that begins the synthesis of a model.

Within the modelling and simulation context, dynamic behaviour is described in terms of the interactions (over time) among some collection of entities that populates the space that the SUI embraces. The feature about these interactions that is of particular interest is the fact that they produce change. The entities in question typically fall into two broad categories, one endogenous (intrinsic to the SUI itself) and the other exogenous.

With respect to the latter, it needs to be recognised that the SUI, like all systems, is a piece of a larger 'universe'; in other words, it functions within an environment. However, not every aspect of this environment has an impact upon the behaviour that is of interest. Those aspects that do have an influence need to be identified and these become the exogenous entities. Examples here are the ships that arrive at a maritime port within the context of a port model developed to evaluate strategies for improving the port's operating efficiency or alternately, the features of a roadway (curves, bumps) within the context of an automobile model being used to evaluate high-speed handling and stability properties. Note that any particular ship that arrives at the port usually exists as an integral part of the model only over some portion of the observation interval. When the service which it seeks has been provided the ship moves outside the realm of the model's behaviour.

More generally (and without regard to category) the entities within the specification fall into classes; for example, sets of customers, messages, orders, machines, vehicles, manufactured widgets, shipments, predators, bacteria, pollutants, forces, and so on. Interaction among entities can occur in many ways. Frequently this interaction occurs because the entities compete for some set of limited resources (a type of entity) such as servers (in banks, gas stations, restaurants, call-centres), transport services (cranes, trucks, tugboats), or health services (ambulances, operating rooms, doctors/nurses). This competition can give rise to yet another class of entity (called queues) in which some entities are obliged to wait for their respective turns to access the resources (sometimes there are priorities that need to be accommodated). On the other hand, entities may exert influence upon other entities in a manner that alters such things as acceleration, direction of motion, energy loss, and so on. Some examples of this breadth of possible entity types are provided in Table 2.1.

TABLE 2.1. Examples of entities.

System Under Investigation (SUI)	General Entities	Special Entities	
		Resource Entities	Queue Entities
Gas station	Cars Trucks	Pumps Attendants	Queue of cars at each pump
Widget manufacturing	Parts Broken machines	Machines Repair technicians	List of component parts List of broken machines
Restaurant	Customers	Tables Servers Kitchen Cooks	Queue of customers waiting for tables Customers at tables waiting for service
Hospital emergency room	Patients Ambulances	Doctors Nurses Examination rooms	Waiting room queue Critical patient queue List of patients in examination rooms waiting for doctor
Ecological system	Predator population Prey population		

We have indicated above that the specification of dynamic behaviour that we seek to develop begins with a description of the change-producing interactions among some set of entities within the SUI. The nature of these interactions is unrestricted and this 'inclusiveness' is one of the outstanding features of the modelling and simulation approach to problem solving. In fact, because of the complexity of the interactions that often need to be accommodated, alternate solution procedures (e.g., analytic) are usually infeasible.

Some entities within the formulation are distinctive inasmuch as they give rise to data requirements. Although these naturally enter into the dialogue about the interactions that need to be identified in model formulation, this occurs only at a relatively generic level. This abstract view is adequate up to a point, but actual behaviour generation cannot take place until the data requirements are accommodated. In effect the data serve to 'energise' the overall model specification.

Such data requirements can exist in a variety of forms. Consider, for example, a customer entity. Typically the data requirement here is the characterisation of the customer arrival rate or equivalently, the time between successive arrivals. This commonly is a random phenomenon and consequently gives rise to the need to identify an appropriate probability distribution function. A similar example can be found in a manufacturing process where a machine entity is subject to failure. The characterisation of such failure is typically in terms of rate of occurrence and repair duration which are again random phenomena and hence require appropriate specification in terms of statistical distributions. Or, alternately, consider the characterisation of the flow through an entity called 'pipe' in a chemical process. A data requirement here could be a single scalar value representing the coefficient of friction associated with the flow. As yet another alternative, consider the two-dimensional array of intercity flight times that would likely be necessary in a study of an airline's operational efficiency. In this case this data object would probably be shared by all 'flight' entities. These examples demonstrate that data requirements can be associated with both exogenous entities (e.g., the customers) and endogenous entities (e.g., the machines).

The detailed specifications for each such data requirement can be viewed as a *data model.* Each of these plays the role of a specialised submodel that has localised relevance. Their elaboration can be separately undertaken and even temporarily deferred without compromising the main thread of model development. In this sense, their development can be associated with the software engineering notion of 'encapsulation'. Each data model is an accessory to the bigger picture of characterising the relevant interactions that exist within the SUI.

The correct formulation of a data model can be a challenging task and its correctness is essential to the quality of the results flowing from the modelling and simulation project. The task is of particular relevance in the context of DEDS (discrete event dynamic system) models because of the central role that is played by the random phenomena that are always present in such models. We explore the topic in Chapter 3.

To illustrate the various notions introduced above, let's return to the gas station example introduced in Chapter 1. The endogenous entities include the attendants and the queue in front of each of the four pumps. There is only one exogenous entity, namely, the vehicles that enter the gas station. Notice that a customer entity is redundant because its role would be indistinguishable from the vehicle entity. Notice also that the pumps themselves are likely of no consequence. They would have to be included among the entities only if it were deemed that the possibility of failure was sufficiently high to have relevance to the credibility of the model. Data models would have to be developed to deal with the characterisation of the arrival rate of the vehicles and their separation into vehicle types and also the service times for each of the three phases of the service function.

The vehicles that enter (and subsequently leave) the gas station, as discussed above, provide an example of a distinctive and important feature of most models that emerge from the DEDS domain. Associated with almost all such models is at least one set of exogenous entities whose members have a transient existence within the scope of the model's behaviour, for example, the vehicles in the gas station model or the ships entering the port in an earlier example. Such entities introduce a variety of specialised requirements upon the way the model is formulated as will become apparent in the discussions that follow. We refer to any such set of exogenous entities which flow through the model as an 'input entity stream'.

2.2.3 Constants and Parameters

The constants and parameters of a model have much in common. In particular, constants and parameters both serve simply as names for the values of features or properties within a model which remain invariant over the course of any particular experiment with the model; for example, g could represent the force of gravity or NCkOut could represent the number of checkout counters in a supermarket. In the case of a constant, the assigned value remains invariant over all experiments associated with the modelling and simulation project. On the other hand, in the case of a parameter, there normally is an intent (usually embedded in the project goals) to explore the effect upon behaviour of a range of different values for the parameter, for example, the parameter C_f that represents the friction coefficient associated with a tire rolling over a road surface.

Often a parameter serves to characterise some 'size attribute' of the SUI, such as the number of berths at a seaport or the number of (identical) generators at a hydroelectric power generating station. In many cases such a size parameter might be associated with a facet of a design goal and the determination of the most appropriate value for it may be one of the reasons for the modelling and simulation project. Consider, for example, a parameter denoted by L_C which represents the passenger load capacity of each elevator in a proposed elevator system for an office tower. A goal of the project could be to determine a 'best' value for L_C that is consistent with specified requirements on the performance of the elevator system.

2.2.4 Time and Other Variables

The endpoint of the modelling process is a computer program that embraces the specification of the dynamic behaviour that we seek to study, that is, the simulation program. A prerequisite for the correct development of any computer program is a high degree of precision in the statement of the requirements that the program must meet. Generally speaking, this corresponds to raising the level of abstraction of the presentation of these requirements. A variety of means is available but perhaps the most fundamental is the incorporation of carefully defined variables that enable the formulation of the requirements in a precise and unambiguous way.

Within the modelling and simulation context, the particular purpose of these variables is to facilitate the formulation of those facets of the SUI's behaviour that are relevant to the goals of the project. In fact, meaningful dialogue about most aspects of a modelling and simulation project is severely hampered without the introduction of variables. They provide the means for elaborating, clarifying, and abstracting not only the dynamic behaviour that is of interest but, as well, the goals of the project.

Variables provide an abstraction for features of the model whose values typically change as the model evolves over the course of the observation interval. Variables fall into a number of broad categories and each of these is examined in the discussion below. Time itself is a very special variable that is common to all dynamic models.

2.2.4.1 Time

Because our interest is exclusively with dynamic systems, there is one variable that is common to all models that we consider, namely, time (which we generally denote with the symbol t). Apart from its pervasiveness, time is a special variable for two additional reasons. First of all, it is a 'primitive' variable in the sense that its value is never dependent upon any other variable. Secondly, and in direct contrast, most other variables are dependent on time; that is, they are functions of time.

It needs to be emphasised here that the variable t represents 'virtual time' as it evolves within the model. This (except for certain special cases) has no relation to 'wall clock' (i.e., real) time

2.2.4.2 Time Variables

As indicated above many of the variables within the context of our discussion of dynamic models are functions of time; that is, they are time-dependent variables or simply *time variables*. If V is designated as a time variable then this is usually made explicit by writing $V(t)$ rather than simply V. Within the context of our modelling and simulation discussions, time variables are frequently regarded as representing 'time trajectories'. Standard mathematical convention associates with $V(t)$ a statement about the set of values of t for which there is a defined value for V. This set is called the domain of V and we denote it $\boldsymbol{D}[V]$. In our case, the most comprehensive domain for any time variable is the observation interval I_o, and often the time variables that we discuss have this domain.

However, because of the underlying nature of the computational process, a time variable $V(t)$, which is the outgrowth of behaviour generation, will have defined values at only a finite subset of I_o; that is, the domain set $\boldsymbol{D}[V]$ has finite size. The underlying implication here is that from the perspective of observed behaviour, the value of a time variable $V(t)$ can be associated with a finite set of ordered pairs; that is,

$$\boldsymbol{T}[V] = \{(\hat{t}, \hat{v}): \hat{t} \in \boldsymbol{D}[V]\} \tag{2.1}$$

where $\hat{v} = V(\hat{t})$. We call the set $\boldsymbol{T}[V]$ the *trajectory set* for V. The reader is cautioned not to attempt an incorrect generalisation of the above. The assertion is simply that time variables reflected in the observed behaviour of a dynamic model have a finite domain set or can be represented with a finite domain set. We do not claim that all time variables have a finite domain set.

2.2.4.3 Input, State, and Output Variables

There are three important categories of variables normally associated with any model. We examine each in turn.

Input Variables: Earlier we introduced the notion of exogenous entities being associated with the model to reflect the impact of the SUI's environment upon its behaviour. This environmental impact can be viewed as the input to the model. The representation of the input is provided by some suitably defined set of input variables. It follows therefore that a model's behaviour cannot be generated without a specification for the values of its input variables.

It is important to appreciate that there can be a relatively subtle but nevertheless fundamental difference between the generic notion of an 'input' to a model and the variable chosen to characterise it. Consider, for example, an input variable $f_a = f_a(t)$, which represents the force of air friction upon the motion of a high-performance aircraft. The magnitude of this force necessarily depends on the aircraft's altitude because the origin of the force is air density, which in turn varies with altitude. This example illustrates that it is possible for an aspect of the model's behaviour (in this case, the aircraft's altitude) to directly influence the value of an input variable (the air friction). Such a mutual interrelationship is, in fact, not uncommon. On the other hand, we need to recognise that in this example the input (i.e., environmental influence itself) is the force that air density exerts on any object that moves through it. This property is fundamental to the environment in which the aircraft is operating and certainly cannot be altered by the aircraft's behaviour. The fact that its value varies with altitude (as reflected in the value acquired by $f_a(t)$) is an essential aspect of this fundamental physical property.

The notion of an 'input entity stream' was introduced earlier. Such a collection of exogenous entities flowing into a DEDS model most certainly corresponds to an input and consequently needs to have a characterisation in terms of an input variable. The two key features of any input entity stream are the time of the first arrival and the interarrival times, or more simply, the arrival times of the entities within the stream. The formulation of a suitable input variable that captures these features is explored in Chapter 4.

State Variables: The set of state variables for a dynamic model is very special for three reasons. One of these relates to a 'constructive property' which is intrinsic to the state variables. This property simply means that the model's dynamic behaviour can be entirely defined in terms of the state variables together with (not surprisingly) the model's input variables and parameters. We note also that the 'state of the model' at time t is a reference to the value of its state variables at time t.

The state variables for a model are not unique; that is, many alternate sets of variables that have this constructive property can usually be identified. Strictly speaking, the selection of any set of state variables includes the condition that it is a minimal set in the sense that no variable can be removed without destroying the constructive property. In our discussions we are generally very flexible with this requirement and this may result in acceptance of a degree of redundancy among the variables designated as the set of state variables. In such situations, it is possible to infer the values of one or more variables within this set from the values of other variables. Achievement of enhanced clarity is usually the motivation for allowing such redundancy.

A second reason why the set of state variables of a dynamic model is very special is that this collection of variables completely captures the effects of 'past' behaviour insofar as it affects future behaviour (both observed and unobserved, that is, from the perspective of model output). A characterisation of this important property can be developed in the following way: suppose $\boldsymbol{X}(t)$, $\boldsymbol{U}(t)$, and $\boldsymbol{P}$,[1] respectively, denote the state variables, the input variables, and the parameters of a model M, whose behaviour is to be observed over the observation interval $I_O = [T_A, T_B]$. Let T be a point in time between T_A and T_B and suppose the behaviour of the model has evolved up to $t = T$ and is stopped. Restarting the model with knowledge of only $\boldsymbol{X}(T)$, $\boldsymbol{U}(t)$, for $t \geq T$, and $\boldsymbol{P}$ would produce the same observed behaviour over $[T, T_B]$ as if the behaviour generating process had not been stopped at $t = T$. In fact this property, which we call Property Σ, has very special relevance because it provides the basis for identifying a set of state variables; in particular, the set of variables represented by the vector $\boldsymbol{X}(t)$ qualifies as a set of state variables for a model if and only if it has Property Σ.

[1] A collection of variables organized as a linear array is called a *vector variable.* We use bold font to indicate vector variables. The number of variables represented is called the dimension of the vector. Sometimes the actual size of the vector is not germane to the discussion and is omitted.

In the highly structured models that typically arise from the realm of continuous-time dynamic systems (CTDS), the state variables can usually be readily identified. In general however, identification of the state variables is not always an easy task. It is therefore useful to observe here that Property Σ, provides a useful means for testing whether a particular set of variables qualifies as a set of state variables. The fact that values must be specified for all state variables before a unique experiment with the model can be initiated provides a key identification criterion. Viewed from an inverse perspective, if an experiment can be duplicated without any need to record the initial value of some variable, then that variable is not a state variable.

Typically there is some collection of entities that flows within the scope of the behaviour of those models that have their origin within the DEDS domain. Consider, in particular, the 'transient entities' associated with an input entity stream. Linked to these entities are attributes and some of these condition the way in which interaction with other components of the model takes place. At any point in time both the existence of these flowing entities as well as their attribute values must be captured within the state variables of the model in order that the requirements of Property Σ be duly respected. Particular care must therefore be taken in identifying the model's state variables in these situations.

Note also that these transient entities typically contribute to data that are relevant to output requirements, and hence to observed behaviour. Assurance of the integrity of this particular category of data is clearly essential and this task is usually reflected in the formulation of state variables for the model. More specifically, this task must be duly recognised when any candidate set of state variables is assessed with respect to the requirements of Property Σ.

Our discussion here of the essential features of state variables has been carried out with the tacit assumption that state variables are simply time functions. In many circumstances this is indeed the case. However, as becomes apparent in the discussions of Chapter 3, the requirements of Property Σ can give rise to the need to formulate state variables as more complex data structures, for example, lists whose entries are themselves lists.

The third reason why the set of state variables of a dynamic model is special relates to the output variables of the model and we defer the examination of that relationship to the discussion of output variables that follows.

Output Variables: The requirements that are embedded within the project goals give rise to the model's output variables. The output variables of a model typically reflect those features of the SUI's behaviour that originally motivated the modelling and simulation project. Because of this, it is not unreasonable to even regard a model as simply an outgrowth of its output variables.

In effect therefore, an output variable serves as a conduit for transferring information about the model's behaviour that is of interest from the perspective of the project goals. The output variable Y could, for example, represent the velocity of a missile over the course of its trajectory from the moment of firing to the moment of impact with its target or Y might represent the number of messages waiting to be processed at a node of a communications network viewed over the duration of the observation interval.

Output variables fall into various categories that have relevance at different stages of the discussion. The initial separation is into two groups called *point-set output variables* (PSOVs) and *derived scalar output variables* (DSOVs), respectively. PSOVs, in turn, can be separated into two subcategories called *time variables* and *sample variables*. An output variable Y that is a time variable is, in fact, simply a function of time; that is, $Y = Y(t)$. The need to recognise output variables of this type should not be surprising inasmuch as any dynamic model evolves over time and consequently it is reasonable to expect behaviours of interest that are functions of time. Any output variable that is a time variable is always expressible in terms of the model's state variables. This represents the third reason why the state variables of a model are special.

An important issue that needs to be examined relates to the manner in which the values of a time variable are recorded. In this regard we make the simple assumption that these are recorded as they become available (at least from the conceptual perspective of our present discussion). We refer to this assumption in the sequel as *Assumption R*. It is important because it eliminates the need for concern about the data recovery issue that is the focus of Property Σ. It is interesting nevertheless to observe that Assumption R imposes no particular constraint because there really is no other option available except to record relevant data as they become available over the course of the behaviour-generating process.

By way of illustrating the implications of Assumption R, let's suppose that $Y(t)$ is an output variable. As was pointed out earlier, because Y is a time variable, an essential aspect of its specification is its domain set $\boldsymbol{D}[Y]$. Furthermore, the 'value of Y' can be regarded as a set of ordered pairs given by the trajectory set:

$$\boldsymbol{T}[Y] = \{(\hat{t}, \hat{y}): \hat{t} \in \boldsymbol{D}[Y] \}, \tag{2.2}$$

where $\hat{y} = Y(\hat{t})$. The behaviour-generating process ensures that as the time variable t traverses the observation interval the successive pairs $(\hat{t}, \hat{y})$ are generated. The consequence of Assumption R is that these pairs are immediately deposited into $\boldsymbol{T}[Y]$.

The second subcategory of PSOVs is called sample variables. These are restricted to the family of DEDS models. In Chapter 4 we introduce a conceptual modelling framework for such models that includes the notion of consumer entities flowing within (and often, through) the model. These entities are instances of a class and the class has an assigned collection of attributes which are inherited by each instance of the class. A sample variable corresponds to such an attribute, that is, to an attribute of a consumer entity class. The entity instances that flow within the model typically leave a 'data trail' that is comprised of the acquired values for one or more designated attributes. The value of a sample variable is associated with those data.

Consider, for example, the messages that flow through a communications network. Suppose m_k is a particular message (an instance of a class that might be called Message). At each of the switching nodes through which m_k passes, it generally encounters a delay and the total delay d_k, encountered by m_k during its existence within the realm of the model would likely be stored in an attribute for this entity class, possibly called D. We regard the 'value of D' at the end of the experiment to be the collection of values $\{d_1, d_2, \ldots, d_N\}$ (where N is the total number of messages passing through the network over the course of the observation interval). The attribute D is an example of a sample variable and its value is a *sample set* which we denote $\Psi[D]$; that is,

$$\Psi[D] = \{d_1, d_2, \ldots, d_N\}.$$

In general, then, a sample set is a set of data values that are 'deposited' by a particular collection of entity instances flowing in a DEDS model. Each such deposited value is the value of a specific attribute (the sample variable) that is common to each instance of the class.

We need to acknowledge here an inherent ambiguity in our presentation of a sample variable. Although we have referred to its 'value' as a collection of values (as contained in a sample set) it is, in general, a random variable and the sample set is simply a collection of samples (observations) of that random variable.[2] The specific observations contained in a sample set $\Psi[\ ]$ are rarely of interest. Generally they serve only as a means for acquiring information for determining the parameters of the probability distribution of the sample variable in question. This small anomaly in terminology provides the benefit of permitting a parallel treatment of trajectory sets and sample sets in our discussions. However, in some circumstances, this irregularity is relevant and needs to be recognised.

[2] In the terminology of stochastic processes (see Section A1.8 of Annex 1) a sample set can be viewed as an observation of a stochastic process.

The anomaly in question is readily apparent in our communications example. The sample variable D is, in fact, a random variable that represents the waiting time of messages. The specific observations of D as contained in the sample set $\Psi[D]$ are rarely of interest. Generally they serve only as a means for acquiring information needed for the estimation of probability distribution parameters relating to D. It is these parameters that are typically relevant to resolving the project goals.

It is not always the case that the values in an output set (i.e., a trajectory set $\mathcal{T}[\ ]$ or a sample set $\Psi[\ ]$) are of direct relevance to the resolution of the project goals. Often it is some feature of the data in such an output set that is of primary interest. This gives rise to another category of output variable that we call a *derived scalar output variable* (DSOV) that is derived from the data in an output set (either $\mathcal{T}[\ ]$ or $\Psi[\ ]$). For example we could have $y_a = \mathrm{AVG}(\mathcal{T}[\ ])$ or $y_{mx} = \mathrm{MAX}(\Psi[\ ])$ where AVG and MAX are set operators that yield the average and maximum values, respectively, of the data within the referenced set. Many such operators can be identified, for example, MIN, NUMBER, AREA, and so on. Notice that the values of DSOVs such as y_a and y_{mx} may be established either by a postprocessing step at the end of a simulation experiment or possibly concurrently with the execution of the simulation experiment.

DSOVs play an especially significant role in modelling and simulation projects in the discrete event dynamic system domain. In that context they are often referred to as *performance measures* and they are always random variables. This, in particular, means that a specific trajectory set or a sample set has no particular significance except to the extent that it serves to help determine the properties of the random phenomenon that underlies the behaviour that is being observed. It is those parameters that are usually important to the project goals. This important topic is explored in detail in the discussions of Chapters 3 and 6.

Our perspective is that data are deposited into a sample set as they become available. This perspective is consistent with Assumption R and also parallels the manner in which time variable data are accumulated. The value of the DSOV that is linked to a particular sample set can be obtained as a separate step at the completion of the model execution or concurrently during the model's execution.

Various DSOVs can be associated any particular sample set. For example, in our communications network there may be a requirement for the value of the scalar variable A_d which represents the average delay taken over all the messages processed over the course of the observation interval. More specifically:

$$A_d = \mathrm{AVG}(\Psi[D]) = [\sum_{k=1}^{N} d_k] / N \ . \tag{2.3}$$

Or consider the scalar variable $\max_d$ that represents the maximum delay encountered by the set of messages m_k, $k = 1, 2, \ldots, N$. Then:

$$\max_d = \text{MAX}(\Psi\,[D])\,, \tag{2.4}$$

where MAX is an operator that provides the largest value contained in a set of numerical values. Clearly the both scalar variables A_d and $\max_d$ are derived from the same sample set. Note also that both A_d and $\max_d$ are in fact random variables and any experiment with the model simply yields a single observation of those random variables.

We conclude this section by including Figure 2.2 that illustrates how the parameters and the various types of variables interact to generate behaviour and how the observation of behaviour is provided via the output variables of the model.

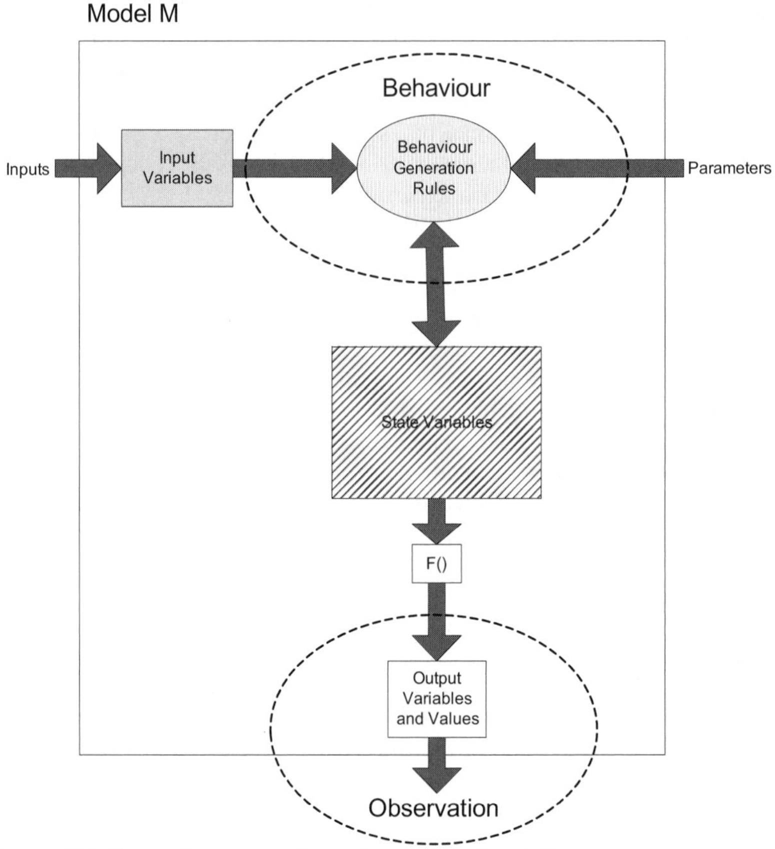

FIGURE 2.2. Interaction among the various types of variables.

2.2.5 An Example – The Bouncing Ball

The example we consider in the discussion that follows illustrates how the notions of parameters and variables play a crucial role in enabling a clear and succinct formulation of a model for dynamic system.

A boy is standing on the ice surface of a frozen pond and (at $t = 0$) throws a ball into the air with an initial velocity of V_0. When the ball leaves the boy's hand it is a distance of y_0 above the ice surface. The initial velocity vector makes an angle of θ_0 with the horizontal. The boy's objective is to have the ball bounce at least once and then fall through a hole that has been cut in the ice. The hole is located at a distance H from the point where the boy is standing. There is a wind blowing horizontally. The general configuration is shown in Figure 2.3.

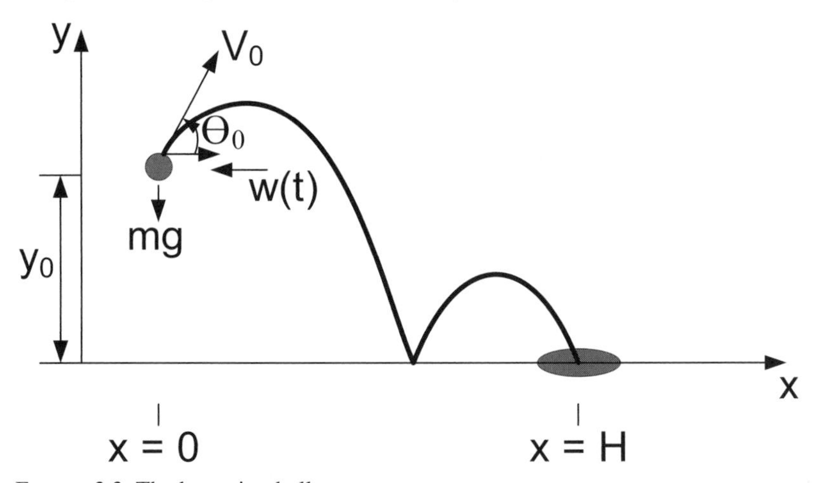

FIGURE 2.3. The bouncing ball.

The goal of our modelling and simulation project is to determine a value for the release angle θ_0 which will result in a trajectory for the ball that satisfies the boy's objective.

A model for the underlying dynamic system can be formulated by a straightforward application of Newton's second law (the familiar $F = ma$). We begin by introducing four state variables to characterise the ball's flight trajectory; namely:

$x_1(t)$: The ball's horizontal position at time t
$x_2(t)$: The ball's horizontal velocity at time t
$y_1(t)$: The ball's vertical position at time t
$y_2(t)$: The ball's vertical velocity at time t

There are two forces acting on the ball. The first is gravity and the second is the force resulting from the wind. In order to proceed to a next level of refinement, two assumptions are in order:

- We assume that when the ball is released from the boy's hand, the velocity vector V_0 lies in a vertical plane that passes through the boy's position and the location of the hole in the ice (this ensures that the ball is heading, at least initially, in the direction of the hole).
- We assume that the horizontal wind velocity is parallel to the plane specified above (this ensures that the wind will not alter the 'correct' direction of the ball's initial motion, i.e., towards the hole).

With these two assumptions the ball's initial motion can be described by the following set of differential equations:

$$\begin{aligned}\frac{dx_1(t)}{dt} &= x_2(t)\\ \frac{dx_2(t)}{dt} &= -W(t)/m\\ \frac{dy_1(t)}{dt} &= y_2(t)\\ \frac{dy_2(t)}{dt} &= -g\end{aligned} \tag{2.5}$$

where $W(t)$ represents the force of the wind acting on the ball's horizontal motion and g represents the gravity force acting on the ball. Each of these four first order differential equations needs to have a specified initial condition. These are: $x_1(0) = 0$, $x_2(0) = V_0 * \cos(\theta_0)$, $y_1(0) = y_0$, $y_2(0) = V_0 * \sin(\theta_0)$ where y_0 is the height above the ice surface of the boy's hand when he releases the ball. (The value assigned to $x_1(0)$ is arbitrary and zero is a convenient choice.)

In view of the boy's objective, it is reasonable to assume that the ball leaves the boy's hand with an upward trajectory (in other words, $\theta_0 > 0$). Sooner or later, however, gravity will cause the ball to arc downwards and strike the ice surface (hence $y_1 = 0$). At this moment (let's denote it T_C) the ball 'bounces' and this represents a significant discontinuity in the ball's trajectory. A number of additional assumptions must now be introduced to deal with the subsidiary modelling requirement that characterises this bounce. These are as follows.

- We assume that the bounce takes place in a symmetric way in the sense that if the angle of the velocity vector (with respect to the horizontal) at the moment prior to the collision is θ_C, then immediately after the collision the velocity vector has an orientation of $-\theta_C$.

- Energy is lost during the collision and we take this to be reflected in a reduction in the magnitude of the velocity vector (loss in kinetic energy). More specifically, if $|V_C|$ is the magnitude of the velocity vector immediately prior to the collision, then we assume that $\alpha\,|V_C|$ is the magnitude after the collision, where $0 < \alpha < 1$.

In effect, the two above assumptions provide a specification for the behaviour that characterises the dynamics of the ball at the point of collision with the ice surface. More specifically, we have formulated a model of the bounce dynamics which is:

$$\begin{aligned} \theta_C^+ &= -\theta_C \\ |V_C^+| &= \alpha\,|V_C| \,. \end{aligned} \tag{2.6}$$

Here θ_C^+ is the angle of the velocity vector at time T_C^+ which is the moment of time that is incrementally beyond the moment of contact, T_C. Similarly $|V_C^+|$ is the magnitude of the velocity vector at time T_C^+.

Although the underlying equations of motion remain unchanged following the bounce, there is a requirement to initiate a new trajectory segment that reflects the changes that occur due to the collision. This new segment of the ball's trajectory begins with 'initial' conditions that incorporate the assumptions outlined above; namely,

$$\begin{aligned} x_1(T_C^+) &= x_1(T_C) \\ x_2(T_C^+) &= \alpha\, x_2(T_C) \\ y_1(T_C^+) &= 0 \\ y_2(T_C^+) &= -\,\alpha\, y_2(T_C) \,. \end{aligned} \tag{2.7}$$

Our model for the trajectory of the ball is given by Equations (2.5) and (2.7). In this model, $W(t)$ (the force exerted by the wind) represents an input variable and α could be regarded as a parameter if, for example, a secondary project goal were to explore the relationship between α and the problem solution θ_0^*. The state variables are x_1, x_2, y_1, and y_2. This is clearly reflected in their essential role in reinitialising the ball's trajectory following each collision with the ice surface.

Before leaving this example, it is of some interest to revisit the stated goal of this modelling and simulation project. In particular, consider the fundamental issue of whether the underlying problem has a solution. It is important to recognise here that the existence of an initial release angle of the ball (θ_0) that will cause the ball to fall through the hole in the ice is not guaranteed. This is not a deficiency in the model but is simply a consequence of the underlying physics. The boy's throwing action gives the ball kinetic energy which is dependent on the release velocity (V_0). This energy may simply be insufficient to accommodate the energy losses encountered by the ball over the course of its trajectory and the ball may not even be able to travel the distance H where the hole is located.

2.3 The Modelling and Simulation Process

An outline of the essential steps involved in carrying out a modelling and simulation study is provided in the discussion that follows. Although the initial steps can be effectively presented using various notions that have been previously introduced, there are several aspects of the latter stages that require extensive elaboration. This is provided in the discussions that follow. An overview of the process is provided in Figure 2.4.

The overview of Figure 2.4 does not include a preliminary phase during which solution alternatives for the problem are explored and a decision is made to adopt a modelling and simulation approach. We note that the option of also carrying out other complementary approaches is entirely reasonable and is often prudent for some portions of the problem. Although this preliminary phase is not explicitly represented in Figure 2.4 its existence and importance must nevertheless be recognised.

It should be emphasised that a modelling and simulation project of even modest size is often carried out by a team of professionals where each member of the team typically contributes some special expertise. There is, therefore, a need for communication among team members. Some facets of the discussion have their basis in this communication requirement.

2.3.1 The Project Description

The process begins with the preparation of a document called the *project description*. This document includes a statement of the project goal(s) and a description of those behavioural features of the SUI that have relevance to the goals. These behaviour features are typically formulated in terms of the various entities that populate the space that the SUI embraces with particular focus on the interactions among these entities. It is, more or less, an informal description inasmuch as it relies mainly on the descriptive power of natural language. The language is, furthermore, often heavily coloured with jargon associated with the SUI. This jargon may not be fully transparent to all members of the project team and this can contribute to both an inherent lack of precision and, as well, to communication problems.

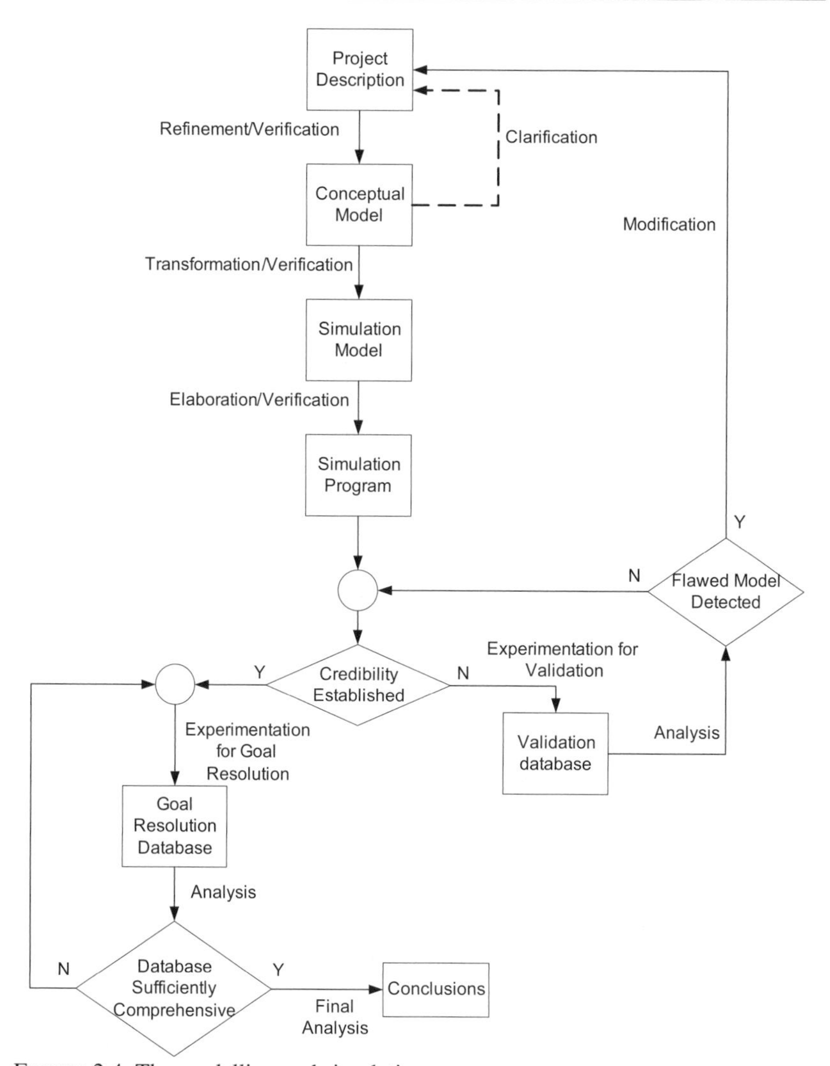

FIGURE 2.4. The modelling and simulation process.

With few exceptions, the SUI also has structural features that provide the context for the interactions among the entities (e.g., the layout of the pumps at a gas station or the topology of the network of streets being serviced by a taxi company). Informal sketches are often the best means of representing these structural features (see, e.g., Figure 2.3). These are an important part of the presentation because they provide a contextual elaboration that can both facilitate a more precise statement of the project goals and as well, help to clarify the nature of the interaction among the entities. Because of these contributions to understanding, such sketches are often a valuable component of the project description.

2.3.2 The Conceptual Model

The information provided by the project description is, for the most part, unstructured and relatively informal. Because of this informality it is generally inadequate to support the high degree of precision that is required in achieving the objective of a credible model embedded within a computer program. A refinement phase must be carried out in order to add detail where necessary, incorporate formalisms wherever helpful, and generally enhance the precision and completeness of the accumulated information. Enhanced precision is achieved by moving to a higher level of abstraction than that provided by the project description. The reformulation of the information within the project description in terms of parameters and variables is an initial step because these notions provide a fundamental means for removing ambiguity and enhancing precision. They provide the basis for the development of the simulation model that is required for the experimentation phase.

There is a variety of formalisms that can be effectively used in the refinement process. Included here are mathematical equations and relationships (e.g., algebraic and/or differential equations), symbolic/graphical formalisms (e.g., Petri nets, finite state machines), rule-based formalisms, structured pseudocode, and combinations of these. The choice depends on suitability for providing clarification and/or precision.

The result of this refinement process is called the *conceptual model* for the modelling and simulation project. The conceptual model may, in reality, be a collection of partial models each capturing some specific aspect of the SUI's behaviour. The representations used in these various partial models need not be uniform.

The conceptual model is a consolidation of all relevant structural and behavioural features of the SUI in a format that is as concise and precise as possible. It provides the common focal point for discussion among the various participants in the modelling and simulation project. In addition, it serves as a bridge between the project description and the simulation model that is essential for the experimentation activity (i.e., the simulation phase). As we point out below, the simulation model is a software product and its development relies on considerable precision in the statement of requirements. One of the important purposes of the conceptual model is to provide the prerequisite guidance for the software development task.

It can frequently happen that the formulation of the conceptual model is interrupted because it becomes apparent that the information flowing from the project description is inadequate. Missing or ambiguous information are the two most common origins of this difficulty. The situation can be corrected only by returning to the project description and incorporating the necessary clarification. This possible 'clarification loop' is indicated with a dashed line in Figure 2.4.

In Figure 2.4, a verification activity is associated with the transition from the project description to the conceptual model. Both verification and the related notion of validation are examined in detail in Section 2.4. As will become apparent from that discussion, verification is included as part of the transition under consideration because it involves a reformulation of the key elements of the model from one form to another and the integrity of this transformation needs to be confirmed.

In the modelling and simulation literature, the phrase 'conceptual model' is frequently reduced simply to 'model'. Our usage of the word 'model' without a modifier generally implies a composite notion that includes a conceptual model and its simulation model and simulation program successors where the latter two notions are described in the discussion that follows.

As a concluding observation in this discussion, it is worth pointing out that there is by no means a common understanding in the modelling and simulation literature of the nature and role of a conceptual model. The overview presented by Robinson [2.15] gives considerable insight into the various perspectives that prevail.

2.3.3 The Simulation Model

The essential requirement for the experimentation phase of a modelling and simulation project is an executable computer program that embodies the conceptual model. It evolves from a transformation of the conceptual model into a representation that is consistent with the syntax and semantic constraints of some programming language. This program is the *simulation model* for the project. It is the execution of this program (or more correctly, an enhanced version of it; see following Section) that generates the 'behaviour' that emulates pertinent aspects of the system under investigation. The solution to the underlying problem that is embedded in the project goal(s) is obtained from the data reflected in this behaviour.

Typically the simulation model is written using the specialised facilities of a programming language that has been designed specifically to support the special requirements of simulation studies. Many such languages have appeared in recent years; some examples are: SIMSCRIPT II.5, MODSIM, GPSS, SIMAN, ACSL, Modelica, Arena, CSIM, and SIMPLE ++. Such languages generally provide features to support the management of time, collection of data, and presentation of required output information. In the case of projects in the DEDS domain, additional features for the generation of random variates, management of queues, and the statistical analysis of data are also provided.

The simulation model is the penultimate stage of a development process that began with the decision to formulate a modelling and simulation project to resolve an identified problem. The simulation model is a software product and as such, the process for its development shares many of the general features that characterise the development of any software product.

Note that in Figure 2.4 the transition from the conceptual model to the simulation model is associated with two activities: namely, transformation and verification. As in the earlier transition from project description to conceptual model, verification is required here to confirm that the transformation has been correctly carried out.

2.3.4 The Simulation Program

The outline of the simulation model provided above is idealised inasmuch as it suggests that the simulation model is directly capable of providing the behaviour-generating mechanism for the simulation activity. In reality this program code segment is never self-sufficient and a variety of auxiliary services must be superimposed. The result of augmenting the simulation model with complementary program infrastructure that provides these essential functional services is the *simulation program.*

The services in question fall into two categories: one relates to fundamental implementation issues whereas the other is very much dependent on the nature of the experiments that are associated with the realisation of the project goals. Included within the first category are such basic tasks as initialisation, control of the observation interval, management of stochastic features (when present), solution of equations (e.g., the differential equations of a continuous system model), data collection, and so on. Convenient programming constructs to deal with these various tasks are normally provided in software environments specifically designed to support the simulation activity. But this is certainly not the case in general-purpose programming environments where considerable additional effort is often required to provide these functional requirements.

The second category of functional services can include such features as data presentation (e.g., visualisation and animation), data analysis, database support, optimisation procedures, and the like. The extent to which any particular modelling and simulation project requires services from this second category can vary widely. Furthermore, modelling and simulation software environments provide these services only to varying degrees and consequently, when they are needed, care must be taken in choosing an environment that is able to deliver the required services at an adequate level.

The manner in which the support services to augment the simulation model are invoked varies significantly among software environments. Almost always there is at least some set of parameters that need to be assigned values in order to choose from available options. Often some explicit programming steps are needed. Considerable care must be taken when developing the simulation program to maintain a clear demarkation between the code of the simulation model and the code required to invoke the ancillary services. Blurring this separation can be detrimental because the resulting simulation program may become difficult to verify, understand, and/or maintain. It has, in fact, been frequently noted (e.g., Oren [2.10]) that an important quality attribute of a simulation software platform is the extent to which it facilitates a clear separation of the code for the simulation model from the infrastructure code required for the experimentation that is required for the achievement of the project goal(s).

Figure 2.4 indicates that a verification activity needs to be carried out in the transition from the simulation model to the simulation program. This need arises because this transition typically involves a variety of decisions relating to the execution of the simulation model and the correctness of these decisions must be confirmed. Consider, for example, a simulation model that incorporates a set of ordinary differential equations. Most modelling and simulation programming environments offer a variety of solution methods for such equations and each has particular strengths and possibly weaknesses as well. If the equations in question have distinctive properties, then there exists a possibility of an improper choice of solution method. The verification process applied at this stage would uncover the existence of such a flaw when it exists.

2.3.5 The Operational Phases

Thus far our outline of the modelling and simulation process has focused on the evolution of a series of interdependent representations of SUI. However, with the existence of the simulation program, the stage is set for two operational phases of the process that we now examine. The first of these is the validation phase whose purpose is to establish the credibility of each of the model realisations, from the perspective of the project goals. The notion of validation is examined in some detail in Section 2.4 which follows below, and hence we defer our discussion of this phase.

The second phase, which can begin only after the model's credibility has been established, is the experimentation phase, or more specifically, the simulation phase. This activity is presented in Figure 2.4 as the task of 'goal resolution'. This is achieved via a sequence of experiments with the simulation program during which an ever-increasing body of data is collected and analysed until it is apparent that a 'goal resolution database' is sufficiently complete and comprehensive to permit conclusions relating to the goal(s) to be confidently formulated.

2.4 Verification and Validation

A simulation model is a software product and like any properly constructed artefact its development must adhere to design specifications. Assuring that it does is a verification task. All software products furthermore have a well-defined purpose (e.g., manage a communications network, or ensure that an optimal air/fuel mixture enters the combustion chamber of an internal combustion engine). In the case of a simulation model the purpose is to provide an adequate emulation of the behavioural features of some SUI, where 'adequate' is assessed from the perspective of the project goals. Assuring that this is achieved is a validation task.

Both verification and validation are concerned with ensuring the credibility of the conclusions that are reached as a consequence of the experiments carried out with the simulation program. They can be reasonably regarded as part of the general thrust of quality assurance (the topic of the following section). However the central role of these notions within the modelling and simulation process, as presented in Figure 2.4, suggests that they merit special treatment. The range of perspectives relating to the processes associated with these notions further justifies an examination that extends beyond their obvious contribution to quality assurance.

The terms verification and validation are used in a variety of disciplines, notably software engineering. By and large, the distinction in the meaning of these two notions is often poorly understood. In the software engineering context, however, a remarkably concise and revealing presentation of the essential difference can be expressed in terms of two closely related questions. These (originally formulated by Boehm [2.4]) are:

Verification: Are we building the product right?
Validation: Are we building the right product?

The product referred to here is the software product being developed. Reinterpretation within a modelling and simulation context is straightforward. The 'product' is the model and the notion of 'building the right product' corresponds to developing a model that has credibility from the perspective of the project goals. On the other hand, 'building the product right' corresponds to ensuring that the artefact that begins as a meaningful and correct problem description and then undergoes various transformations that culminate in a simulation program is never compromised during these various transformations.

Verification is concerned with ensuring that features that should (by design) be clearly apparent in each manifestation of the model are indeed present. Whether these features properly reflect required or expected model behaviour (always from the perspective of the project goals) is an issue that falls in the realm of validation.

By way of illustration, consider a modeling and simulation project whose primary purpose is to provide support for the design of the various control systems that are to be incorporated into a new, highly automated, thermoplastics manufacturing plant. Certain thermodynamics principles have been identified as being best suited as the basis for formulating the model of one particular aspect of the chemical kinetics that is involved in the process. The approach will, in all likelihood, give rise to a conceptual model that incorporates a set of partial differential equations. The task of ensuring that these differential equations are correctly formulated on the basis of the principles involved and ensuring that they are correctly transformed into the format required by the simulation software environment to be used, falls in the realm of verification. Confirmation that the selected principles are indeed an adequate means of representing the relevant behaviour of the chemical process is a validation task.

Consider an alternate but similar example where a modeling and simulation project is concerned with exploring alternatives for enhancing the operating efficiency of a large metropolitan hospital. The model is to be organised as a number of interacting components. One of these will focus on the operation of the elevator system which has frequently been observed to be a point of congestion. The behaviour of any elevator system is described by a relatively complex set of rules. Ensuring that these rules are correctly represented in each of several realisations (e.g., natural language in the initial statement, then an intermediate and more formal representation such as a flow chart and finally in the program code of the simulation model) is part of the verification activity. Confirmation of the correctness of the rules themselves is a validation task.

A simulation model always functions in some software environment and assumptions about the integrity of the environment are often made without any particular basis. Confirmation of this integrity is a verification task. Consider, for example, the adequacy of the numerical software for the solution of the differential equations in a continuous-time dynamic system model or the adequacy of the mechanism used for generating a random number stream required in a DEDS model. Confirmation that such essential tools are not only available but are sufficiently robust for the project requirements is part of the verification activity.

It has been rightly observed (e.g., Neelamkavil [2.12] that 'complete validation' of a model is an objective that is beyond the realm of attainability; the best that can be hoped for is 'failure to invalidate'. A related idea is contained in one of a collection of verification and validation principles suggested by Balci [2.1], namely, that the outcome of the validation activity is not binary valued. Degrees of success must be recognised and accepted and the credibility of the conclusions derived from the experiments with the model treated accordingly. The practical reality for accepting less than total success in the validation endeavour originates in the significant overhead involved. Carrying out validation to a level that totally satisfies all concerned parties can be both expensive and time consuming. A point of diminishing returns is invariably reached and compromises, together with acceptance of the attendant risk, often become unavoidable.

Validation must necessarily begin at the earliest possible stage of the modelling and simulation project, namely, at the stage of problem definition. Here the task is simply to ensure that the statement of the problem is consistent with the problem that the project originator wants to have solved. This is of fundamental importance because, for the members of the project team that will carry out the project, the problem statement is the problem. The documented problem statement is the only reference available for guidance. All relevant facets must therefore be included and confirmation of this is a validation task.

The problem definition has many facets and most have direct relevance to the validation task. One which is especially relevant is the statement of the project goals. These have a profound impact that ranges from the required level of granularity for the model to the nature of the output data that need to be generated. Consider, for example, the model of an airliner. A model that has been validated within the context of a project that is evaluating a business plan for a commercial airline will most likely not qualify as an adequate model within the context of a project that seeks to determine the aircraft's aerodynamic characteristics. In effect then, one of the most fundamental guiding principles of any validation activity is that it must be guided by the goals of the study.

One (essentially naïve) perspective that might be adopted for validating a model is simply to ensure that its observed behaviour 'appears correct', as reflected by animation, graphical displays, or simply the values of some set of designated variables. The assessment here is clearly entirely qualitative rather than quantitative and hence is very subjective. A far more serious shortcoming of this approach is that it makes no reference to the goals of the modelling and simulation study. As noted above, this is a very serious flaw. The absence of this context carries the naïve (and most certainly incorrect) implication that the model has 'universal applicability'. However, it is rare indeed that a model of 'anything' is appropriate for all possible modelling and simulation projects to which it might be linked.

Nevertheless, the relatively superficial approach given above does have a recognised status in the validation toolkit when it is refined by including the understanding that the observers are 'domain experts' and that their judgement is being given with full understanding of the model's purpose. With these qualifiers, the approach is referred to as *face validation*.

It is reasonable to assume that within the framework of the project goals, a collection of (more or less) quantifiable anticipated behaviours for the model can be identified. These will usually be expressed in terms of input–output relationships or more generally in terms of cause/effect relations. Consider, for example, a model developed to investigate the aerodynamic characteristics of an aircraft. The occurrence of an engine failure during a simulation experiment should (after a short time interval) lead to a decrease in the aircraft's altitude. If this causal event does not take place, then there is a basis for suspicion about the model's adequacy.

Or consider introducing the occurrence of disruptive storms in a harbour model. It is reasonable to expect that this would result in a decrease in the operating efficiency of the harbour as measured by the average number of ships per day passing through the loading/unloading facilities.

As a final example, consider doubling the arrival rate of tourists/convention attendees in an economic model of an urban area. This should result in an approximate doubling in the occupancy rate of the hotels in the area. Furthermore, an occupancy rate increase of more than a factor of two should be cause for some reflection about possible flaws in the model's specification.

The general approach outlined above is often called *behaviour validation*. An implicit assumption in the approach is that a verified simulation program is available for experimentation. The approach has been examined in some detail by Birta and Ozmizrak [2.3] who incorporate the notion of a validation knowledge base that holds the collection of behavioural features that need to be confirmed. The investigation includes a discussion of a procedure for formulating a set of experiments that efficiently covers the tests that are implied in the knowledge base. An accessory question that does need resolution prior to the implementation of the process relates to the level of 'accuracy' that will be expected in achieving the designated responses. Behaviour validation has several noteworthy aspects; for example, the knowledge base can conveniently accommodate insights provided by a domain expert and as well, it can accommodate data acquired from an observable system when such an option exists.

In fact this latter feature is closely related to a notion called *replicative validation*, that is, confirming that the simulation program is capable of reproducing all available instances of the SUI's input–output behaviour. This notion is clearly restricted to the case where the SUI actually exists and behavioural data have been collected. But even in such circumstances there remain open questions; for example, could there not be 'too much' data available and if so, how can they be organised into meaningful (nonredundant classes), and how is the impact of project goals accommodated?

Validation in the modelling and simulation context must also embrace the data modelling task. For example, suppose that once ordered, the arrival time for a replacement part for a machine in a manufacturing process is random. There are at least two possible choices here for representing this situation. One is simply to use the mean delay time (assumed to be known) and an alternative is to use successive samples drawn from a correctly specified stochastic distribution. Ensuring that a satisfactory choice is made (with due regard to project goals) can be regarded as a validation task.

Accreditation is a notion that is closely related to validation. Accreditation refers to the acceptance, by a designated accreditation authority, of some particular simulation model for use within the context of a particular modelling and simulation project. Several important issues are associated with this notion: for example, what guidelines are followed in the designation of the accreditation authority and how is the decision-making procedure with respect to acceptance carried out. These are clearly matters of critical importance but they are, for the most part, very situation-dependent and for this reason we regard the topic of accreditation as being beyond the scope of our discussions. Certification is an equivalent issue which is explored in some detail by Balci [2.2].

We conclude this section by observing that the importance of model credibility has been recognised even at legislative levels of government because of the substantial government funding that is often provided in support of large-scale modelling and simulation projects. In 1976 the American government's General Accounting Office presented to the U.S. Congress the first of three reports that explored serious concerns about the management, evaluation, and credibility of government-sponsored simulation models (see [2.7] through [2.9]). For the most part these concerns were related to verification and validation issues in the context of modelling and simulation projects carried out by, or on behalf of, the U.S. Department of Defense. This latter organisation is possibly the world's largest modelling and simulation user community and a comprehensive presentation of its perspective about verification and validation can be found in [2.6]. An overview of some of the material contained in [2.6] together with a discussion of verification and validation issues in some specialised circumstances (e.g., hardware-in-the-loop, human-in-the-loop, distributed environments) can be found in Pace [2.13].

2.5 Quality Assurance

Quality assurance within the framework of modelling and simulation is a reference to a broad array of activities and methodologies that share the common objective of ensuring that the goals of the simulation project are not only achieved but are achieved in a timely, efficient, and cost-effective manner. An interesting overview of these can be found in Ören [2.10]. As we have previously noted, a significant thrust of the quality assurance effort necessarily deals with ensuring the credibility of the simulation model (where credibility here must always be interpreted from the perspective of the goals of the project.) Nevertheless, there is a variety of other important issues that fall within the realm of quality assurance. We examine some of these in the discussion below.

Documentation

A modelling and simulation project of even modest complexity can require many months to complete and will likely be carried out by a team having several members. Information about the project (e.g., assumptions, data sources, credibility assessment) is typically distributed among several individuals. Personnel changes can occur and in the absence of documentation, there is a possibility that important fragments of information may vanish. Likewise the reasons for any particular decision made during the course of the project may be completely obvious when it is made, but may not be so obvious several months later. Only with proper documentation can the emergence of this unsettling uncertainty be avoided. Comprehensive documentation also facilitates the process of 'coming-up-to-speed' for new members joining the team.

Project documentation must not only be comprehensive but must also be current. Deferring updates that reflect recent changes is a prescription for rapid deterioration in the value of the documentation because prospective users will become wary of its accuracy and hence will avoid reliance on it. In the extreme, documentation that is deferred until the end of the project essentially belies the intent of the effort.

Program Development Standards

Premature initiation of the program development phase must be avoided. Often there is an urge to begin the coding task before it is entirely clear what problem needs to be solved. This can result a simulation program that is poorly organised because it is continually being 'retrofitted' to accommodate newly emerging requirements. Any computer program developed in this manner is highly prone to error.

Testing

Testing is the activity of carrying out focused experiments with the simulation program with a view towards uncovering specific properties. For the most part, testing is concerned with establishing credibility and consequently considerable care needs to be taken in developing, and documenting, the test cases. Testing activity that is flawed or inadequate can have the unfortunate consequence of undermining confidence in the results flowing from the simulation project.

Testing can have a variety of objectives. For example, regression testing is undertaken when changes have taken place in the simulation program. In such circumstances it is necessary to confirm not only that any anticipated behavioural properties of the simulation model do actually occur but also that improper side-effects have not been introduced. This implies carrying out some suite of carefully designed experiments before and after the modifications.

Another testing perspective is concerned with trying to acquire some insight into the boundaries of usefulness of the simulation program relative to the goals of the project. This can be undertaken using a process called stress testing whereby the model is subjected to extreme conditions. For example, in the context of a manufacturing process the effect of extremely high machine failure rates could be explored or alternately, in a communication system context the impact of data rates that cause severe congestion could be explored. The intent of such testing is to create circumstances that provide insight into the limits of the model's plausibility in terms of an adequate representation of the SUI's behaviour.

Experiment Design

We use the phrase 'experiment design' to refer to a whole range of planning activities that focus on the manner in which the simulation program will be used to achieve the project goals. The success of the project is very much dependent on the care taken in this planning stage. Poor experiment design can seriously compromise the conclusions of the study and in the extreme case may even cast suspicion on the reliability of the conclusions.

Some examples of typical matters of concern are:

- What data need to be collected (with due regard to the analysis requirements and tools)?
- How will initialisation and transient effects be handled?
- Are there particular operational scenarios that are especially well suited to providing the desired insight when the relative merits of a number of specified system design alternatives need to be examined?
- Is there a useful role for special graphics and/or animation and if so, what should be displayed?

Frequently the project goals include an optimisation requirement and the difficulty of this task is often underestimated. Care is required both in the formulation of the optimisation problem itself and in the formulation of a procedure for its solution. Problem formulation usually corresponds to the specification of a scalar-valued criterion function whose value needs to be either minimised or maximised. The nature of this function is often clearly apparent from the goals of the project. Care must be taken to avoid attempting to embed in the project goals several such functions whose simultaneous optimisation is in conflict. The parameters available for adjustment in the search for an extreme value for the criterion function need to be identified. Frequently there are constraints in the admissible values for these parameters and such constraints must be clearly identified. Alternately there may be prescribed requirements on certain features of the model's behaviour that have to be incorporated.

The identification of an appropriate procedure for solving the problem then has to be carefully considered. Numerous techniques can be found in the classical optimisation literature and these are generally directly applicable within the modelling and simulation context provided stochastic effects are not present. Their applicability is, however, seriously undermined when stochastic behaviour is an integral part of the criterion function as is often the case in the modelling and simulation context. Although true optimality may be infeasible, sometimes a suboptimal solution can be a reasonable expectation. In such circumstances it is highly desirable to have available a means for estimating the likely deviation from optimality of the accepted solution. In practice, when stochastic effects are present, the search for optimality may simply have to be abandoned and replaced with the relatively straightforward task of selecting the best alternative from among a finite collection of options.

Presentation/Interpretation of Results

Often the person/organisation that has commissioned the modelling and simulation project remains remote from the development stage and periodic presentations are normally necessary. Unless explicitly requested, great detail about the simulation model's features should not dominate these presentations. The focus must be on results obtained from the simulation experiments that relate directly to the goals of the project. This is not to suggest that additional information that appears pertinent should not be presented but its possibly tangential relevance should be clearly pointed out. Wide availability of increasingly more sophisticated computer graphics and animation tools can be creatively incorporated but the visual effects they provide should serve only to complement, but not replace, comprehensive quantitative analysis.

2.6 The Dynamic Model Landscape

Models of dynamic systems can be characterised by a number of features. For the most part these are inherited from the underlying system (i.e., the SUI) that the model represents. We examine some of these characterising features in the discussion that follows.

Deterministic and Stochastic

The system context for a large class of modelling and simulation projects includes random elements. Models that emerge from such contexts are called stochastic models which are very distinct from deterministic models that have no random aspects. Values taken from any particular experiment with a stochastic model must be regarded as observations of some collection of random variables. The need to deal with random aspects of stochastic models (and the underlying SUI) has a very substantial impact on essentially all facets of both the modelling and the simulation phases of a project. A whole range of related considerations must be carefully handled in order to ensure that correct conclusions are drawn from the study. A few of these are listed below.

- Only aggregated results are meaningful, hence many simulation experiments need to be carried out.
- Dealing with the start-up issue may require ignoring data from the initial portion of a simulation experiment.
- The need to formulate a collection of data models that capture the various random phenomena that are embedded in the model.

Discrete and Continuous

In models for discrete event dynamic systems (i.e., DEDS models) state changes occur at particular points in time whose values are not known a priori. As a direct consequence, (simulated) time advances in discrete 'jumps' that have unequal length.

In contrast, with models that emerge from the domain of continuous time dynamic systems (i.e., CTDS models), state changes occur continuously (at least in principle) as time advances in a continuous fashion over the length of the observation interval. It must, however, be stressed that this is an idealised perspective that ignores the realities introduced by the computational process. It is simply infeasible for any practical procedure to actually yield data at every value of time within the continuum of the observation interval. Thus, from the perspective of the observer, state changes do apparently occur with discrete 'jumps' as the solution unfolds over the observation interval.

Our presentation in this textbook may give the erroneous impression that models neatly separate into the two broad categories that we refer to as DEDS models and CTDS models. This is an oversimplification. There is, in fact a third category of models that are usually called combined models where the name reflects the combination of elements from both the discrete and continuous domains. As an illustration consider the parts in a manufacturing plant that move from one workstation to another on the way to assembly into a final product. At these workstations, queues form and the service function provided by the workstation may have random aspects (or may become inoperative at random points in time). Thus the basic elements of a DEDS model are present. At some workstations the operation may involve heating the part to a high temperature in a furnace. This heating operation and the control of it would best fall in the realm of a CTDS model. Hence the overall model that is needed has components from the two basic domains.

Work on the development of modelling formalisms and tools for handling this third category of combined models has a long history. The interested reader wishing to explore this topic in more detail will find relevant discussions in Cellier [2.5], Ören [2.11], and Praehofer [2.14].

Linear and Nonlinear

The properties of linearity and nonlinearity of systems are basic considerations in many areas of analysis, e.g., mathematics and system theory. However, because the experiments that are inherent. within the modelling and simulation context are always assumed to be carried out by a computer the information that is 'delivered' evolves entirely from numerical computation. Hence the inherent simplifications to the analysis process that are introduced by the linearity property have no particular consequence in the modelling and simulation realm. This absence of any need to distinguish between linear and nonlinear systems and models is most certainly one of the noteworthy features of the modelling and simulation approach to problem solving.

2.7 Exercises and Projects

2.1 Technical papers in the modelling and simulation applications literature are sometimes lacking in the clarity with which they deal with such fundamentals as

a) The goals of the modelling and simulation study
b) Outline of the conceptual model
c) Identification of input and output variables
d) Model validation efforts
e) Evaluation of success in achieving a solution to the problem that motivated the study

Choose two papers in some particular application area of modelling and simulation and compare the effectiveness with which the authors have addressed the items listed above. Some application areas that could be considered are:

a) Network management and control
b) Ecological and environmental systems
c) Biomedicine and biomechanics
d) Power generation and distribution
e) Automated manufacturing
f) Robotics and autonomous systems
g) Transportation and traffic
h) New product development

Technical papers and/or pointers to technical papers in these areas can be found at Web sites such as www.scs.org and www.informs-cs.org/wscpapers.html.

2.8 References

2.1. Balci, O.,(1994), Validation, verification, and testing techniques throughout the life cycle of a simulation study, *Annals of Operations Research*, **53**: 121–173.

2.2. Balci, O., (2001), A methodology for certification of modeling and simulation applications, *ACM Transactions on Modeling and Computer Simulation*, **11**: 352–377.

2.3. Birta, L.G. and Ozmizrak, N.F., (1996), A knowledge-based approach for the validation of simulation models: The foundation, *ACM Transactions on Modeling and Computer Simulation*, **6**: 67–98.

2.4. Boehm, B.W., (1979), Software engineering: R&D trends and defence needs, in: P. Wegner (Ed.), *Research Directions in Software Technology*, MIT Press, Cambridge, MA.

2.5. Cellier, F.E., (1986), Combined discrete/continuous system simulation – aplication, techniques and tools, in *Proceedings of the 1986 Winter Simulation Conference*.

2.6. Department of Defense (DoD) Recommended Practices Guide (RPG) for Modeling and Simulation VV&A, Millennium Edition (available at http://vva.dmso.mil).

2.7. General Accounting Office, (1976), Report to the Congress: Ways to improve management of federally funded computerized models, report LCD-75-111, U.S. General Accounting Office, Washington, DC.

2.8. General Accounting Office, (1979), Guidelines for model evaluation, report PAD-79-17, U.S. General Accounting Office, Washington, DC.

2.9. General Accounting Office, (1987), DOD simulations: Improved assessment procedures would increase the credibility of results, report GAO/PEMD-88-3, U.S. General Accounting Office, Washington, DC.

2.10. Ören, T.I., (1981), Concepts and criteria to access acceptability of simulation studies, *Communications of the ACM*, **24**: 180–189.

2.11. Ören, T.I., (1971), GEST: General system theory implementor, a combined digital simulation language, PhD Dissertation, University of Arizona, Tucson.

2.12. Neelamkavil, F., (1987), *Computer Simulation and Modeling*, John Wiley and Sons, Chichester, UK.

2.13. Pace, D.K., (2003), Verification, validation and accreditation of simulation models, in M.S. Obaidat and G.I. Papadimitriou (Eds.), *Applied System Simulation: Methodologies and Applications*, Kluwer Academic, Boston.

2.14. Praehofer, H., (1991), System theoretic formalisms for combined discrete continuous system simulation, *International Journal of General Systems*, **19**: 219–240.

2.15. Robinson, S., (2006), Issues in conceptual modelling for simulation: Setting a research agenda, in *Proceedings of 2006 Operations Research Society Simulation Workshop*, March, Lexington, England.
2.16. Shannon, R.E., (1975), *Systems Simulation: The Art and Science*, Prentice-Hall, Englewood Cliffs, NJ.

PART 2
DEDS Modelling and Simulation

In the second part of this book we examine the modelling and simulation process within the discrete event dynamic systems (DEDS) domain. The presentation is, for the most part, guided by the general process presented in Figure 2.4.

We have previously emphasised (Section 2.3.1), that the project description should provide a clear statement of the project goals. From these it should be possible to identify those behavioural features of the system under investigation (SUI) that have relevance to the model development process. One behavioural feature that is specific to DEDS is the central role played by random phenomena. The interarrival times between messages entering a communication network and the time to service customers at the checkout counter of a grocery store are examples of such phenomena.

Data modelling is an essential subtask of conceptual modelling phase in the DEDS domain. It is concerned, in part, with correctly representing the features of the SUI's environment that have an impact on its behaviour and this can be a demanding and time-consuming task. The project goals guide the identification of the data models that are required. The data modelling task is considerably facilitated when the SUI currently exists and is accessible because then data collection is possible. When the SUI does not yet exist (or indeed, may never actually 'exist') data modelling becomes a very uncertain undertaking and essentially depends on insight and intuition.

There are three main world views or frameworks for building a DEDS simulation model: the activity scanning world view, the event scheduling world view, and the process-oriented world view. With the superposition of a variety of operational features (e.g., generation and management of random variates for handling random phenomena, management of a predefined time-advance mechanism) the simulation model evolves into a simulation program.

Chapters 3 and 4 deal with conceptual modelling. Chapter 3 provides an overview of some key aspects of random behaviour and discusses data modelling. These data models are then integrated into a conceptual modelling framework that captures the relevant structural and behavioural features of the SUI; this is presented in Chapter 4. The framework presented in Chapter 4 is based on the activity scanning world view. Chapter 5 shows how a conceptual model that has been formulated in the framework described in Chapter 4 can be transformed into either an event scheduling simulation model or a process-oriented simulation model. The programming environments used in these two cases are Java and GPSS, respectively.

The project goals have a direct impact on the way in which experimentation is carried out with the simulation program. A basic objective of experimentation is to produce values for the performance measures stipulated in the project goals. In Chapter 6, we examine how the experimentation activity has to be organised in order to acquire meaningful values for these performance measures.

Chapter 3 DEDS Stochastic Behaviour and Data Modelling

3.1 The Stochastic Nature of DEDS

This section explores some fundamental aspects of the random (stochastic) nature of DEDS and introduces several assumptions that are typically made about it.

Consider a simple view of the operation of a delicatessen counter which has one server. Customers arrive at the counter, wait in a queue until the server is available, and then select and purchase items at the counter.

Two important random phenomena drive this SUI. The first is the arrival of customers which is usually expressed in terms of interarrival times, that is, the time between successive customer arrivals. The second is the time it takes to select and purchase items at the counter which is referred to as the service time. Both of these phenomena can be represented by discrete stochastic processes (an overview of stochastic processes is provided in Section A1.8 of Annex 1). The arrival process, for example, can be represented by $X = (X_1, X_2, X_3, \ldots, X_n)$, where X_j is the time between the arrival of the $(j-1)$th customer and the jth customer, $j = 1, 2, \ldots, n$ (X_1 is measured with respect to the left boundary of the observation interval). A convenient assumption here is that the observation interval corresponds to the boundaries of a business day and that n customers are processed over the course of a day. The service time of each of the n customers can likewise be represented by a stochastic process; that is, $Y = (Y_1, Y_2, Y_3, \ldots, Y_n)$. In our simple model, both of these random phenomena are entirely independent of any other phenomena or interactions in the system. Throughout our discussions we refer to stochastic processes with this independent attribute as *autonomous stochastic processes*.

As an alternate example, consider the status of a machine in a manufacturing system which we can represent with the continuous stochastic process *Status(t)*. A simple approach is to assign *Status* a value of 1 when the machine is operational and a value of 0 when it is not; for example, it has malfunctioned and is being repaired (see Figure 3.1). Continuous stochastic processes in a DEDS model are piecewise-constant time functions because their values change only at discrete points in time, as is the case with *Status(t)*. Notice that the stochastic process, *Status(t)*, could be viewed as having two independent component parts; namely a discrete stochastic processes for the 'uptime' durations, $U = (U_1, U_2, U_3, \ldots, U_n)$ and another for 'downtime' durations, $D = (D_1, D_2, D_3, \ldots, D_n)$. *Status(t)* could in fact, be constructed from these two component data models.

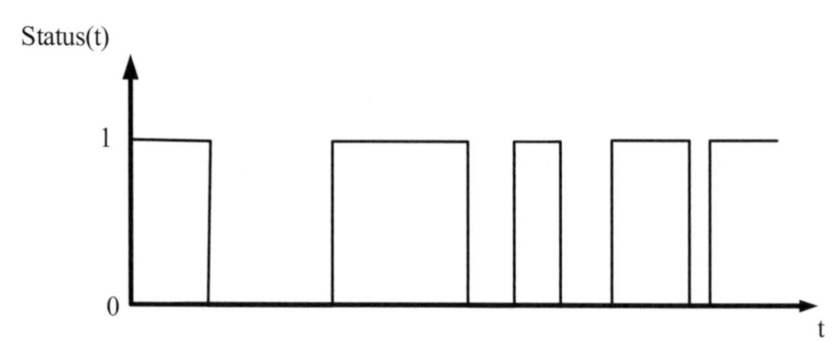

FIGURE 3.1. Continuous stochastic process representing machine status.

In our delicatessen example, the autonomous stochastic processes X and Y give rise to dependent stochastic processes via the inherent behaviour properties of the model. Consider, for example, the waiting times of customers in the deli queue and the length of the queue. The waiting times can be represented as a discrete stochastic process: $W = (W_1, W_2, W_3, \ldots, W_n)$, where W_j is the waiting time of the jth customer. The length of the queue, *L(t)*, is likewise a dependent random phenomenon which is, furthermore, piecewise constant.

Properties of the dependent random phenomena are typically of interest in resolving the project goals and these properties are usually called performance measures. Often interest focuses on the change in value of performance measures that results from some predetermined change in system operation or system structure. For example, the service time could be reduced by restricting customer choices to some collection of prepackaged items. A change in the structure of the system would result if an additional person were hired to serve behind the counter and a two-queue service protocol was established. The goal of a modelling and simulation project could then be to evaluate the effect of such changes on customer waiting time.

When stochastic processes are used to represent autonomous random phenomena simplifying assumptions are typically incorporated. For example, a common assumption is that customer interarrival times can be represented with a *homogeneous* stochastic process, that is, the sequence of random variables that constitute the stochastic process are independent and identically distributed (IID).

Unfortunately there are many cases where such assumptions are simply not realistic. Consider our deli counter example. There are 'busy periods' over the course of a business day, during which customer arrivals occur more frequently. This implies that the mean of the customer interarrival time distribution will be shorter during these busy periods. It is reasonable to assume that dependent stochastic processes such as waiting times will be affected and will also be nonstationary. When appropriate, this issue can be circumvented by redefining the observation interval so that the study is restricted, for example, to the busy period. Then the validity of a homogeneous stochastic process assumption for the customer interarrival time can be reasonably assured.

Even in cases where the autonomous stochastic processes within a DEDS model are stationary over the observation interval, dependent stochastic processes can still exhibit transient behaviour. These transient effects are a consequence of initial conditions whose impact needs to dissipate before stationary behaviour evolves. Consider, for example, waiting times when the deli of our example first opens in the morning. The first customer will experience no waiting time and receive service immediately. Subsequent customers will likewise experience short waiting times. As time progresses, more customers will enter the queue and waiting times could start to lengthen but in any event, the exceptional circumstances immediately following the opening will disappear.

This behaviour will occur even when we assume that the customer interarrival times are represented by a homogeneous stochastic process. If the mean of the interarrival times changes to a new value at the start of the busy period, the waiting times will again go through a transient period prior to eventually reaching steady-state behaviour. Note that it can be reasonably assumed that the waiting time stochastic process is positively correlated.

With very few exceptions, the variables in a DEDS model can be regarded as stochastic processes: some as autonomous stochastic processes and others as dependent stochastic processes. The values acquired by a variable that is regarded as a dependent stochastic process are an outgrowth of the autonomous stochastic processes coupled with the model's behaviour specifications; hence consideration of data models in this circumstance is not meaningful. Such values, however, are generally of considerable importance because they provide the sample set and/or trajectory set output that is required for the achievement of the project goals. These issues are examined in Section 3.2 below.

A data model is required for each autonomous stochastic process that is identified in a conceptual model. Such models provide the basis for generating the values that are associated with these processes. Such a process might, for example, represent the interarrival times of an input entity stream as introduced in Chapter 2.

Creating a data model consists of determining appropriate probability distributions for the constituent random variables of the stochastic process. Data models can be very complex. Such models could be required to represent nonstationary stochastic processes or even multivariate stochastic processes where one stochastic process is correlated to another. However, our treatment of data modelling in this textbook is restricted in scope. In particular, we limit our considerations to autonomous stochastic processes that are piecewise homogeneous (see Section A1.8 of Annex 1).

The key feature of a homogeneous stochastic process is that the data modelling task reduces to the identification of a single underlying distribution function (because all constituent random variables have the same distribution). The situation is somewhat more demanding in the general piecewise homogeneous case (with m segments) inasmuch as there is a distribution required for each of the m segments.

A useful overview of data modelling is provided by Biller and Nelson [3.3]. More comprehensive discussions can be found in a variety of references such as: Banks et al. [3.2], Law and Kelton [3.13], and Leemis and Park [3.14].

3.2 DEDS Modelling and Simulation Studies

The goals of a modelling and simulation project implicitly define one of two possible types of study. The differences between them are easy to appreciate. However, as becomes apparent in later discussions (in particular, in Chapter 6) these differences have a significant impact on the nature of the experimentation procedure that needs to be carried out and in one case it requires considerably more effort. The main differences arise from two interdependent features; one relates to specifications on the right-hand boundary of the observation interval and the other relates to possible constraints imposed on the acquired data. We refer to the two alternatives as bounded horizon studies and steady-state studies. They[1] are summarised below:

[1] These two types of study are often referred to as 'terminating simulations' and 'nonterminating simulations', respectively

Bounded Horizon Study:

- The right-hand boundary of the observation interval is specified in the problem statement either explicitly by a given value of time or implicitly by some combination of values acquired by the model's state variables.
- There are no restrictions on the properties of the dependent stochastic processes that are of interest. Often transient behaviour dominates.

Steady-State Study:

- The right-hand boundary of the observation interval is not provided in the problem statement. Its value emerges in an indirect fashion because the observation interval extends to a point where the times series of acquired data is long enough to reflect steady-state behaviour.
- Steady-state behaviour of the dependent stochastic processes that are of interest is essential. In other words, the focus is on behaviour patterns in the absence of transient effects.

3.3 Data Modelling

Data models (expressed in terms of theoretical probability distributions) are provided for the various autonomous stochastic processes that appear in the DEDS examples presented in this textbook. This might give the erroneous impression that defining such data models is simple and straightforward. In reality much time and effort is required for carrying out the data modelling task that gives rise to these data models. Furthermore, it must be strongly emphasised that improper data models can destroy the value of the results that flow from a simulation study. Data models play the role of input data to a computer program and a long established principle in software engineering is that 'garbage in equals garbage out'!

When the SUI exists it may be possible to obtain insights about its various autonomous stochastic processes by observing the existing system. Data can be gathered and analysed to obtain information necessary for the formulation of suitable data models. In other cases, when such data are not available (e.g., the SUI does not exist or the collection of data is impossible or too costly), data models may have to be constructed on the basis of the insight provided by domain specialists, that is, 'educated guesses'.

In this section we first consider the case when data can be obtained from the SUI and we provide an overview of steps required to develop, from the collected data, the underlying distributions that serve as data models. Some general guidelines are given for formulating data models when no data exist.

Data models are integrated, in a modular fashion, into the conceptual modelling framework that is presented in Chapter 4. Data modelling is usually carried out in parallel with the conceptual modelling task. Both of these modelling exercises can give rise to the need for refinements to the problem description and/or project goals.

3.3.1 Defining Data Models Using Collected Data

Our introduction to data modelling is restricted to modelling autonomous stochastic processes that are (piecewise) homogeneous. The objective is to formulate either an appropriate theoretical or empirical distribution derived from the collected data.

Collected data can be used to formulate a data model that is specified in terms of a cumulative distribution function that is called an empirical CDF. Section 3.3.4 discusses this approach. When the empirical distribution is continuous, an inherent disadvantage of the approach is that it will not generate values outside the limits of the observed values. It can nevertheless yield values other than those that have been observed. An alternate (and generally preferred) approach is to use statistical techniques to fit a theoretical distribution to the collected data. A theoretical distribution provides a number of advantages; for example, it smooths out irregularities that could arise with the empirical alternative and it can yield values outside the boundaries of the observed values. Theoretical distributions always have embedded parameters (see Sections A1.3.6 and A1.4.4 of Annex 1) which provide a simple means for adjusting the distribution to best fit the collected data.

Our first task is to determine if the collected data do indeed belong to a homogeneous stochastic process. This requires two tests: one to determine if the stochastic process is identically distributed and a second to determine if the constituent random variables are independent. A number of analysis techniques exist for testing for these properties and a few are presented in Section 3.3.2. The final task is to fit a theoretical distribution to the collected data. Software is available for analysing collected data and fitting theoretical distributions. Such software is available in standalone form, for example, ExpertFit [3.12] and Stat::Fit [3.7]. This functionality is often integrated directly into simulation packages; e.g., Arena [3.9] or ProModel [3.7] (which includes Stat::Fit).

3.3.2 Do the Collected Data Belong to a Homogeneous Stochastic Process?

This section presents two techniques for evaluating independence of collected data and a technique for evaluating stationarity.

3.3.2.1 Testing for Independence

Two graphical methods for evaluating independence are presented here: scatter plots and autocorrelation plots. In both cases the objective is to investigate possible dependencies among the values in a times series obtained as an observation of a stochastic process (see Section A1.8 of Annex 1). More specifically, we assume that our collected data are the time series $x = (x_1, x_2, x_3, \ldots, x_n)$ which is an observation of a stochastic process, $X = (X_1, X_2, \ldots, X_n)$.

A scatter plot is a display of the points $P_i = (x_i, x_{i+1})$, $i = 1, 2, \ldots, (n-1)$. If little or no dependence exists, the points should be scattered in a random fashion. If, on the other hand, a trend line becomes apparent then dependence does exist. For positively correlated data, the line will have a positive slope; that is, both coordinates of the points P_i will be either large or small. If data are negatively correlated the trend line will have a negative slope; that is, a small value of x_i will be associated with a large value of x_{i+1} and vice versa.

We illustrate the method with two separate time series. The first consists of 300 data values generated from a gamma distribution (with $\alpha = 2$, $\lambda = 1/3$)[2]. The second has 365 values representing the daily maximum temperatures in Ottawa, Ontario, Canada[3] between May 20, 2005 and May 20, 2006.

Figure 3.2 shows the scatter plot for the first case. Clearly there is no apparent trend and consequently independence can be assumed. The scatter plot shown in Figure 3.3 for the temperature data shows a trend line with a positive slope. The implied positive correlation is to be expected inasmuch as there is a great likelihood that the temperature on successive days will be similar.

---

[2] Generated using the Microsoft ® Office Excel 2003 Application.

[3] Source: http://ottawa.weatherstats.ca.

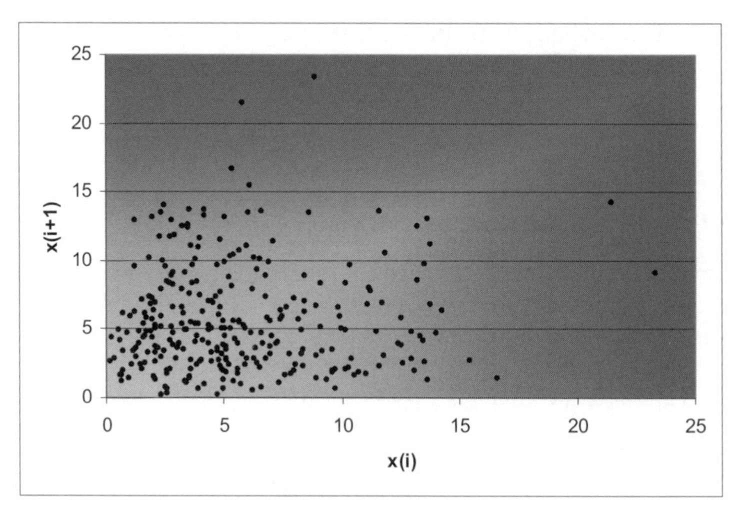

FIGURE 3.2. Scatter plot showing uncorrelated data.

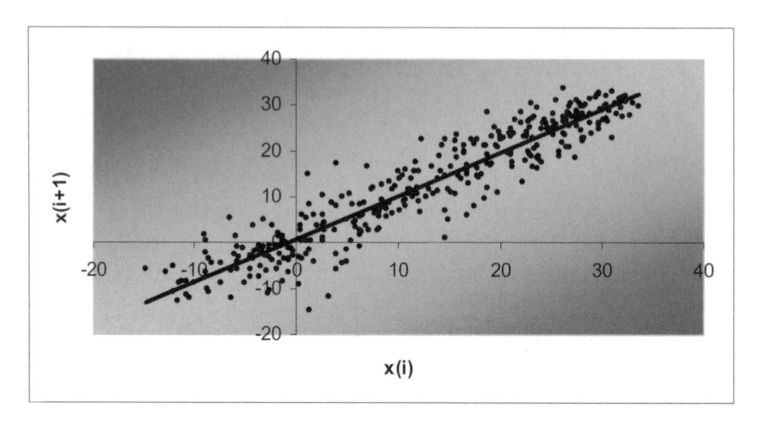

FIGURE 3.3. Scatter plot showing correlated data.

A scatter plot is a presentation of data values that are immediately adjacent, that is, that have a lag of 1. An autocorrelation plot on the other hand is more comprehensive because it evaluates possible dependence for a range of lag values. An autocorrelation plot is a graph of the sample autocorrelation $\hat{\rho}(k)$ for a range of lag values k, where:

$$\hat{\rho}(k) = \frac{\sum_{i=1}^{n-k}(x_i - \bar{x}(n))(x_{i+k} - \bar{x}(n))}{(n-k)s^2(n)} .$$

Here $\bar{x}(n)$ and $s^2(n)$ are estimates of the sample mean and sample variance respectively for the time series (some elaboration can be found in Section A1.8 of Annex 1).

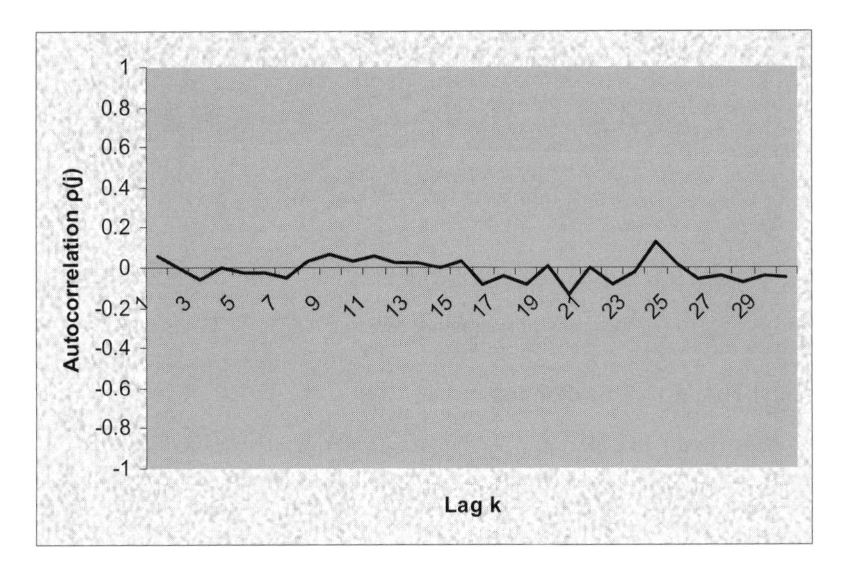

FIGURE 3.4. Autocorrelation plot showing uncorrelated data.

Figure 3.4 shows the autocorrelation plots for the time series obtained from the gamma distribution. The graph shows that the sample autocorrelation is low for all lag values which reinforces the earlier conclusion that the data are independent. For the temperature time series the autocorrelation plot in Figure 3.5 shows high values for the sample autocorrelation for all lag values between 1 and 30, indicating a high level of correlation between temperatures over the first month.

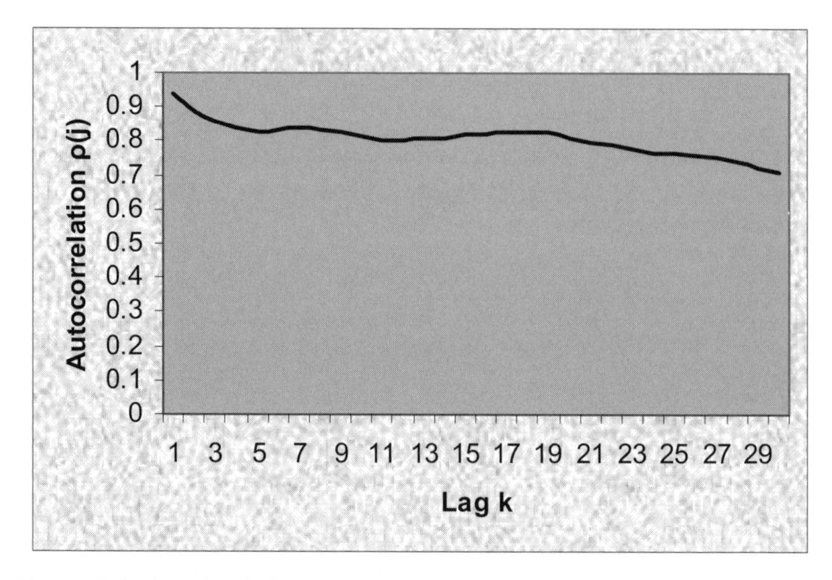

FIGURE 3.5. Autocorrelation plot showing correlated data.

3.3.2.2 Testing for Stationarity

We begin with the assumption that a collection of values $\{\hat{x}_i: i = 1, 2, \ldots, m\}$ has been acquired for the stochastic process X. Each $\hat{x}_i$ is an $n(i)$-tuple of values obtained for the constituent random variables within X; that is, $\hat{x}_i = (x_{i,1}, x_{i,2}, \ldots, x_{i,n(i)})$. The process X could, for example, represent the interarrival times of customers at the deli counter introduced earlier. The $n(i)$-tuple (or time series) $\hat{x}_i$, could be interpreted as an observation of X on the ith day of an m-day collection period. Testing for stationarity can be a relatively elaborate process but as a minimum it requires the assessment of the degree to which average values within the collected data remain invariant over time.

We outline below a graphical method that provides insight into the variation over time of the average value of the collected data hence an approach for carrying out a fundamental test for stationarity. Assume that the collected data extend over a time interval of length T. The procedure begins by dividing T into a set of L time cells of length Δt. Recall that each data value (say $x_{i,j}$) is necessarily time indexed (either explicitly or implicitly) and consequently falls into one of the L time cells. The $n(i)$ values in the time series x_i can then be separated into disjoint subsets according to the time cell to which each value belongs. The average value of the data in each cell is computed for each of the m time series and then a composite average over the m time series is determined and plotted on a

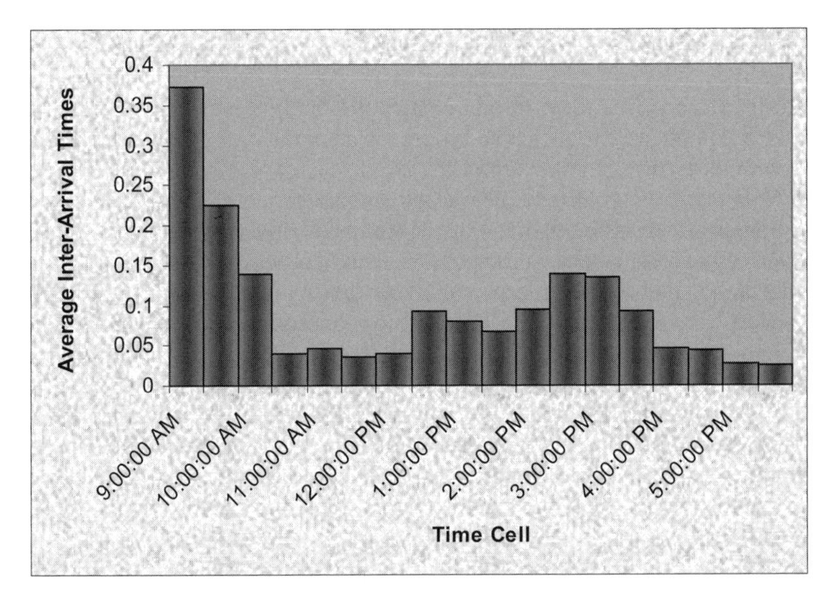

FIGURE 3.6. Nonstationary interarrival times.

time axis that is similarly divided into L cells. The resulting graph therefore displays the time behaviour of average values within the cells.

Within the context of our deli example, the recorded data could represent observations of the customer interarrival times. For day i, we denote by $\overline{a}_{ci}$ the average interarrival time in cell c. Then, for each time cell c, we compute an overall average interarrival time $\overline{a}_c$ using data from all m days; that is,

$$\overline{a}_c = \frac{1}{m} \sum_{i=1}^{m} \overline{a}_{ci} .$$

The value $\overline{a}_c$ provides an estimate for the mean of the underlying distribution of those interarrival time random variables whose time index falls in the time cell c.

A plot of $\overline{a}_c$ versus time cell c provides a visual aid for evaluating how the mean of the distributions vary over time. Figure 3.6 shows such a plot computed from three days of observed interarrival times in the deli between 9 AM and 6 PM. An interval Δt of 30 minutes was used. The plot clearly shows that the mean does vary because smaller averages occur around noon and at the end of the day, that is, during rush-hour periods.

3.3.3 Fitting a Distribution to Data

Fitting a theoretical distribution that matches time series data obtained from a homogeneous stochastic process is a trial-and-error procedure. The procedure usually begins with a histogram developed from the collection of N values belonging to some particular time series. If the objective is a continuous distribution then the histogram provides a representation whose shape approximates the underlying probability density function. On the other hand, if the objective is a discrete distribution, then the histogram provides a representation whose shape approximates the underlying probability mass function. A plot of the associated cumulative distribution function can also be helpful for specifying empirical distributions.

The general shape of the histogram serves to suggest possible theoretical distribution candidates. Parameters that are associated with theoretical distributions then need to be estimated. Goodness-of-fit tests are generally used to determine how well the parameterised distribution candidates fit the data. A selection is made based on the results from this analysis.

As an example, consider a time series obtained by observing the 'group sizes' that enter a restaurant over the course of a particular business day. The distribution of interest here is discrete and the histogram shows the number of occurrences of each of the group sizes as contained in the available time series data. The histogram shown in Figure 3.7 illustrates a possible outcome. A group size of 4 clearly occurs most frequently. The associated cumulative distribution is also provided in Figure 3.7; and it shows, for example, that just over 70% of the group sizes are equal to or less than 4.

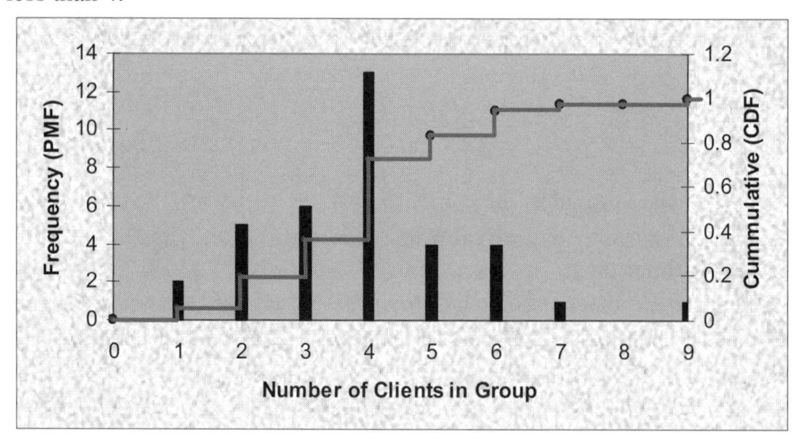

FIGURE 3.7. Histogram for discrete valued data.

An approximation process is required to handle the continuous case, that is, the case where observed data values can assume any value in a prescribed interval of the real line. In this circumstance, a histogram is constructed by dividing the interval into subintervals called *bins*. The number of values that fall into each bin is counted and plotted as the frequency for that bin. The 'smoothness' of the graph that results is very much dependent on the bin size. If the bin size is too small, the resulting plot can be ragged. If the bin size is too large, the graph's value for inferring a candidate distribution can be compromised. Banks et al. [3.2] suggest choosing the number of bins to be $\sqrt{n}$, where n is the number of values in the time series observation. On the other hand, Stat::fit [3.7] recommends $\sqrt[3]{2n}$ for the number of bins.

Figure 3.8 shows a histogram created using 100 data values (n = 100) generated from an exponential distribution using 22 bins. It illustrates how a ragged plot can occur when the bin size is too small. Figure 3.9 shows the histogram using 10 bins (value recommended by Banks et al.). Figure 3.10 shows the histogram using 6 bins (recommended value used in Stat::Fit).

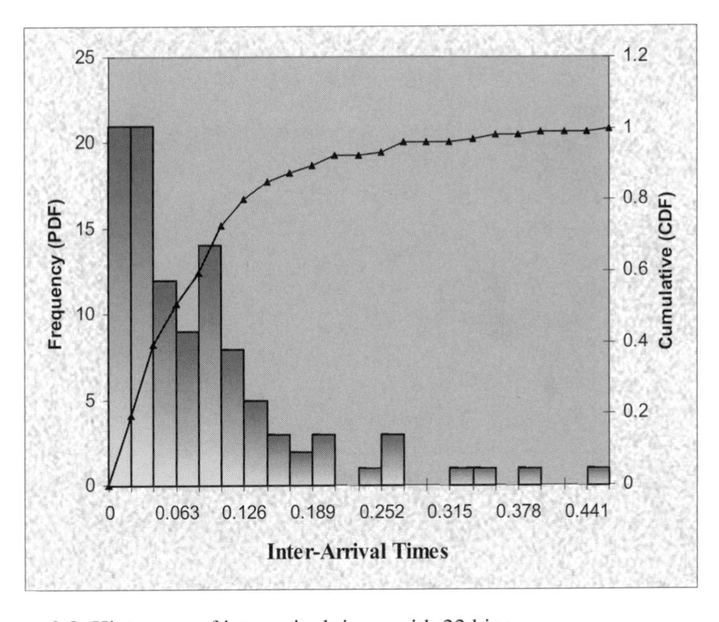

FIGURE 3.8. Histogram of interarrival times with 22 bins.

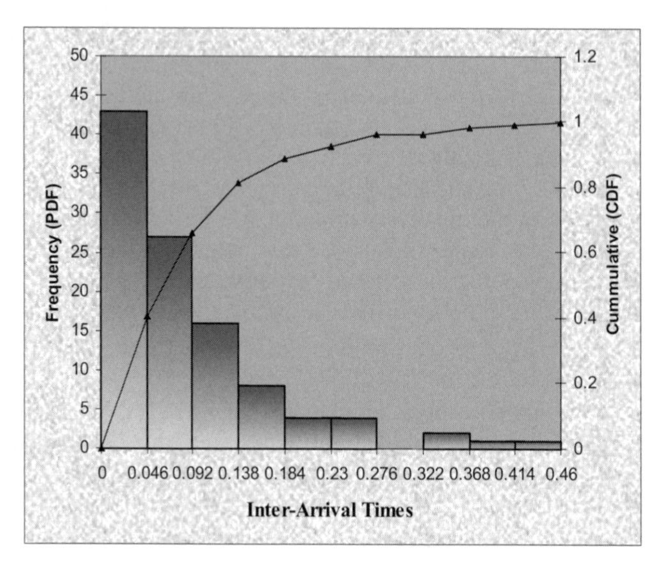

FIGURE 3.9. Histogram of interarrival times with 10 bins.

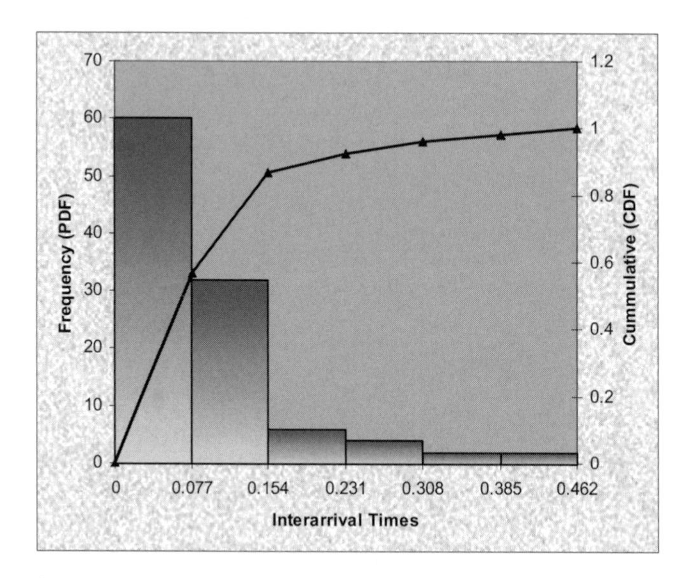

FIGURE 3.10. Histogram of interarrival times with 6 bins.

Once the histogram has been created, the shape of the histogram is used to select one or more theoretical distributions as possible candidates for the data model. Estimated values for the parameters of each of these candidates must be obtained and a number of estimation methods developed for this purpose are available. In the discussion that follows we briefly examine the category of estimators called maximum likelihood estimators (further details can be found in [3.13]).

The sample mean $\bar{x}(n)$ and the sample variance $s^2(n)$ of the time series observation play a key role in the maximum likelihood parameter estimation procedure. Table 3.1 shows how maximum likelihood estimators for the parameters of several distributions are computed. Estimators for other distributions can be found in [3.2] and [3.13].

Once parameters for a candidate distribution have been estimated, a goodness-of-fit test needs to be used to determine how well the selected theoretical distribution (with assigned parameters) fits the collected data. Various such tests are available and an overview can be found in Law and Kelton [3.13] or Banks et al. [3.2]. Among the options is the chi-square test[4] which we summarise below.

TABLE 3.1. Maximum likelihood estimators.

Distribution	**Parameters**	**Estimators**
Exponential	λ	$\hat{\lambda} = \frac{1}{\bar{x}(n)}$
Normal	μ, σ^2	$\hat{\mu} = \bar{x}(n)$ $\hat{\sigma}^2 = s^2(n)$
Gamma	α, λ	Compute $T = \left[\ln(\bar{x}(n)) - \frac{1}{n} \sum_{i=1}^{n} \ln(x_i) \right]^{-1}$ and find $\hat{\alpha}$ from Table 3.4 using linear interpolation. $\hat{\lambda} = \frac{\hat{\alpha}}{\bar{x}(n)}$

Suppose D_c is the parameterised candidate distribution to be tested. The objective is to determine if there is a basis for rejecting D_c because it provides an inadequate match to the collected data. The first step in the test is to determine a value for an adequacy measure A_m that essentially compares the frequencies in the histogram formulated from the collected data to expected frequency values provided by D_c.The definition of A_m is:

[4] The test is shown for the continuous distribution. For discrete distributions, each value in the distribution corresponds to a class interval and $p_i = P(X = x_i)$. Class intervals are combined when necessary to meet the minimum-expected-interval-frequency requirement (E_i is less than 5).

$$A_m = \sum_{i=1}^{k} \frac{(O_i - E_i)^2}{E_i} , \tag{3.1}$$

where:

- k is the number of class intervals which initially is equal to the number of bins in the data histogram. In other words, a class interval is initially associated with each bin.
- E_i is the expected frequency for the ith class interval based on D_c. It is defined as $E_i = np_i$, where p_i is the probability that a value falls into the interval; that is, $p_i = \mathrm{P}[x_i < x \le x_{i+1}]$ where x_i and x_{i+1} are the boundaries of the ith class interval and n is the number of values in the available time series data. The probability p_i can be computed using the cumulative density function $F(x)$ of D_c; that is,

  $$p_i = F(x_{i+1}) - F(x_i).$$

- When E_i is less than 5, the interval is combined with an adjacent one (thus the new interval contains multiple adjacent bins) and the value of k is appropriately reduced. The E_i for the new enlarged interval is then re-evaluated. This step is repeated until all E_i values are greater than 5.
- The value O_i corresponds to the frequency observed in the histogram bin that corresponds to the ith class interval. For a class interval that contains more than one bin, the frequencies from the corresponding histogram bins are summed to provide the value for O_i.

Clearly the intent is to have A_m as small as possible. The practical question, however, is whether A_m is sufficiently small. The decision is made by comparing the value of A_m with a 'critical value' χ^*, obtained from the chi-square distribution table as given in Table 3.5. The selection of χ^* depends on two parameters; the first is the degrees of freedom, $\nu = k - g - 1$, where g is the number of parameters embedded in D_c and the second is α (the level of significance) for which 0.05 is a commonly used value. If $A_m > \chi^*$, then A_m is not sufficiently small and D_c should be rejected as a distribution option.

The procedure outlined above for fitting a theoretical distribution to a time series observation is illustrated with the gamma distributed data used in Section 3.3.2.1. Seventeen (the approximate square root of 300) bins were used to generate the histogram shown in Figure 3.11. Two theoretical distributions are selected for consideration as candidates, first an exponential distribution and then a gamma distribution.

Values for the sample mean and sample variance of the data are $\bar{x}(300) = 5.58$ and $s^2(300) = 14.26$ respectively. The estimator of the exponential distribution's single parameter $\hat{\lambda}$ is equal to 1/5.58 = 0.179. Table 3.2 shows the values for O_i derived from the data histogram and E_i derived from the exponential distribution candidate (with mean 5.58). Notice how the bins 12 to 17 are collapsed into two class intervals. Using the data in Table 3.2 it follows (from Equation (3.1)) that A_m = 37.30. With $\nu = 11$ (k = 13 and g = 1) and $\alpha = 0.05$, the critical value (from Table 3.5) is $\chi^* = 19.68$. Because $A_m > \chi^*$ the exponential distribution candidate is rejected by the test.

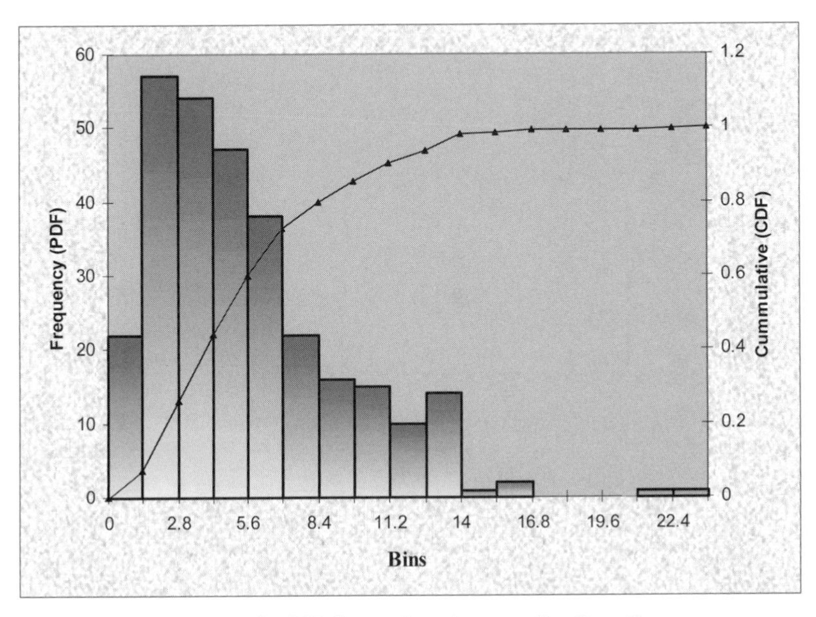

FIGURE 3.11. Histogram for 300 data values (gamma distributed).

From Table 3.1, the estimators for the gamma distribution's two parameters are $\hat{\alpha} = 2.141$ and $\hat{\lambda} = 0.384$ (notice that these do not equal the parameters used in generating the data: namely $\alpha = 2$ and $\lambda = 1/3$). Table 3.3 shows the values for O_i derived from the data histogram and E_i derived from the gamma distribution candidate. In this case bins 10 to 17 are collapsed into two class intervals. From Equation (3.1), it follows that A_m = 10.96. Now $\nu = 8$ ($k = 11$, $g = 2$) and with $\alpha = 0.05$, the critical value χ^* = 15.51. Because $A_m < \chi^*$, the gamma distribution candidate is not rejected by the chi-square test.

TABLE 3.2. Observed and expected frequency data for exponential distribution candidate.

	Class Interval		O_i		E_i	
1	0 - 1.4		22		66.56	
2	1.4 - 2.8		57		51.79	
3	2.8 - 4.2		54		40.30	
4	4.2 - 5.6		47		31.36	
5	5.6 - 7		38		24.40	
6	7 - 8.4		22		18.99	
7	8.4 - 9.8		16		14.78	
8	9.8 - 11.2		15		11.50	
9	11.2 - 12.6		10		8.95	
10	12.6 - 14		14		6.96	
11	14 - 15.4		1		5.42	
12	15.4 - 16.8		2	2	4.22	7.50
	16.8 - 18.2		0		3.28	
13	18.2 - 19.6		0	1	2.55	7.29
	19.6 - 21		0		1.99	
	21 - 22.4		1		1.55	
	22.4 - 23.8		1		1.20	

TABLE 3.3. Observed and expected frequency data for gamma distribution candidate.

	Class Interval	O_i		E_i	
1	0 - 1.4	22		26.11	
2	1.4 - 2.8	57		52.17	
3	2.8 - 4.2	54		53.96	
4	4.2 - 5.6	47		46.17	
5	5.6 - 7	38		36.05	
6	7 - 8.4	22		26.66	
7	8.4 - 9.8	16		19.02	
8	9.8 - 11.2	15		13.23	
9	11.2 - 12.6	10		9.028	
10	12.6 - 14	14	15	6.07	10.10
	14 - 15.4	1		4.034	
11	15.4 - 16.8	2	4	2.655	11.03
	16.8 - 18.2	0		1.733	
	18.2 - 19.6	0		1.124	
	19.6 - 21	0		0.725	
	21 - 22.4	1		0.465	
	22.4 - 23.8	1		0.297	

TABLE 3.4. Estimating the α parameter for the gamma distribution.

T	$\hat{\alpha}$	T	$\hat{\alpha}$	T	$\hat{\alpha}$
0.0	0.0187	2.7	1.4940	10.3	5.3110
0.0	0.0275	2.8	1.5450	10.6	5.4610
0.0	0.0360	2.9	1.5960	10.9	5.6110
0.1	0.0442	3.0	1.6460	11.2	5.7610
0.1	0.0523	3.2	1.7480	11.5	5.9110
0.1	0.0602	3.4	1.8490	11.8	6.0610
0.1	0.0679	3.6	1.9500	12.1	6.2110
0.1	0.0756	3.8	2.0510	12.4	6.3620
0.1	0.0831	4.0	2.1510	12.7	6.5120
0.2	0.1532	4.2	2.2520	13.0	6.6620
0.3	0.2178	4.4	2.3530	13.3	6.8120
0.4	0.2790	4.6	2.4530	13.6	6.9620
0.5	0.3381	4.8	2.5540	13.9	7.1120
0.6	0.3955	5.0	2.6540	14.2	7.2620
0.7	0.4517	5.2	2.7550	14.5	7.4120
0.8	0.5070	5.4	2.8550	14.8	7.5620
0.9	0.5615	5.6	2.9560	15.1	7.7120
1.0	0.6155	5.8	3.0560	15.4	7.8620
1.1	0.6690	6.0	3.1560	15.7	8.0130
1.2	0.7220	6.2	3.2570	16.0	8.1630
1.3	0.7748	6.4	3.3570	16.3	8.3130
1.4	0.8272	6.6	3.4570	16.6	8.4630
1.5	0.8794	6.8	3.5580	16.9	8.6130
1.6	0.9314	7.0	3.6580	17.2	8.7630
1.7	0.9832	7.3	3.8080	17.5	8.9130
1.8	1.0340	7.6	3.9580	17.8	9.0630
1.9	1.0860	7.9	4.1090	18.1	9.2130
2.0	1.1370	8.2	4.2590	18.4	9.3630
2.1	1.1880	8.5	4.4090	18.7	9.5130
2.2	1.2400	8.8	4.5600	19.0	9.6630
2.3	1.2910	9.1	4.7100	19.3	9.8130
2.4	1.3420	9.4	4.8600	19.6	9.9630
2.5	1.3930	9.7	5.0100	20.0	10.1600
2.6	1.4440	10.0	5.1600		

Derived from table provided by Choi and Wette [3.4].

TABLE 3.5. Chi-square distribution.

Degrees of freedom ν	Level of Significance (α)				
	0.005	0.01	0.025	0.05	0.1
1	7.88	6.63	5.02	3.84	2.71
2	10.60	9.21	7.38	5.99	4.61
3	12.84	11.34	9.35	7.81	6.25
4	14.86	13.28	11.14	9.49	7.78
5	16.75	15.09	12.83	11.07	9.24
6	18.55	16.81	14.45	12.59	10.64
7	20.28	18.48	16.01	14.07	12.02
8	21.95	20.09	17.53	15.51	13.36
9	23.59	21.67	19.02	16.92	14.68
10	25.19	23.21	20.48	18.31	15.99
11	26.76	24.72	21.92	19.68	17.28
12	28.30	26.22	23.34	21.03	18.55
13	29.82	27.69	24.74	22.36	19.81
14	31.32	29.14	26.12	23.68	21.06
15	32.80	30.58	27.49	25.00	22.31
16	34.27	32.00	28.85	26.30	23.54
17	35.72	33.41	30.19	27.59	24.77
18	37.16	34.81	31.53	28.87	25.99
19	38.58	36.19	32.85	30.14	27.20
20	40.00	37.57	34.17	31.41	28.41
21	41.40	38.93	35.48	32.67	29.62
22	42.80	40.29	36.78	33.92	30.81
23	44.18	41.64	38.08	35.17	32.01
24	45.56	42.98	39.36	36.42	33.20
25	46.93	44.31	40.65	37.65	34.38
26	48.29	45.64	41.92	38.89	35.56
27	49.64	46.96	43.19	40.11	36.74
28	50.99	48.28	44.46	41.34	37.92
29	52.34	49.59	45.72	42.56	39.09
30	53.67	50.89	46.98	43.77	40.26
40	66.77	63.69	59.34	55.76	51.81
50	79.49	76.15	71.42	67.50	63.17
60	91.95	88.38	83.30	79.08	74.40
70	104.21	100.43	95.02	90.53	85.53
80	116.32	112.33	106.63	101.88	96.58
90	128.30	124.12	118.14	113.15	107.57
100	140.17	135.81	129.56	124.34	118.50

Generated using Microsoft Excel function CHIINV

3.3.4 Empirical Distributions

When it is difficult to fit a theoretical distribution to the collected data, an empirical distribution can usually be formulated to serve as a data model. The procedure requires a cumulative distribution function (CDF) and this can be easily developed from the histogram of the collected data.[5] Observe, for example, that in Figure 3.11 the CDF is already present and defined by a series of points. Values between these points can be obtained by interpolation (e.g., linear interpolation).

We use the Java Class *Empirical*[6] as a data module. It first creates an empirical distribution and then uses the resulting data model to generate samples. The main steps involved are as follows. An array of histogram frequencies is first provided to an *Empirical* object. When instantiated, the object creates an internal representation of the CDF from these frequencies. The *Empircal* object generates random variates from the distribution using the CDF and the inverse transform method discussed later in Section 3.4.2.

Figure 3.12 shows a short Java method that instantiates an *Empirical* object called *empDM* (empirical data module), extracts the CDF representation from *empDM*, and then invokes *empDM* to generate ten random numbers. The object *empDM* is instantiated with the Class constructor that has the following three arguments.

- The array *histogram* contains the frequency values from a data histogram provided by the user. (Figure 3.12 provides 17 values taken from Figure 3.11.)
- The value *Empirical. LINEAR_INTERPOLATION* for the second argument indicates that the distribution is continuous and linear interpolation is to be used to generate radom samples from the constructed CDF. Otherwise when the value *Empirical. NO_INTERPOLATION* is used for the second argument a discrete CDF is created and used.
- The third argument is a uniform random number generator object. (in Figure 3.12, the *MersenneTwister* uniform random number generator is used).

The object first defines L points $(y_i, F(y_i))$ on the CDF, $F(y)$, in the following way:

where:

$$y_i = i * \text{scaleFactor} \quad with \quad i = 1,2, \; \ldots\ldots \; L \text{ and scaleFactor is the bin width.}$$

$$F(y_i) = F(y_{i-1}) + \frac{histogram[i]}{K}$$

$$F(y_0) = 0 \; and \; K = \sum_{k=1}^{L} histogram[k]$$

---

[5] It is also possible to derive the CDF directly from the collected data.

[6] The Empirical Class is provided as part of the cern.colt Java package provided by CERN (European Organization for Nuclear Research). See Section 5.4.3 for more details on using this package.

The method *cdf(i)* returns the value $F(y_i)$. Table 3.6 shows the points defined on the CDF by *empDM* in the program of Figure 3.12 (the table is generated by the program).

```
public static void main(String[] args)
{
  double randomValue;
  double[] histogram = {
              22, 57, 54, 47, 38, 22, 16,
              15, 10, 14, 1, 2, 0, 0, 0, 1, 1
                };
  double scaleFactor=1.4; // Width of the histogram bin
  double yMax = histogram.length*scaleFactor; // maximum data value

  // Create Empirical Object
  Empirical empDM=new Empirical(histogram,
                              Empirical.LINEAR_INTERPOLATION,
                              new MersenneTwister());

  // Lets get defining points on the CDF from empDM
  for(int i = 0 ; i<=histogram.length ; i++)
  {
    System.out.println(i+", "+(i*scaleFactor)+", "+empDM.cdf(i));
  }

  // Get empDM to generate 10 random numbers
  for(int i=0 ; i<20 ; i++)
  {
    randomValue = yMax*empDM.nextDouble();
    System.out.println(randomValue);
  }
}
```

FIGURE 3.12 Implementing a data module using an empirical distribution.

The random numbers generated by the *Empirical* object *empDM* in Figure 3.12 vary between 0 and 1. Thus random values returned by *empDM* must be multiplied by yMax (23.8) to obtain values that fall into the domain of

TABLE 3.6. The CDF used by the empirical object empDM.

i	y_i	$F(y_i)$
0	0	0
1	1.4	0.073333
2	2.8	0.263333
3	4.2	0.443333
4	5.6	0.6
5	7	0.726667
6	8.4	0.8
7	9.8	0.853333
8	11.2	0.903333
9	12.6	0.936667
10	14	0.983333
11	15.4	0.986667
12	16.8	0.993333
13	18.2	0.993333
14	19.6	0.993333
15	21	0.993333
16	22.4	0.996667
17	23.8	1

the CDF. The code shown in Figure 3.12 produces the following values: 6.3445029, 9.3559343, 1.4540021, 8.0323057, 5.3995883, 6.0610477, 3.8339216, 6.385926, 1.9355035, 9.2322093, 1.5383085, 4.5596041, 3.2674369, 7.8097124, 4.0771284, 5.7625121, 4.9623389, 2.9529618, 3.1037785, 10.877524. These correctly fall in the domain of the CDF, that is, between 0 and yMax.

3.3.5 Data Modelling with No Data

When data cannot be collected or are not available (e.g., the SUI does not exist), then educated guesses provide the means of last resort for formulating data models for the autonomous stochastic processes. These guesses can be based on research material and/or on information obtained from individuals who are particularly familiar with the SUI.

When the only features that can be confidently assumed about a random phenomenon that needs to be modelled are its minimum and maximum values then a reasonable distribution candidate is the uniform distribution (see Section A1.4.4.1). Because of this 'minimum knowledge' feature the uniform distribution is sometimes referred to as the distribution of maximum ignorance.

If the minimum, maximum, and modal values of a distribution can be specified, then the triangular distribution provides a convenient choice (see Section A1.4.4.2). The Beta distribution (see Section A1.4.4.8) can provide a variety of forms over the unit interval [0, 1] and can be easily shifted to accommodate other intervals.

Hundreds of distributions have been created to model many different types of phenomena. The type of phenomena under consideration often suggest a particular group of candidate distributions that are especially relevant. Several examples of this relationship are presented in Annex 1. A comprehensive discussion can be found in Banks et al. [3.2].

3.4 Simulating Random Behaviour

In this section we provide an overview of some of the most common techniques for generating samples from a specified distribution, in other words, techniques for generating random variates. The presentation is not intended to be complete and comprehensive. Its purpose is primarily to provide some insights into the techniques that are widely implemented in simulation software environments and consequently are conveniently accessible. Nevertheless, some appreciation for the nature of the procedures being invoked can provide a basis for ensuring correct usage, understanding potential shortcomings, and dealing with unanticipated results.

3.4.1 Random Number Generation

As will become apparent in the following subsection, the common methods for generating random variates depend on the availability of a stream of random numbers that are uniformly distributed on the unit interval. Although a procedure for generating uniformly distributed random numbers has its own intrinsic importance, this alternate role considerably amplifies this importance. On first glance, the development of such a procedure might seem straightforward but this is far from being so. The subtle complexities that need to be addressed have given rise to a considerable body of research literature. The special journal issue of Reference [3.5] is recommended for readers wishing to explore the topic in more detail within a modeling and simulation context.

From a theoretical point of view the basic requirement is that any value in the [0, 1] interval be equally likely and that there be no interdependence among the values that are generated (e.g., values to the right of the mean and to the left of the mean should not occur in batches or values should not tend to have a pattern of successively diminishing or successively increasing). From a practical point of view, there are implicit requirements for:

- Computational efficiency because many thousands of variates may be needed for any particular simulation experiment.
- Reproducibility because it should be possible to replicate any particular random number stream in order to repeat experiments.
- Hardware independence; that is, the procedure should not be intimately locked into the hardware architecture of any particular computer in order to ensure portability.

The reproducibility requirement might correctly suggest a fundamental contradiction to the reader. The important implication here is that we are obliged to abandon our original quest and be satisfied with the generation of pseudorandom numbers which provide reproducibility at the expense of genuine 'randomness.' We note furthermore that any implementation via an algorithmic process will intrinsically provide reproducibility. Thus the challenge reduces to the search for a 'good' algorithm, that is, one that yields a random number stream that has satisfactory statistical properties and also provides the efficiency and hardware independence that we seek.

The most widely used technique for generating streams of pseudorandom numbers is an approach called the *linear congruential* method. It is the remarkably simple iterative formula:[7]

$$K_i = (a\, K_{i-1} + c) \bmod m; \text{ with } i = 1, 2, \ldots,$$

where a and m are positive integers and c is a nonnegative integer. The initial value K_0 (likewise a positive integer) is called the *seed*. It was first proposed by Lehmer [3.15] with $c = 0$ in which case the method is called the *multiplicative congruential* method. The case where $c \neq 0$ (which is called the *mixed congruential* method) was suggested by Rotenberg [3.17] and Coveyou [3.6]. As might be expected, the values chosen for the parameters a, c, and m have a significant impact on the statistical quality of the sequence of numbers that are generated.

[7] Mod is the modulo operator; p mod q yields the remainder when p is divided by q where both p and q are positive integers.

Several basic features of the formula should be noted:

- The integer values that are generated fall in the [0, $m - 1$] interval. They can be shifted into the [0, 1) interval simply by dividing by m. In other words, the values $u_i = K_i/m$ fall in the range [0, 1).
- Suppose that the qth value K_q in the sequence $K_1, K_2, \ldots, K_p, \ldots, K_q$ is the first occurrence of equality to a previously generated value K_p, then the subsequence between K_p and K_q will be continually recycled as the sequence continues. Such an occurrence must happen sooner or later because there are at most m distinct values that can be generated. In other words, the process has a maximum *period* of m.
- Because at most m distinct values can be generated, there is an immediate divergence from the properties of a 'genuine' continuous random variable U that is uniformly distributed on the [0, 1] interval. For example, suppose $m > 3$; then the probability that U falls between $2.5/m$ and $2.6/m$ is $0.1/m$. However, there is zero probability that the linear congruential method will yield a value in this range.

These apparent shortcomings of the approach can, to a large extent, be overcome by suitable choices for the available parameters, a, c, and m. Selection of a very large value for m has obvious advantages. In fact, m is typically chosen to be of the form 2^b where b is the word length of the computer being used. That choice has the added advantage of simplifying the modulo calculation which can be carried out with a shift or mask operation.

As noted earlier, the longest possible period for the linear congruential method is m. An obvious question then is whether there exist parameter selections that will yield this limiting period. The following result due to Hull and Dobell [3.8] answers the question.

> The linear congruential method has full period (a period of m) if and only if the following conditions hold (throughout, divides means exactly divides; i.e., zero remainder):
> – The only positive integer that divides both m and c is 1.
> – If q is a prime number (divisible only by itself and 1) that divides m, then q divides $(a - 1)$.
> – If 4 divides m, then 4 divides $(a - 1)$ (i.e., $a = 4k - 1$ for some positive integer k).

Notice that the first of these conditions precludes the existence of a full period multiplicative congruential method. If m is chosen to be a power of 2, for example, $m = 2^b$ ($b > 0$) then a full period method results if and only if c is an odd integer and $a = 4k - 1$ for some positive integer k.

Although a full period multiplicative congruential generator is not possible (because the first condition listed above fails), the statistical properties of this method have generally proved to be superior to those of the mixed generators. Consequently it is widely used. It can be shown (see, for example, Knuth [3.11]) that if m is chosen to be 2^b then the maximum period is 2^{b-2} and this period is achieved if K_0 is odd and $a = 8k + h$, where k is a nonnegative integer and h is either 3 or 5.

A variety of tests has been formulated for assessing the acceptability of any random number stream, for example, frequency test, runs test, and poker test. These have been designed to detect behaviour that is inconsistent with the statistical integrity of the stream. Some details of these tests can be found in Banks et al. [3.2] and Law and Kelton [3.13].

3.4.2 Random Variate Generation

Our concern in this section is with generating samples from arbitrary, but specified, distributions. The techniques we outline depend on the availability of random samples that are uniformly distributed. In practice these random samples will originate from an algorithmic procedure which provides samples whose statistical properties only approximate those of a uniform distribution, for example, the techniques outlined in the previous section. Consequently we are obliged to acknowledge that the results generated will likewise fall short of being ideal.

One of the most common techniques for generating variates from a specified distribution is the *inverse transform method.* Its application depends on the availability of the CDF, $F(x)$, of the distribution of interest. The method is equally applicable for both the case where the distribution is continuous or discrete. We consider first the continuous case.

Application of the method in the continuous case requires the assumption that $F(x)$ is strictly increasing; that is, $F(x_1) < F(x_2)$ if and only if $x_1 < x_2$. A representative case is shown in Figure 3.13. Because of this assumption it follows that $P[X \leq x] = P[F(X) \leq F(x)]$. The procedure is illustrated in Figure 3.13 and is as follows (F^{-1} denotes the inverse of the CDF).

- Generate a sample u from the uniform distribution on [0, 1] (recall that the CDF for this distribution is $F_U(u) = P[U \leq u] = u$).
- Take $y = F^{-1}(u)$ to be the generated value (note that $F^{-1}(u)$ is defined because u falls in the range 0 and 1, which corresponds to the range of F).

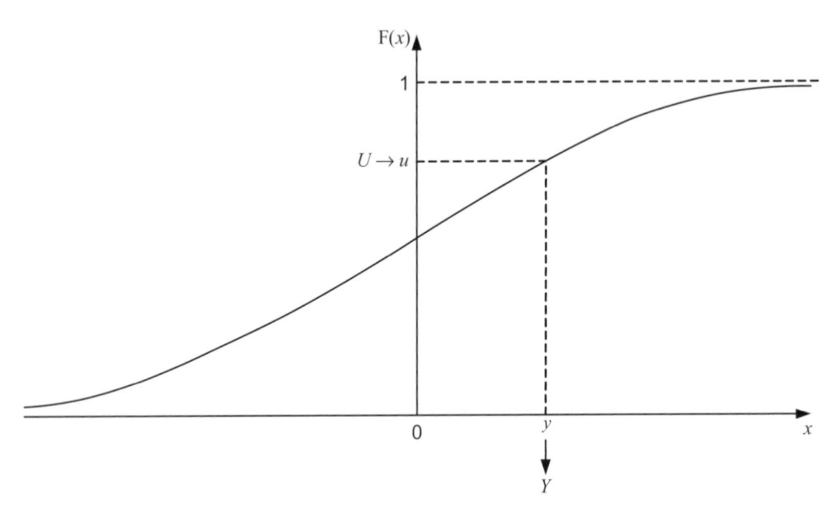

FIGURE 3.13. Illustration of the inverse transform method for the case of a continuous distribution.

The procedure, in effect, creates a random variable $Y = F^{-1}(U)$. To confirm that the procedure is doing what we hope it is doing, we need to demonstrate that $P[Y \le y] = F(y)$, that is, that the CFD for Y is the one of interest. This demonstration is straightforward:

$$
\begin{aligned}
P[Y \le y] &= P[F^{-1}(U) \le y] \\
&= P[F(F^{-1}(U) \le F(y)] \quad \text{(because } F \text{ is strictly increasing)} \\
&= P[U \le F(y)] \\
&= F(y)
\end{aligned}
$$

When the CDF of interest can be explicitly 'inverted,' the procedure becomes especially straightforward. Consider the case of the exponential distribution with mean $1/\lambda$; the CDF is $F(x) = 1 - \exp(-\lambda\, x)$. We begin by setting $u = F(x) = 1 - \exp(-\lambda\, x)$ and then obtaining an expression for x in terms of u. This can be readily achieved by taking the natural logarithm which yields:

$$x = -\ln(u - 1)\,/\lambda. \qquad (3.2)$$

The implication here is that if we have a sequence of u's that are uniformly distributed random values on the interval [0, 1] then the corresponding values x given by Equation (3.2) will be a sequence of samples from the exponential distribution having mean $1/\lambda$.

The inverse transform method is equally applicable when the requirement is for random variates from a discrete distribution. Suppose that X is such a random variable whose range of values is $x_1, x_2, \ldots, x_n$. Recall that in this case the CFD, $F(x)$ is:

$$F(x) = \mathrm{P}[X \leq x] = \sum_{x_i \leq x} p(x_i) \ ,$$

where $p(x_i) = \mathrm{P}[X = x_i]$. The procedure (which is illustrated in Figure 3.14) is as follows.

- Generate a sample u from the uniform distribution on [0, 1] (i.e., from the distribution whose CDF is $F_U(u) = \mathrm{P}[U \leq u] = u$).
- Determine the smallest integer K such that $u \leq F(x_K)$ and take x_K to be the generated value.

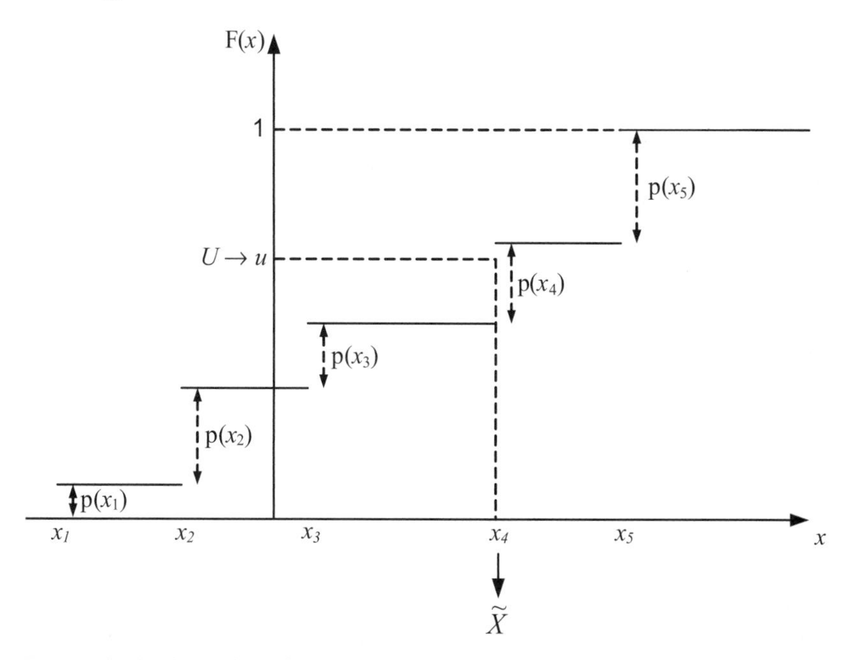

FIGURE 3.14. Illustration of the inverse transform method for the case of a discrete distribution.

Repetition of the procedure generates a stream of values for a random variable that we can represent by $\widetilde{X}$. To verify the correctness of the procedure, we need to demonstrate that the values x_i that are generated satisfy the condition $\mathrm{P}[\widetilde{X} = x_i] = p(x_i)$, for $i = 1, 2, \ldots, n$. Observe first that $\widetilde{X} = x_1$ if and only if $U \leq F(x_1)$. Thus:

$$\mathrm{P}[\widetilde{X} = x_1] = \mathrm{P}[U \leq F(x_1)] = F(x_1) = p(x_1) \ .$$

Assume now that $i > 1$. The procedure ensures that:

$$\widetilde{X} = x_i \quad \text{if and only if } F(x_{i-1}) < U \leq F(x_i).$$

Consequently:

$$\begin{aligned} \mathrm{P}[\overline{X} = x_i] &= \mathrm{P}[F(x_{i-1}) < U \leq F(x_i)] \\ &= F_U(F(x_i)) - F_U(F(x_{i-1})) \\ &= F(x_i) - F(x_{i-1}) \\ &= p(x_i)\,, \end{aligned}$$

which completes the demonstration. Note that the above relies on the fact that for any discrete CDF, $F_Y(y)$, it is true that $\mathrm{P}[a < Y \leq b] = F_Y(b) - F_Y(a)$.

Because the CDF for the normal distribution cannot be written in closed form, the inverse transform method is not conveniently applicable for generating samples from that distribution. Fortunately there are a number of alternative general methods available which can be used. In fact, there is one technique that is specifically tailored to the normal distribution. This is the *polar method* and a description can be found in Ross [3.16].

An alternate general technique is the *rejection–acceptance method.* Its implementation depends on the availability of the probability density function of the distribution of interest (hence it can be used for generating samples from the normal distribution). It is equally applicable for both discrete and continuous distributions. In its simplest form, the method involves generating samples from a uniform distribution and discarding some samples in a manner which ensures that the ones that are retained have the desired distribution. The method shares some features with the Monte Carlo method for evaluating integrals (see Section 1.6). A more comprehensive treatment of the method can be found in Reference [3.16] where the presentation includes a more general approach for the underlying procedure than we give in the following discussion.

In the simplest form described here, the implementation depends on the assumption that the probability density function $f(x)$ (or probability mass function) of interest is bounded on the left (by a) and on the right (by b). Consequently if a long 'tail' exists, it needs to be truncated to create the values a and b. We assume also that the maximum value of $f(x)$ on the interval $[a, b]$ is c.

The procedure (which is illustrated in Figure 3.15) is as follows.

- Generate two samples u_1 and u_2 from the uniform distribution on [0,1]. (i.e., from the distribution whose CDF is $F_U(u) = \mathrm{P}[U \leq u] = u$).
- Compute $\widetilde{x} = a + u_1 (b - a)$.
- Compute $\widetilde{y} = c\, u_2$.

- If $\tilde{y} \leq f(\tilde{x})$ accept $\tilde{x}$ as a valid sample; otherwise repeat the process with the generation of two new samples from *U*.

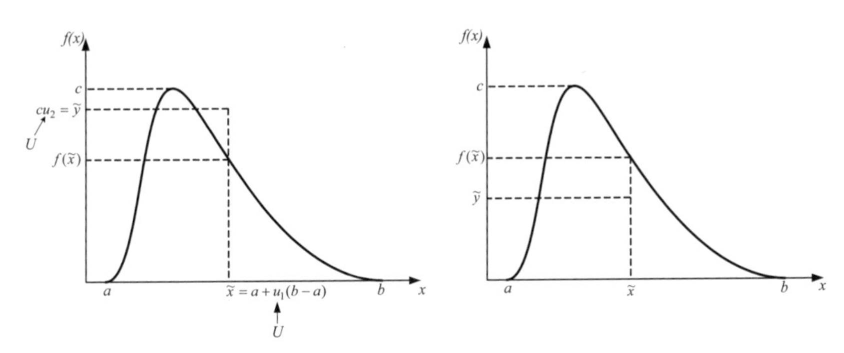

FIGURE 3.15. Illustration of the rejection–acceptance method.

Although by no means a formal proof, the following observations provide some intuitive confirmation of the procedure's correctness. Note that the tuple $(\tilde{x}, \tilde{y})$ is a point in the *abc* rectangle. Because the values u_1 and u_2 are independent samples from the uniform distribution each point in the rectangle is equally likely to occur. Suppose x_1 and x_2 are two distinct points in the $[a, b]$ interval with $f(x_1) > f(x_2)$ (see Figure 3.16). Over a large number of repetitions of the procedure $\tilde{x}$ will coincide with x_1 and x_2 the same number of times. However the occurrence of x_1 is far more likely to be output by the procedure than x_2. More specifically,

- Given that x_1 has occurred the probability of it being output is:

$$\mathrm{P}[c\,U \leq f(x_1)] = \mathrm{P}[U \leq f(x_1)/c\,] = f(x_1)/c.$$

- Given that x_2 has occurred the probability of it being output is:

$$\mathrm{P}[c\,U \leq f(x_2)] = \mathrm{P}[U \leq f(x_2)/c\,] = f(x_2)/c$$

Although the occurrence of x_1 and x_2 are equally likely, the relative proportion of x_1 outputs to x_2 outputs is proportional to $f(x_1)/f(x_2)$ which is consistent with an intuitive perspective of the distribution of interest.

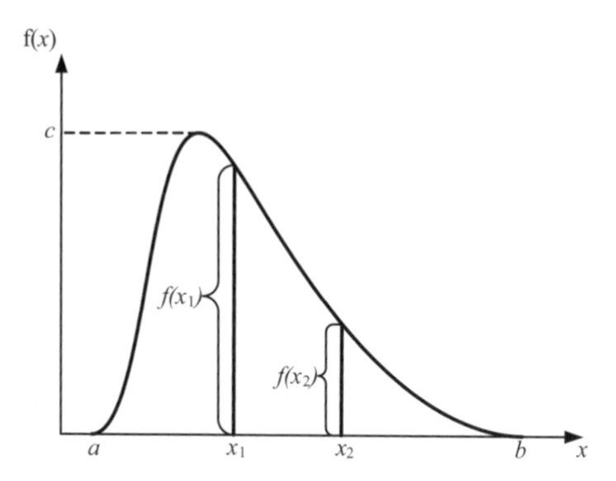

FIGURE 3.16. Illustration of relative output frequency.

One shortcoming of the method is the uncertainty that any particular execution of the procedure will be successful (i.e., yield an acceptable sample). The probability of success is equal to the relative portion of the *abc* rectangle that is filled by the density function of interest. Because the area of the rectangle is $c(b - a)$ and the area of the density function is 1, the probability of success is the ratio $1/[c(b - a)]$. Two representative cases are shown in Figure 3.17: the occurrence of rejections will, on average, be more frequent in the case of $f_2(x)$ than in the case of $f_1(x)$.

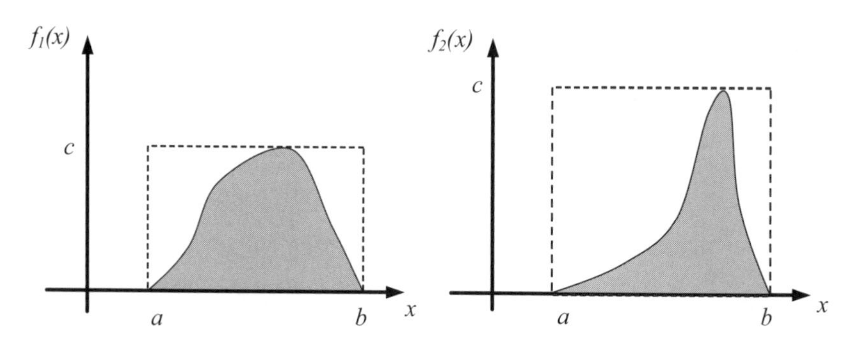

FIGURE 3.17. illustration of relationship between rejection probability and shape.

3.5 References

3.1. Balci O., (1998), Validation, verification and testing, in *The Handbook of Simulation*, Chapter 10, John Wiley & Sons, New York, pp. 335–393.

3.2. Banks, J., Carson II, J.S., Nelson, B.L., and Nicol, D.M., (2005), *Discrete-Event System Simulation*, 4th edn., Pearson Prentice Hall, Upper Saddle River, NJ.

3.3. Biller, B. and Nelson, B.L., (2002), Answers to the top ten input modeling questions, in *Proceedings of the 2002 Winter Simulation Conference*.

3.4. Choi S.C. and Wette R., (1969), Maximum likelihood estimation of the paramters of the gamma distribution and their bias, *Technometrics*, **11**: 683–690.

3.5. Couture, R. and L'Ecuyer, P. (Eds.), (1998), Special Issue on Random Variate Generation, *ACM Transactions on Modeling and Simulation*, **8**(1).

3.6. Coveyou, R.R., (1960), Serial correlation in the generation of psuedo-random numbers, *Journal of the ACM*, **7**: 72–74.

3.7. Harrell, C., Ghosh, B.K., and Bowden, Jr., R.O., (2004), *Simulation Using ProModel*, 2nd ed., McGraw-Hill, New York.

3.8. Hull, T.E. and Dobell, A.R., (1962), Random number generators, *SIAM Review*, **4**: 230–254.

3.9. Kelton, D.W., Sadowski, R.P., and Sturrock, D.T., (2004), *Simulation with Arena*, 3rd ed., McGraw-Hill, New York.

3.10. Kleijnen, J.P.C., (1987), *Statistical Tools for Simulation Practitioners*, Marcel Dekker, New York.

3.11. Knuth, D.E., (1998), *The Art of Computer Programming, Vol. 2: Seminumerical Algorithms*, 3rd ed. , Addison-Wesley, Reading, MA.

3.12. Law A.M. and McComas, M.G., (1997), Expertfit: Total support for simulation input modeling, in *Proceedings of the 1997 Winter Simulation Conference*, pp. 668–673.

3.13. Law, A.M. and Kelton, D.W., (2000), *Simulation Modeling and Analysis*, 3rd ed., McGraw-Hill, New York.

3.14. Leemis, L.M. and Park, S.K., (2006), *Discrete Event Simulation: A First Course*, Pearson Prentice Hall, Upper Saddle River, NJ.

3.15. Lehmer, D.H., (1949, 1951), Mathematical methods in large scale computing units, in *Proceedings of the Second Symposium on Large-Scale Digital Calculating Machinery*, 1949 and *Annals Comput. Lab*, 1951, vol. 26, pp. 141–146, Harvard University Press, Cambridge, MA.

3.16. Ross, S.M., (1990), *A Course in Simulation*, Macmillan, New York.

3.17. Rotenberg, A., (1960), A new pseudo-random number generator, *Journal of the ACM*, **7**: 75–77.

Chapter 4 A Conceptual Modelling Framework for DEDS

4.1 Need for a Conceptual Modelling Framework

A key requirement for carrying out a meaningful discussion of any complex topic is a collection of clearly defined concepts that map onto the various facets of the domain of interest. Within the context of formulating models of discrete-event dynamic systems, this requirement translates into the need for a framework that provides a consistent and coherent way of viewing and describing the mechanisms that give rise to behaviour. In this chapter we present such a framework. It serves as an essential aspect of our goal of exploring the issues that surround the successful completion of any modelling and simulation project in the DEDS domain.

Because of the diversity and the unbounded complexity that characterises the DEDS domain, no standardised and generally accepted framework for representing systems in this class has yet emerged. A variety of existing formalisms, such as finite state machines or Petri nets ([4.4] and [4.6]) can be useful in particular cases but these lack sufficient 'modelling power' (i.e., generality) to be universally applicable. The DEVS approach (Zeigler [4.7] and Zeigler et al. [4.8]) on the other hand certainly has the requisite generality but has restricted accessibility because of its underlying mathematical formality.

Various options are possible. The one we present is informal in nature and has considerable intuitive appeal. But at the same time, it has a high level of both generality and adaptability. Although it is directly applicable to a wide range of project descriptions, it can also be easily extended on an ad hoc basis when specialised needs arise.

4.2 Constituents of the Conceptual Modelling Framework

4.2.1 Overview

Our conceptual modelling framework for DEDS is formulated from a small number of basic components. The first is a collection of *entities* that interact over the course of the observation interval by reacting to, and giving rise to, the occurrence of *events* which are the second important constituent of our framework. Entities and events represent basic building blocks. In addition, however, there is a higher-level construct called an *activity* that is fundamental to our conceptual modelling framework. The set of activity constructs within the model provides the means for capturing relevant dynamic behaviour. Each activity, in fact, represents a specific unit of behaviour that needs to be recognised. It can be viewed as a relationship among some of the entities within the model. An activity is formulated from the basic building blocks of entities and events which are, in turn, characterised in terms of some collection of constants, parameters, and variables that have been chosen to enable this characterisation in a meaningful way (see Section 2.2.4).

The basic concepts underlying our approach are not new. They can be traced back to the activity scanning paradigm that is usually identified as one of the modelling and simulation 'world views'. A comprehensive presentation of activity scanning from a programming perspective can be found in Kreutzer [4.2]. Examples of the utilisation of this paradigm can be found in Martinez [4.3], Shi [4.5], and Gershwin [4.1].

One particular aspect of our activity-based conceptual modelling framework needs to be emphasised; namely, that it must not be interpreted as a programming environment. Its purpose instead is to provide a meaningful foundation for program development or, stated alternately, its purpose is to serve as a vehicle for making the transition from a project description to a simulation program.

4.2.2 Entities and Model Structure

We recognise three sets of entities in our conceptual modelling framework. These are:

- A set of consumer entity instances (cei's), $\hat{C}$, where:

  $$\hat{C} = \bigcup_{1 \le k \le n} C_k$$

  The sets C_k are disjoint and n is SUI-dependent. The cei's in each consumer entity set C_k are all members of the same consumer entity class which we denote by Θ_k.

- A set $\check{R}$ of *resource entities*. Each resource entity has relevance to the cei's from some of the consumer entity classes represented in $\hat{C}$.
- A set $\hat{A}$ of *aggregate entities*. These have real-world counterparts that include queues and groups where the latter is simply a collection without the ordering protocol that is associated with a queue.

The cei's typically interact directly with the resource and the aggregate entities within the model. This interaction gives rise to changes in the state of the model and it is these changes that represent the dynamic behaviour that is of interest. The sets $\hat{C}$, $\check{R}$, and $\hat{A}$ are dependent on the specific nature of the SUI; however, the conceptual model for even the simplest DEDS will normally include at least one consumer entity class, one resource, and one aggregate.

The resource entities can generally be regarded as providing a service that is sought by the cei's. Consequently the cei's within a model can often be viewed as being 'in motion' inasmuch as they disperse throughout the space embraced by the model in search of the services that are available from the various resource entities. Because the rate at which these services can be dispensed is constrained, the service-seeking cei's are frequently obliged to wait until they can be accommodated. Waiting takes place in an aggregate that is connected (perhaps only implicitly) to the resource. In effect, our conceptual modelling framework generally gives rise to a network structure in which the nodes are resources or aggregates and the links are paths that the cei's can traverse.

Implicit in the description given above is the important feature that no cei can exist except by virtue of being connected to either an aggregate or a resource. In effect then, the resource and aggregate entities in a model share the important common feature of providing a 'location' (possibly virtual) where cei's can temporarily reside. This 'residency' typically has finite duration and both its establishment and termination are associated with the occurrence of events. As becomes apparent in the following section, this has important consequences on the manner in which we develop a meaningful characterisation for the various entity types within any particular model. The state variables for the model emerge from this characterisation.

Note also that any particular cei in $\hat{C}$ is permitted to have only a transient existence within the model over the course of the observation interval. In other words the consumer entity sets C_k may be time-dependent; that is, $C_k = C_k(t)$. Transient existence for members of $\hat{A}$ and $\check{R}$ is not precluded but, at the same time, is not typical.

To illustrate these essential notions of our conceptual modelling framework, consider the operation of a department store. Customers arrive and generally intend to make one or more purchases at various shopping areas of the store. At each chosen area a customer browses, makes a selection, and then pays for it at the closest service desk before moving on to the next shopping area. Upon completion of the shopping task the customer leaves the store.

For this fragment of a project description, the customers correspond to a consumer entity class and those instances that are in the store at any particular moment correspond to a consumer entity set.

Each of the service desks at the various shopping areas corresponds to a resource. Because the service function at a service desk has a finite duration there is a likelihood that some cei's may not receive immediate attention upon arrival at the service desk. Hence it is reasonable to associate a queue entity with each service desk. Note also that the browsing phase at each shopping area introduces the special case of an aggregate that we call a group. The network structure of the underlying dynamics is apparent from the representation given in Figure 4.1.

The cei's within a group are not organised in a disciplined way as in the case of a queue but rather simply form an identifiable 'grouping'. Note furthermore that the discipline that is inherent in a queue introduces two important features. Both of these arise from the fact that there is a natural exit mechanism for cei's within a queue, namely, availability of access to the resource that is associated with the queue. As a consequence:

- the duration of a cei's membership in a queue is implicitly established.
- the destination of a cei that departs from a queue is likewise implicitly established.

In contrast, neither the duration of membership nor the subsequent destination of a cei within a group is implicit in the membership property and both of these therefore need to be explicitly specified.

Several data models are necessarily associated with the above formulation. Included here would be the characterisation of customer arrival rates and service times at the service desks, allocation of the shopping areas to be visited by the arriving customers and the characterisation of the duration of the browsing phase at each area, and so on. It is especially important to observe that these various data models will provide the basis for generating events that give rise to change. For example, the event associated with the end of a particular customer's browsing phase will generally (but not necessarily) result in that customer's relocation into the queue associated with the service desk of that service area.

Many details have yet to be clarified; for example, how is the set of shopping areas that a particular customer visits selected? What is the order of the visitations? And how many servers are assigned to the service desks? Can a particular customer balk, that is, not make any purchase at one or more of the assigned shopping areas and if so, then under what circumstances? The information for dealing with these questions is not provided in the given project description fragment but would most certainly be necessary before a meaningful conceptual model could be formulated. Indeed one of the important functions of the conceptual modelling process is to reveal the absence of such essential details.

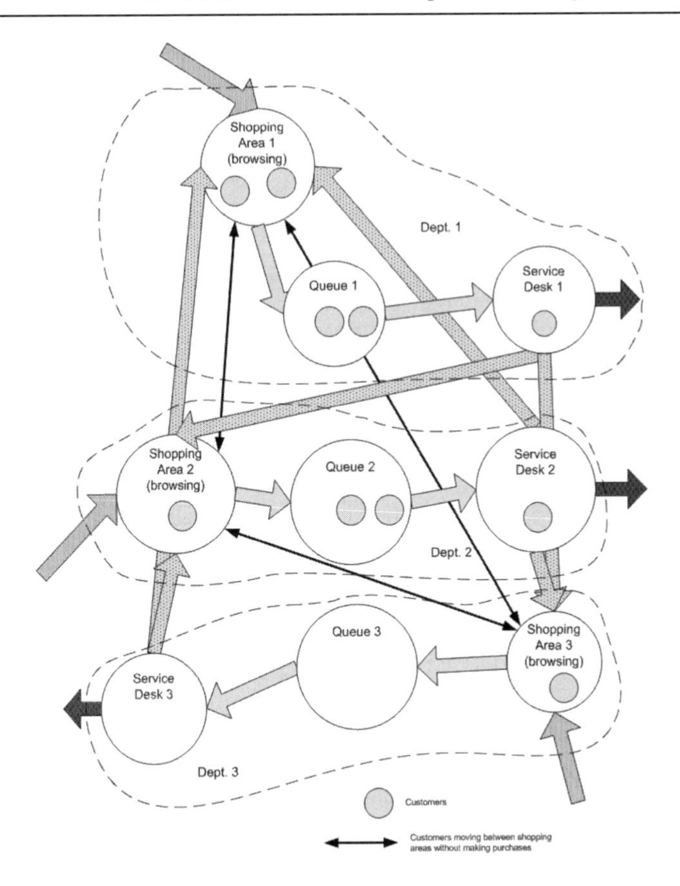

FIGURE 4.1. Structural view of the conceptual model for department store shoppers.

We refer to the conceptual modelling perspective for discrete event dynamic systems that is outlined above as the *ABCmod framework* (Activity-Based Conceptual modelling) and we refer to the conceptual model that emerges as an *ABCmod conceptual model.* As becomes apparent in the discussion that follows, an ABCmod conceptual model can be represented as a structure with three components; that is,

$$\text{ABCmod} = < \Theta, \Omega, \Phi > ,$$

where:

Θ is a set of consumer entity classes.

Ω is a set of service entities (the union of a set of aggregates and a set of resources).

Φ is a set of activity constructs.

There are two types of activity constructs, the Activity and the Action. An Activity can be viewed as a unit of behaviour which represents an abstraction of some purposeful task that takes place within the SUI. Generally the specification for an Activity references one of the service entities and one or more consumer entity instances. An Activity has a duration and results in changes in the value of some state variables within the model. An Action resembles an Activity but with the important difference that an Action has no duration; that is, it unfolds at one particular instant of (simulated) time.

The specifications for the sets Θ, Ω, and Φ are derived from the information provided in the project description. They are presented as a set of tables. The templates for these tables are included with the discussion of these various sets which follows in Sections 4.2.3 and 4.2.4.

It needs to be stressed that a flexible interpretation of the sets Θ and Ω is essential. For example, our suggestion above that the resources provide a service to the cei's should not be rigidly interpreted at the exclusion of other possibilities. Consider, for example, a SUI that includes a collection of supermarkets (a set of resources) and two particular consumer entity classes called 'shoppers' and 'delivery trucks'. Two distinct relationships can be identified between these classes and the set of resources: namely, a 'shopping' Activity and a 'delivery' Activity, respectively. In the first case the resources (the supermarkets) do provide a service for the shoppers whereas in the second case, the resources receive a service from the delivery trucks. Thus references in the sequel to the notion of resources providing a service to cei's should be loosely interpreted and regarded simply as a semantic convenience.

Likewise note that the distinction between membership of entities in the sets Θ and Ω can become blurred. Consider, for example, a set of machines within a manufacturing environment that are subject to failure. A team of maintenance personnel is available to carry out repairs. While the machines are operating, they clearly serve as a resource in the manufacturing operation but when they fail they become consumer entity instances that need the service function of the maintenance team. How the model builder chooses to view such situations is not especially critical; the choice is typically governed by the perspective that seems most natural to the modeller.

4.2.3 Characterising the Entity Types

The discussion in Section 4.2.2 is primarily concerned with presenting a structural perspective of our conceptual modelling framework. We now undertake a more detailed examination of the constituents of this structure.

We begin by exploring how members of the sets Θ and Ω can be characterised in a way that facilitates the specification of dynamic behaviour (namely, the specification of the activity constructs in the set Φ) which we consider in Section 4.2.4. The basis for identifying the state variables for the model emerges from this characterisation. Our approach is to propose for each entity category a set of attributes that provide the basis for a meaningful characterisation. It is important to appreciate that the choice of attributes depends very much on the nature of the SUI and on the project goals (e.g., its output requirements).

As noted earlier, the cei's that belong to $\hat{C}$ can be viewed as flowing among the aggregates and the resources in $\hat{A}$ and $\check{R}$, respectively. There is therefore an essential requirement here to track both the existence and the status of these entities to ensure that they can be processed correctly by the rules that govern the model's behaviour. In addition, there may be a particular trail of data produced by the cei's that is relevant to the output requirements that are implicit in the project goals. These various requirements demonstrate the need to characterise each cei in some meaningful way. This can be achieved by associating with each consumer entity class Θ_k a set of m_k attributes that has been chosen in a way that satisfies the underlying requirements. There are three common categories of attributes that we call property reflectors, path reflectors, and elapsed-time reflectors. They are described below.

- Property reflectors: The cei's from any particular class Θ_k may have a variety of essential properties or features that have direct relevance to the manner in which they are treated in the rules of behaviour, for example, a 'size' which may have one of three values (small, medium, or large) or a 'priority' which may have one of two values (high or low). The value assigned to an attribute that falls in the category of a property reflector remains invariant over the course of a consumer entity instance's existence within the scope of the model.
- Path reflectors: In view of the inherent network structure of our conceptual modelling framework, it is sometimes necessary to maintain an explicit record of what nodes a cei has already visited or alternately, what nodes remain to be visited. Attributes that have this function fall in the category of path reflectors. The values of these attributes naturally change as a particular consumer entity instance flows through the model. The values of attributes that are path reflectors play a key role in initiating the transfer of the cei's to their next destination.
- Elapsed-time reflectors: Output requirements arising from the project goals often need data that must be collected about the way that cei's have progressed through the model. Frequently this requirement is for

some type of elapsed time measurement. The values assigned to attributes that fall in the elapsed-time category serve to provide a basis for the determining the required output data. For example, it may be required to determine the average time spent by cei's from some particular class Θ_k waiting for service at a particular resource entity. The principal attribute introduced in this context could function as a timestamp storing the value of time t when the waiting period begins. A data value for the required data set would then be produced as the difference between the value of time when the waiting period ends and the timestamp.

Suppose *Con* is a cei belonging to the consumer entity class Θ_k. The collection of attributes associated with Θ_k provides the basis for characterising *Con* using a variable that we write as *C.Con*. Specifically,

$$C.Con = (\alpha_1, \alpha_2, \text{----} \ \alpha_{m_k}),$$

where α_j is the jth attribute for the class Θ_k. The value of the variable *C.Con* at any moment in time t corresponds to the value of the collection of attributes, α_j, $j = 1, 2, \ldots, m_k$ as they exist at time t. The notation $C.Con.\alpha_j$ is used to reference the value of the jth attribute of consumer entity instance *Con*. The template for a consumer entity class is shown in Table 4.1.

When a specific cei belonging to Θ_k first appears within the scope of the model's behaviour (and hence becomes a member of C_k) values are assigned to its attributes (if, in fact, a meaningful value is not yet available, we assume for convenience that a temporary value of 'nil' is implicitly assigned). The result is an m_k-tuple of data values that we refer to as an *attribute-tuple*. In effect, such an attribute-tuple serves as a surrogate for that specific cei.

TABLE 4.1. Template for summarising a consumer entity class.

Consumer Entity Class: *EntityClassName*	
A description of the consumer entity class called EntityClassName.	
Attributes	**Description**
AttributeName1	*Description of the attribute called AttributeName1.*
AttributeName2	*Description of the attribute called AttributeName2.*
.	.
.	.
AtributeNamen.	*Description of the attribute called AttributeNamen*

A perspective that is frequently appropriate is one where cei's (or more precisely, their attribute-tuple surrogates) flow from resource to resource accessing the services associated with these resources. At any particular point in time, however, access to a particular resource may not be possible because the resource is already engaged (busy) or is otherwise not available (e.g., out of service because of a temporary failure). Such circumstances are normally handled by connecting the cei to an aggregate entity associated with the resource where it can wait until access to the resource becomes possible.

The most common aggregate entity is a queue. Connecting a cei to a queue corresponds to placing the entity's attribute-tuple in that queue. From this observation it is reasonable to suggest two particular attributes for any queue entity within the model, namely, *List* and *N*. Here *List* serves to store the attribute-tuples for the cei's that are enqueued in the queue and *N* is the number of entries in that list. Thus we might associate with a queue entity called *Q*, the variable

$$A.Q = (List, N) .$$

Note that the attribute-tuples in *List* need not all be of uniform structure because cei's belonging to a variety of consumer classes may be interleaved in the queue. Also note that it may be appropriate in some circumstances to assume that the attribute-tuples are ordered in *List* according to the queuing discipline specified in the project description. In any event we do adopt the convention that the first entry in *List* is the attribute-tuple that will next be removed from the queue. This particular attribute-tuple is referenced as *A.Q.List[0]*.

It needs to be stressed that the above selection of attributes for characterising a queue is not necessarily adequate for all situations. In some cases, for example, it may be appropriate to include a reference to the specific resource (or even resources) with which the queue is associated.

It is interesting to observe that the queue in which a cei awaits service may only be virtual inasmuch as it may be the service providing entities (resources) that are in motion whereas the service requesting cei's are stationary. The resources may simply cycle through a list of cei's providing the appropriate service function. The queue, although virtual, still remains an effective way of viewing the interaction between these two categories of entity. As an example, consider a manufacturing plant in which there are a large number of machines participating in the manufacturing process. These machines are subject to failure and a repair crew (the resource) moves around the plant carrying out repairs.

The characterisation of a group entity is similar to that of a queue but there are important differences. Consumer entity instances are placed into a group as in the case of a queue, however, there is no intrinsic ordering discipline. In many cases the duration of the connection of a cei to a group is established via a data model. The time T_F when a connected cei is disconnected from a group can be obtained as $(T_S + D)$ where T_S is the value of time when the connection occurred and D is the duration provided by the data model. In such circumstances it is natural to assume that the value T_F is assigned to some attribute (possibly called *TerminationTime*) of each connected cei as part of the connection step. In this situation the set of *TerminationTime* values could provide a basis for imposing an ordering for the cei's connected to the group.

On the basis of the observations above, the attributes for a group could reasonably include *List* and *N* where *List* is the list of attribute-tuples of the cei's connected to the group and *N* is the number of entries in this list. In this circumstance, the characterising variable associated with a group entity called *G* would be:

$$A.G = (List, N) .$$

In some situations it may be useful to extend this characterisation with a reference to the destination of consumer entities following their connection to a group. This is very much context-dependent and provides a further illustration of the need to tailor the characterisation variables of entity types to the specific requirements of a project. The template for an aggregate entity is shown in Table 4.2.

Consider now a resource entity. In our perspective, we choose to regard the cei being serviced as being incorporated into the resource. In other words we assume that each resource entity has an attribute allocated for this purpose (possibly called *Client*) whose value is the attribute-tuple of the cei currently being serviced by that resource entity. As in the case of a group aggregate, the time T_F when a connected cei is disconnected from a resource would be determined as $(T_S + D)$ where T_S is the value of time when the connection to the resource occurred and D is the duration assigned to that cei (typically obtained from a data model). The value T_F would then be assigned to the attribute *TerminationTime* of the connected cei. It is also usually relevant to incorporate a status indicator reflecting whether the resource entity is 'busy'. This implies a binary-valued attribute whose name might be *Busy*.

TABLE 4.2. Template for summarising an aggregate entity.

Aggregate Entity: ***EntityName***	
A description of the aggregate entity called EntityName.	
Attributes	**Description**
AttributeName1	*Description of the attribute called AttributeName1.*
AttributeName2	*Description of the attribute called AttributeName2.*
.	.
.	.
AtributeNamen.	*Description of the attribute called AttributeNamen*

Thus, a typical variable associated with a resource entity *Res* in the model, might be:

$$R.Res= (Client, Busy)\,.$$

Depending on the context, a possibly useful extension of this characterisation could be the addition of another attribute called Broken, whose value would reflect whether the resource entity is in working order. The template for an aggregate entity is shown in Table 4.3.

TABLE 4.3. Template for summarising a resource entity.

Resource Entity: ***EntityName***	
A description of the resource entity called EntityName.	
Attributes	**Description**
AttributeName1	*Description of the attribute called AttributeName1.*
AttributeName2	*Description of the attribute called AttributeName2.*
.	
.	
AtributeNamen.	*Description of the attribute called AttributeNamen*

References to the state of a model are an important and integral part of the discussions surrounding the conceptual model development process. Inasmuch as the model's state at time *t* is simply the value of its state variables at time *t*, a prerequisite for such discussions is a clear understanding of what constitutes the set of state variables for the model. If the model's state variables are incorrectly identified then aspects of the model's development can become muddled and vague and hence errorprone.

There are, unfortunately, very few rules for dealing with the task of state variable identification that have general applicability. The one fundamental notion we rely on is Property Σ as outlined in Section 2.2.4.3. Property Σ is a specification of what must hold true if some set of variables is to serve as the state variables for a model. The requirements of Property Σ can provide the justification for augmenting a candidate set with additional variables and this approach is used below.

The tracking information for all instances of consumer entities within the model at time t is part of the model's state at time t and hence this information needs to be captured in the state variables defined for the model. This has been reflected in our choice of the characterising variables for both aggregate entities and resource entities, namely, the variables, $A.Q$, $A.G$, and $R.Res$ which relate to a queue, a group, and a resource called Q, G, and Res, respectively. These variables can, in fact, be taken to be the model's state variables. This follows from the observation that the information embedded in these variables is needed in order to satisfy the requirements of Property Σ.

4.2.4 Activity Constructs and Model Behaviour

As we have previously stressed, the only models of interest to us are dynamic and hence evolve over time. The variable t which we use to represent time naturally plays a key role in the exploration of this behaviour. Furthermore, the formulation of the specification for the model's behaviour depends on a key assumption, namely, the existence of a mechanism that moves t across the observation interval, beginning at its left boundary. How this is carried out is not relevant to the current discussion but this traversal assumption is essential and it is implicit in the discussions that follow. In concert with the traversal of t across the observation interval, state changes occur within the model and these, in effect, represent the model's behaviour. Another important but implicit assumption relating to the variable t is the assumption that within all sections of any model, the units associated with t are the same for example, seconds, days, years, and the like.

Our concern now is with developing a framework within which the model's behaviour can be formulated in a consistent and coherent manner. Our main constructs for characterising change, or equivalently, behaviour, is the Activity construct which, in turn, depends on the notion of events. A re-examination of the department store example introduced in Section 4.2.2 provides useful insight for developing these notions.

In Figure 4.2 we illustrate a possible manner in which three shoppers (i.e., three instances of the consumer entity class called Shopper) in the department store example might interact. The three shoppers (called A, B,

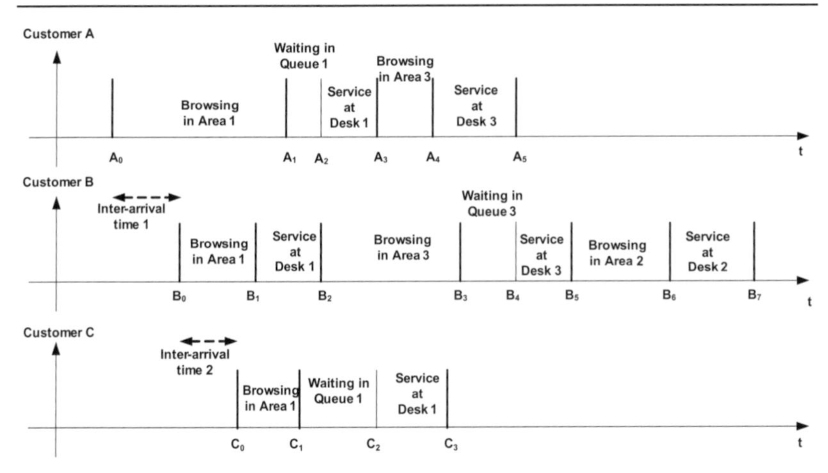

FIGURE 4.2. Behaviour of three department store shoppers.

and C) arrive at times A_0, B_0, and C_0, respectively, and leave the store at times A_5, B_7, and C_3, respectively.

There are a number of important observations that can be made about Figure 4.2. Notice, in particular, that some type of transition occurs at each of the time points A_0 through A_5, B_0 through B_7, and C_0 through C_3. These transitions, in fact, correspond to changes in the state of the model. Notice also that some of these time points are coincident; for example, $A_2 = B_2$, $A_3 = C_2$, and $A_5 = B_4$, suggesting that several different changes can occur at the same moment in time. It is also clear from Figure 4.2 that there are intervals of time during which at least some of these three shoppers are engaged in the same activity; for example, between B_0 and B_1 all three customers are browsing in Area 1 and between C_1 and A_2 customers A and C are in Queue 1.

Each of the service desks corresponds to a resource entity and shoppers need to acquire ('seize') this resource in order to pay for items being purchased before moving on to another shopping area. The payment activity at a service desk is highly structured and is, in fact, representative of a wide class of activities that can take place in DEDS models. Notice several key features:

1. There is a precondition that must be true before the service activity can begin (the server must be available and there must be a shopper seeking to carry out a payment transaction).
2. The service activity has a duration; that is, it extends over an interval of time.
3. One or more state variables change value when the service function is completed (e.g., at time $A_3 = C_2$ the number of shoppers in browsing

Area 3 increases by one and the number in the queue in front of service desk 1 decreases by one).

The payment procedure which shoppers carry out in this example maps onto one of the basic constructs in our conceptual modelling framework called an *Activity*. An Activity represents a unit of behaviour. Its role is to encapsulate some aspect of the interaction among the various entities that exist within the model. An essential feature of this interaction is a collection of changes in the value of some of the state variables within the model. The notion of 'unit' here is intended to suggest minimality; in other words, an Activity should be viewed as an atomic construct in the sense that it captures an aspect of the model's behaviour that is not amenable to subdivision (at least from the perspective taken by the model builder).

In general, an Activity has an initial phase, a duration, and a terminal phase. Both the initial phase and the terminal phase unfold instantaneously (i.e., they consume no (simulated) time). The duration aspect of an Activity carries the important implication that once it has become 'energised', an Activity cannot end until there has been an elapse of some number of time units. This duration need not map onto a contiguous time interval but may instead correspond to a collection of disjoint intervals.

From a structural point of view, an Activity is assembled from more primitive constituents called events. When an event occurs, a construct that we call a *status change specification (SCS)* captures its impact on the model. There is an event associated with both the initial phase and the terminal phase of an Activity construct.

The event is a primitive construct in the sense that it can only appear as a constituent within an Activity construct. Its associated SCS includes, as a minimum, the designation of changes to some set of state variables within the model. The state variables that are referenced generally have some affinity in terms of their relevance in characterising some particular aspect of the model's status.

An event begins and ends at the same point in (simulated) time and consequently all the changes specified in its associated SCS occur simultaneously. An event can be either conditional or scheduled. The distinction reflects the manner in which the event is 'activated'. If it is conditional, its activation depends on the value of one or more state and/or input variables. On the other hand, if it is scheduled, then its activation will occur at some predefined value of time t, and independent of the state of the model.

The event that corresponds to the initial phase of an Activity (i.e., its starting event) occurs when a prescribed logical expression associated with the Activity (its precondition) acquires a TRUE value. Although there are some important exceptions, the precondition is generally formulated in terms of the various state variables and/or input variables within the model. Hence the starting event of an Activity is a conditional event. Furthermore, it always includes a state variable change that inhibits an immediate reactivation of the Activity (in other words, a change which gives the precondition a FALSE value). Notice that the implication here is that when a precondition is present, a starting event is a mandatory component for an Activity.

Activation of the event that corresponds to the terminal phase of an Activity (i.e., its terminating event) takes place immediately upon the completion of the Activity's duration (hence it can be regarded as a scheduled event). Although typically present, a terminating event is not a mandatory feature of an Activity. The state changes resulting from the activation of a terminating event may cause preconditions of multiple activity constructs to become TRUE thereby enabling those constructs. This demonstrates that multiple Activities within the model can be simultaneously 'in progress'.

When an Activity starts it has a tentative duration whose length is either already known before the Activity begins, or else is an intrinsic part of the Activity's specification. This length is frequently established via a data model which, therefore, implies that a data modelling stage is often embedded in the Activity's formulation. In the most common circumstance, the duration Δ of an Activity does not change once the Activity is initiated. Furthermore it typically maps onto a continuous time interval. In these circumstances the termination time t_{end} of an Activity is predetermined when the Activity begins; that is,

$$t_{end} = (t_{start} + \Delta),$$

where t_{start} is the value of time t when the Activity's precondition acquired a TRUE value and hence the Activity was initiated. The terminating event (if present) occurs at time at $t = t_{end}$.

The Activity, as described thus far, should be regarded simply as a generic concept. In reality, its realisation in our ABCmod conceptual modelling framework maps onto a collection of closely related constructs whose properties are outlined in the discussion that follows. Each of these constructs has a predetermined format that can be conveniently presented in terms of a template. These templates serve to provide a convenient means of summarising salient features.

Activity: This is the most fundamental member of the collection of constructs. Each occurrence of this construct in the model has a name and is organised according to a template whose format is given in Table 4.4. Each status change specification (SCS) usually includes (but is not restricted to) the identification of changes in value to some collection of state variables.

TABLE 4.4. Template for an Activity.

	Activity: ***ActivityName***
Precondition	*Boolean expression that specifies the conditions which initiate the Activity*
Event	*SCS associated with Activity initiation*
Duration	*The length of the duration (typically acquired from a Data Module that references a data model)*
Event	*SCS associated with Activity completion*

Our convention of regarding an Activity as an atomic unit of behaviour precludes embedding within it a secondary behaviour unit even when it may be closely related. One such situation occurs when one behaviour unit directly follows upon completion of another without the need to 'seize' a further resource. Our notion of a Triggered Activity provides the means for handling such situations.

As an example, consider a port where a tugboat is required to move a freighter from the harbour entrance to a berth where a loading (or unloading) operation immediately begins. Here the berthing and the loading operations both map onto Activities but the latter Activity is distinctive because (by assumption) no additional resource is required and hence it can be immediately initiated upon completion of the berthing operation. It is because of this absence of a precondition that the loading operation maps onto a Triggered Activity in our ABCmod framework.

Triggered Activity: The distinguishing feature of a Triggered Activity is that its initiation is not established by a precondition but rather by an explicit reference to it within the terminating event of some Activity, for example, TA.*TriggeredActivityName*. Note that this demonstrates that an SCS can be more than simply a collection of specifications for state variable changes inasmuch as it can also include a reference to a Triggered Activity which, in turn, serves to initiate that Activity. The template for the Triggered Activity is given in Table 4.5.

As we have previously indicated, an Activity is associated with a unit of behaviour within the model. The flow of this behaviour may, however, be subjected to an intervention which disrupts the manner in which the Activity unfolds. Such an intervention can have a variety of possible effects; for example,

a) The initial (tentative) duration of the Activity may be altered.
b) The duration may no longer map onto a continuous time interval but may instead map onto two or more disjoint intervals.
c) A combination of (a) and (b).
d) The behaviour intrinsic to the Activity may be stopped and may never be resumed.

TABLE 4.5. Template for the Triggered Activity.

Triggered Activity: *ActivityName*	
Event	*SCS associated with Activity initiation*
Duration	*The length of the duration (typically acquired from a Data Module that references a data model).*
Event	*SCS associated with Activity completion*

There are two possible types of intervention; namely, pre-emption and interruption. We examine each of these in turn.

Pre-emption: This typically occurs in a situation where two (or more) Activities require the same resource which cannot be shared. Such a circumstance is commonly resolved by assigning access priorities to the various competing Activities. With this approach, an Activity can disrupt the duration of some lower-priority Activity that is currently accessing the resource. There is, however, an implication here that some consumer entity instance that is 'connected' to the resource will be displaced. When this occurs, the completion of the service function for the displaced cei is suspended and consequently the duration of the Activity, from the perspective of the displaced cei, becomes distributed over at least two disjoint time intervals, or in the extreme case may never even be completed.

Interruption: Changes in the value of an input variable are generally reflected in one or more of the Activities within the model. For example, in response to a change in value of an input variable, an 'in progress' Activity may undergo a change in the manner in which it completes the task that was initially undertaken. Such an intervention is called an *interrupt.* Generally an interrupt is characterised by a set of changes as reflected in an SCS. An interrupt shares some common features with the notion of pre-emption but the range of possible behaviour alterations is broader.

To accommodate the requirements involved in handling an intervention, a more general construct than the Activity is necessary. This construct is called an Extended Activity and its template is given in Table 4.6.

Extended Activity: As its name suggests, this construct can accommodate more general behaviour and is the most comprehensive of the Activity constructs.

TABLE 4.6. Template for the Extended Activity.

Extended Activity: ***ActivityName***	
Precondition	*Boolean expression that specifies the conditions which initiate the Activity*
Event	*SCS associated with Activity initiation*
Duration	*The length of the duration (typically acquired from a Data Module that references a data model).*
Pre-emption Event	*SCS associated with Activity pre-emption*
Interrupt Precondition	*Boolean expression that specifies the conditions under which an interrupt occurs*
Event	*SCS associated with Activity interruption*
Event	*SCS associated with Activity completion*

The notion of intervention is equally relevant to a Triggered Activity. This gives rise to a generalisation of the Triggered Activity construct that we call an Extended Triggered Activity.

Extended Triggered Activity: Like its 'basic' counterpart, the distinguishing feature of an Extended Triggered Activity is that its initiation is not established by a precondition but rather by an explicit reference to it within the terminating event of some Activity. The template for an Extended Triggered Activity is given in Table 4.7.

The final member of the family of activity constructs is called an *Action*. An Action resembles other Activity constructs inasmuch as it also is initiated when a precondition acquires a TRUE value. The principal difference is that an Action does not include a duration, hence it unfolds at a single point in time and has a single SCS. Often the precondition for an Action is formulated exclusively in terms of the variable t (time), that is, without reference to the state variables of the model. In such circumstances, the Action can be regarded as being autonomous. Such a circumstance rarely occurs in the case of any of the other Activity constructs.

TABLE 4.7. Template for the Extended Triggered Activity.

Extended Triggered Activity: ***ActivityName***	
Event	*SCS associated Activity initiation*
Duration	*The length of the duration (typically acquired from Data Module that references a data model).*
Pre-emption Event	*SCS associated with Activity pre-emption*
Interrupt Precondition	*Boolean expression that specifies the conditions under which an interrupt occurs.*
Event	*SCS associated with Activity interruption*
Event	*SCS associated with Activity completion*

An Action may be intrinsically repetitive. This can occur as a consequence of a recurring change in the value of some variable within the precondition in either an explicit or an implicit fashion. In the former case, the change is embedded in the Action's SCS. An Action with such a repetitive property is called an *Action Sequence*. As becomes clear in the discussion of Input in Section. 4.2.5, this particular realisation of an Action (i.e., the Action Sequence) provides the basis for handling input entity streams within the ABCmod framework. The template for an Action (or Action Sequence) is given in Table 4.8.

TABLE 4.8. Template for the Action/Action Sequence.

Action /Action Sequence: ***ActionName***	
Precondition	*Boolean expression that specifies the conditions which initiate the Action*
Event	*The SCS associated with Action initiation*

Another common application of the Action construct is the accommodation of a circumstance where the current state of the model inhibits a particular state change that might otherwise take place. In effect, a delay of uncertain length is thus introduced. An Action can be used to provide a 'sentinel' that awaits the development of the conditions that permit the state change to occur.

Table 4.9 summarises the important features of the various Activity constructs.

TABLE 4.9. Features of the family of Activity constructs.

Feature	Activity	Triggered Activity	Extended Activity	Extended Triggered Activity	Action/Action Sequence
Precondition	Yes	No	Yes	No	Yes
Starting Event	Yes	Yes	Yes	Yes	Yes
Duration	Yes	Yes	Yes	Yes	No
Intervention	No	No	Yes	Yes	No
Terminating Event	Optional	Optional	Optional	Optional	No

4.2.5 Inputs

A model may have many inputs. Each represents a particular aspect of the SUI's environment that affects the behaviour that is of interest. In general a variable u, selected to represent any such input within the model is, in fact, a function of time; that is, $u = u(t)$. In the case of a DEDS model the essential information about an input $u(t)$ is normally available as a sequence of ordered pairs of the form: $<(t_k, u_k): k = 0, 1, 2, \ldots>$ where t_k is a value of time and $u_k = u(t_k)$ and we assume that $t_i < t_j$ for $i < j$. Each of the time values t_k in this sequence identifies a point in time where there is a noteworthy occurrence in the input u (typically a change in its value). We refer to this sequence as the *characterising sequence* for u and denote it CS[u].

For any input u, the specifications that allow the construction of CS[u] are part of the data modelling task associated with model development. In this regard, however, note that there are two separate sequences that can be associated with the characterising sequence CS[u] = $[(t_k, u_k): k = 0, 1, 2, \ldots]$. These are:

$$CS_D[u] = <t_k: k = 0, 1, 2, \ldots>$$

$$CS_R[u] = <u_k: k = 0, 1, 2, \ldots>,$$

which we call, respectively, the *domain sequence* for u and the *range sequence* for u. It is almost always true that the domain sequence for any input u has a stochastic characterisation, that is, a stochastic data model. Generally this implies that if t_j and $t_{j+1} = t_j + \Delta_j$ are successive members of $CS_D[u]$, then the value of Δ_j is provided by a stochastic model. The range sequence may or may not have a stochastic characterisation.

From the perspective of the ABCmod conceptual modelling process we assume that the data modelling task has been completed. This, in particular, means that valid mechanisms for creating the domain sequence and the range sequence for each input variable are available.

Often a variable u that represents an input to a DEDS model falls in the class of piecewise-constant (PWC) time functions. A typical member of this class is shown in Figure 4.3.

The input variable $u(t)$ in Figure 4.3 could, for example, represent the number of electricians, at time t, included in the maintenance team of a large manufacturing plant that operates on a 24 hour basis but with varying levels of production (and hence varying requirements for electricians). The behaviour of the model over the interval $[t_j, t_{j+1})$ likely depends explicitly on the value $u_k = u(t_j)$ hence the representation of $u(t)$ as a PWC function is not only meaningful but is, in fact, essential. The characterising sequence for u is:

$$\text{CS}[u] = <(t_0,1), (t_1,2), (t_2,4), (t_3,4), (t_4,3), (t_5,1), (t_6,2)>.$$

Observe also that with the interpretation given above this particular input is quite distinctive inasmuch as neither its domain sequence nor its range sequence will likely have a stochastic characterisation.

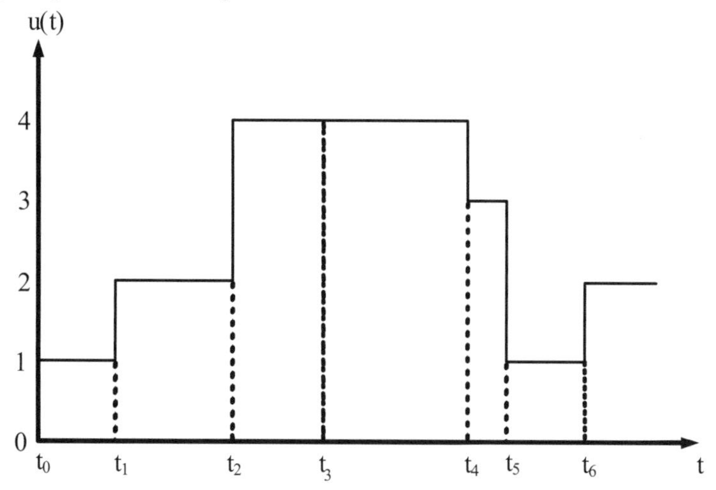

FIGURE 4.3. A piecewise constant time function.

As another example consider a case where an input variable u represents the number of units of a particular product requested on orders received (at times $t_1, t_2, \ldots, t_j, \ldots$) by an Internet-based distributing company. The characterising sequence would be written as

$$\text{CS}[u] = <(t_1, u_1), (t_2, u_2), \ldots, (t_j, u_j) \ldots>.$$

Note however that only the specific values $u_1 = u(t_1)$, $u_2 = u(t_2)$, . . . , $u_j = u(t_j)$ are relevant. In other words, representation of this particular input as a PWC time function is, strictly speaking, not appropriate because the value of u between order times has no meaning. Note also that the data model for this input would need to provide a specification for both the domain sequence $CS_D[u] = <t_1, t_2, \ldots, t_j, \ldots>$ and the range sequence of order values $CS_R[u] = <u_1, u_2, \ldots, u_j, \ldots>$ and both would likely be in terms of specific probability distribution functions.

The notion of an input entity stream was introduced in Chapter 2 as the vehicle for describing the arrival of members of the various consumer entity classes that typically flow into the domain of a DEDS model. Suppose we associate the variable $\tilde{u}_K = \tilde{u}_K(t)$ with the input entity stream corresponding to consumer entity class Θ_K. The characterising sequence for $\tilde{u}_K$ can be written as

$$CS[\tilde{u}_K] = <(t_\eta, 1), (t_{\eta+1}, 1), (t_{\eta+2}, 1), \ldots, (t_j, 1), \ldots>.$$

Here each value in the domain sequence $<t_\eta, t_{\eta+1}, t_{\eta+2}, \ldots, t_j, \ldots>$ is the arrival time of a cei from class Θ_K. Each element of the range sequence has a value of 1 (i.e., $\tilde{u}_K(t_j) = 1$ for all j) because we generally assume that arrivals occur one at a time. Also $\eta = 0$ if the first arrival occurs at $t = t_0$ (the left boundary of the observation interval) and $\eta = 1$ otherwise (i.e., the first occurrence is at $t_1 > t_0$). The domain sequence is constructed from the arrival process associated with consumer entity class Θ_K.

Of particular importance to the handling of input variables in our ABCmod framework is a time variable $M[u](t)$ which we associate with the input u to a DEDS model. This function is called the *timing map* for $u(t)$ and is constructed from the information embedded in the domain sequence $CS_D[u]$ for u. Suppose t_{j-1} and t_j are two successive times in the domain set $CS_D[u]$; then

$$M[u](t) = t_j \quad \text{for} \quad t_{j-1} < t \leq t_j .$$

In other words, at time t, $M[u](t)$ provides the value of the entry in $CS_D[u]$ that most closely follows (or is equal to) t. In more formal terms, $M[u]$ is a mapping from the segment of the real line that corresponds to the observation interval to the set $CS_D[u]$. The timing map for the input function shown in Figure 4.3 is shown in Figure 4.4.

There are generally many inputs to a DEDS model; some correspond to input entity streams and some do not. Nevertheless all inputs have a characterising sequence and hence an associated timing map. The behaviour implicit in each such input is captured in the ABCmod framework by an Action Sequence whose precondition is based on the timing map for the input variable. It should be noted that it is only in limited circumstances that the domain sequence $CS_D[u]$ is explicitly known; generally the values in $CS_D[u]$ evolve in a stochastic manner. The implication here is that the timing map for u can rarely be explicitly drawn. Note also that if the first entry in $CS_D[u]$ is t_0 (the left-hand boundary of I_0) then, by definition, $M[u](t_0) = t_0$.

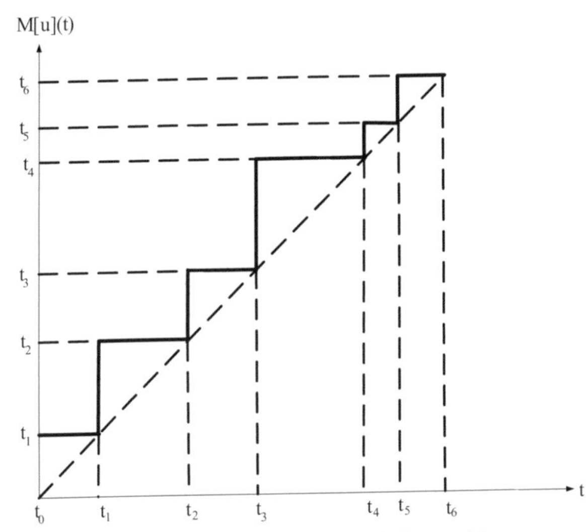

FIGURE 4.4. Timing map for the input function of Figure 4.3.

The salient features of all inputs to a particular DEDS model are summarised in an Inputs template whose format is shown in Table 4.10. The templates for the Action Sequences referenced in Table 4.10 are given in Table 4.11(a) and Table 4.11(b).

TABLE 4.10. Template for Inputs.

Inputs				
Input Variable	**Description**	**Data Models**		**Action Sequence**
		Domain Sequence	**Range Sequence**	
$u(t)$	*Description of the input which the input variable represents*	*Details for* $CS_D[u]$	*Details for* $CS_R[u]$	*Name of the associated Action Sequence*
$\tilde{u}_K(t)$	*Description of the input entity stream which the input variable represents*	*Details for* $CS_D[\tilde{u}_K]$	*Details for* $CS_R[\tilde{u}_K]$	*Name of the associated Action Sequence*

TABLE 4.11. Templates for Action Sequences.

(a) Case where the Action Sequence corresponds to the input variable, $u(t)$.

Action Sequence: *ActionName(u)*	
Precondition	$t = M[u](t)$
Event	*Typically the assignment to u of the value which it acquires at time t (this need not be different from the value prior to time t). The values for* $M[u](t)$ *are derived from the domain sequence* $CS_D[u]$.

(b) Case where the Action Sequence corresponds to an input entity stream.

Action Sequence: *ActionName(C.EntityClassName)*	
Precondition	$t = M[\tilde{u}_K](t)$
Event	*Typically a collection of value assignments to the attributes of the arriving consumer entity instance from the class referenced in the argument together with a state variable change that reflects the initial positioning (connection) of the arriving consumer entity instance. The values for* $M[\tilde{u}_K](t)$ *are derived from the domain sequence* $CS_D[\tilde{u}_K]$.

4.2.6 Outputs

In our discussion of output variables in Section 2.2.4 we indicated that such variables fall into one of two categories based on the nature of the values that they represent. More specifically, the value associated with an output variable may be either:

- A set of data values (either a trajectory set associated with a time variable or a sample set associated with a consumer entity class)
- A scalar value usually obtained via an operation on a data set

Inasmuch as the output variables provide the information that is either explicitly or implicitly required for achieving the goals of the modelling and simulation project, it is important that they be documented in a meaningful way in our ABCmod framework. This documentation is organised in terms of three templates whose general structure is shown in Table 4.12 through Table 4.14.

Recall that both the time variables and the sample variables listed in the Trajectory Set and the Sample Set templates always correspond to attributes defined for entities within the model. (In the interest of facilitating transformation to program code, we have replaced $\mathcal{T}$ with TRJ and Ψ with PHI.)

TABLE 4.12. Template for Summarising Trajectory Sets.

Trajectory Sets	
Name	**Description**
TRJ[y]	*Description of the time variable y(t)*

TABLE 4.13. Template for Summarising Sample Sets.

Sample Sets	
Name	**Description**
PHI[y]	*Description of the sample variable y whose values populate the sample set PHI[y]*

TABLE 4.14. Template for summarising derived scalar output variables.

Derived Scalar Output Variables (DSOV's)			
Name	**Description**	**Output Set Name**	**Operator**
Y	*The meaning of the value acquired by Y*	*The value of Y is obtained by carrying out an operation on the values in this output set*	*The operation that is carried out on the underlying data set to obtain the value assigned to Y*

4.2.7 Data Modules

It is rarely possible to formulate an ABCmod conceptual model without the need to access data. The simplest such requirement is the case where there is a need for a sample from a prescribed distribution function. Alternately, the requirement might be for a sample from one of several specified distributions according to some prescribed rule. The convention we have adopted in our ABCmod framework is to encapsulate any such data delivery requirement within a named data module which serves as a 'wrapper' for the data specification. The rationale here is simply to facilitate modification of the actual source of the data if that need arises. The collection of such data modules that are referenced within an ABCmod conceptual model is summarised in a table whose template is shown in Table 4.15.

TABLE 4.15. Template for summarising data modules.

Data Modules		
Name	**Description**	**Data Model**
ModuleName(parameterlist)	*Description of the purpose of the data module called ModuleName.*	*Details of the mechanism that is invoked in order to generate the data values provided by the data module called ModuleName.*

4.2.8 Standard Modules and User-Defined Modules

A variety of 'standard' operations reoccur in the formulation of the SCSs within the various activity constructs that emerge during the development of any ABCmod conceptual model. We assume the existence of modules to carry out these operations and each of these is briefly outlined below.

- InsertQue(QueueName, item)
  This module carries out an insertion into a queue called QueueName. The second argument holds the item to be inserted. The insertion takes place according to the declared queuing protocol associated with the QueueName.
- InsertQueHead(QueueName, item)
  This module carries out an insertion into a queue called QueueName. The second argument holds the item to be inserted. The item is inserted at the head of the queue.
- RemoveQue(QueueName, item)
  This module removes the item which is at the head of the queue called QueueName. The removed item is placed in the second argument.
- InsertGrp(GroupName, item)
  This module carries out an insertion into a group called GroupName. The second argument holds the item to be inserted.
- RemoveGrp(GroupName, item)
  This module removes an item from the group called GroupName. The item to be removed is identified in the second argument.
- Put(SampleSetName, item)
  This module is used to place items into the Sample Set called

SampleSetName. The second argument holds the item to be placed into the set.

- Leave(cei)

  It frequently occurs that a cei's existence within the model comes to an end. The purpose of this module is to explicitly indicate this occurrence which is typically invoked as part of the SCS of the terminating event of an Activity. The argument serves to provide an explicit reference to the cei in question.

Situations typically arise where modules are needed to carry out specialised operations that are not included in the 'standard set' above. These can be freely defined wherever necessary to augment the ABCmod framework and facilitate the conceptual modeling task. They are called user-defined modules and a template is given in Table 4.16.

Note that references to standard modules and user-defined modules in the formulation of activity constructs have the form SM. *ModuleName* () UM.*ModuleName* (), respectively.

TABLE 4.16. Template for summarising user-defined modules.

User-Defined Modules	
Name	**Description**
ModuleName(parameter list)	*Description of the purpose of the user-defined module called ModuleName.*

4.2.9 Intervention

We have previously pointed out (Section 4.2.4) that an Activity may be subjected to an intervention which disrupts the manner in which it unfolds. This disruption has a direct impact upon the duration of the Activity. In our ABCmod framework we recognise two types of intervention which we call Pre-emption and Interruption.

Pre-emption is most commonly associated with the circumstance where the initiation of one Activity (e.g., called ActP) disrupts the flow of another Activity (e.g., called ActQ) because a resource that is required by both Activities must be taken from ActQ and reassigned to ActP because ActP has higher-priority access to the resource. The ABCmod conceptual model presentation of such a circumstance requires that ActQ be formulated as an Extended Activity (see Table 4.6) with a Pre-emption subsegment within its Duration segment. A directive of the form: 'PE.ActQ' in the starting SCS of ActP initiates the pre-emption. This directive links directly to the Pre-emption subsegment of ActQ where the consequences of the Pre-emption are specified.

In the ABCmod framework we typically use the Interrupt notion as the means for handling an intervention that results from a change in the value of one of the conceptual model's input variables. Such a change may have an impact on several Activities and the Interruption subsegment of the Extended Activity provides the means for formulating the condition that defines the occurrence of an Interrupt as well as the consequent reaction to it.

An intervention of the Interrupt category is illustrated in Example 3 that is presented in the following section. Although no explicit example of the Pre-emption category is given, it is convenient mechanism for handling the circumstances outlined in Problem 4.4 of Section 4.4 (Exercises and Projects). Details are left to the reader.

4.3 Some Examples of Conceptual Model Development in the ABCmod Framework

We have emphasised the importance of the conceptual modelling phase of a modelling and simulation project because of its role in bridging the gap between the generalities of the project description and the precision required for the development of the simulation program. In Section 4.2 we presented a framework for formulating a conceptual model for discrete-event dynamic systems and in this section we present several examples to illustrate its application. These examples have been chosen to illustrate how a variety of features that can arise within the DEDS context are handled in the ABCmod framework.

4.3.1 Example 1

Project Description

Tankers are loaded with crude oil at a port that currently has berth facilities for loading three tankers simultaneously. There is one tugboat available to assist the tankers. Tankers of all sizes require the service of the tug to move from the harbour into a berth and later to move out of the berth back to the harbour, at which point they leave the port. The general configuration is shown in Figure 4.5.

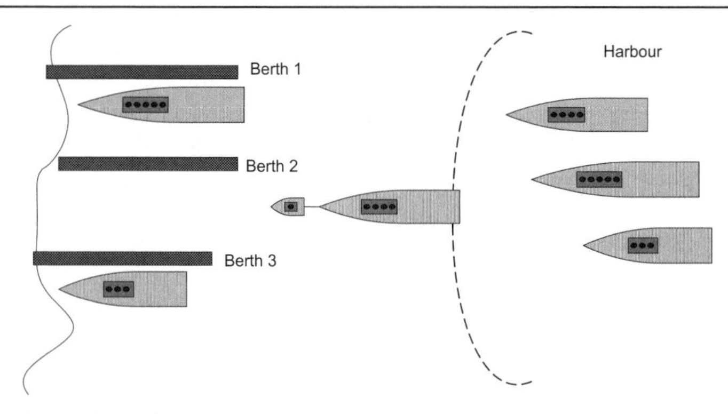

FIGURE 4.5. View of the port's operation.

Project Goals

The port authority is concerned about the increasing number of complaints arising from excessive turnaround time experienced by some tankers. It is therefore considering the construction of a fourth berth. The project goal is to acquire some insight into the likely impact of such an expansion. More specifically, the goal is to compare, for the two cases (three berths and four berths), the average number of berths that are occupied and, as well, the average waiting time of tankers, based on steady-state conditions.

Some clarification is in order with respect to the interpretation of 'waiting time'. For any particular tanker, it is the accumulation of those time intervals during which the tanker is not engaged in some purposeful task. Hence it includes the periods when the tanker is waiting to be towed to berths or waiting to be towed back to the harbour. In the ideal case, these tasks would be initiated immediately and waiting time would be zero.

SUI Details

The interarrival time between tankers can be modelled as a homogeneous stochastic process that has an underlying exponential distribution with a mean of 8 hours. There are three sizes of tanker: small, medium, and large and 25% of the arriving tankers are small, 25% are medium, and 50% are large. The loading time for tankers of each size is a random variable that has a uniform distribution. However specific details differ as follows: small tankers require 9 ± 1 hours to load, medium tankers require 12 ± 2 hours, and large tankers require 18 ± 3 hours. Additional tanker characteristics are summarised in Table 4.17.

The process followed by each tanker upon arrival in the harbour is as follows.

1. It waits in a (virtual) queue for the tugboat to assist it to dock in an available berth. The queue functions on a first-in-first-out (FIFO) basis and is independent of tanker size.

2. The tanker is moved from the harbour into a berth (the berthing operation) by the tug; the loading procedure begins immediately and the tug is released to carry out its next task.
3. When the loading is complete, the tanker once again requests the service of the tug which effectively places it in a queue with other tankers that are waiting for the tug's assistance to depart from a berth.
4. The loaded tanker is moved from the berth back to the harbour entrance by the tug (the deberthing operation).
5. The tanker leaves the harbour and the tug proceeds to its next task.

TABLE 4.17. Tanker characteristics.

Tanker Size	**Proportion of Arrivals (%)**	**Loading Time (hours)**
Small	25	UNIFORM(8,10)
Medium	25	UNIFORM(10,14)
Large	50	UNIFORM(15,21)

The berthing operation (moving an empty tanker from the harbour to an empty berth) takes 2 hours and the deberthing operation (moving a loaded tanker from its berth to the harbour) takes 1 hour. Both these operations are independent of tanker size. When not towing a tanker, the tug requires 0.25 hour to travel from the harbour to the berth area and vice versa. When the tug finishes a berthing operation, it will deberth the first tanker that has completed loading. If no tanker is available for deberthing but tankers are waiting in the harbour and a berth is available, then the tug will travel to the harbour and begin berthing the tanker that has been waiting the longest in the harbour queue. Otherwise, the tug remains idle in the berth area.

When the tug finishes a deberthing operation and there are tankers waiting in the harbour, the tug will berth the one that has been waiting the longest. If there is no tanker waiting to be berthed and the berths are empty, then the tug will simply wait at the harbour entrance for the arrival of a tanker. If, on the other hand, there are no tankers in the harbour but there are tankers in the berths, then the tug will travel back to the berth area without any tanker in tow. Then, if there are tankers ready for deberthing, it will begin deberthing the one that has waited longest. Otherwise the tug will remain idle in the berth area until either a tanker finishes loading and needs to be deberthed or a tanker arrives in the harbour, in which case it will travel back to the harbour without a tanker in tow. The ABCmod conceptual model for example 1 is given in Tables 4.18 through 4.30. For convenience of reference we have assigned the name ABCmod.Port.ver1 to this conceptual model.

ABCmod.Port.ver1

Structural Components

A number of constants are introduced to facilitate the formulation of the conceptual model and these are summarised in Table 4.18. There is one parameter, namely, MaxBerth which has two values of interest, that is, 3 (the 'base' case) and 4.

TABLE 4.18. Summary of constants and parameters for *ABCmod.Port.ver1.*

Constants		
Name	**Role**	**Value**
t_0	Left boundary of the Observation Interval	0 (clock time)
t_f	Right boundary of the Observation Interval	Cannot be predetermined because a steady-state study is required
t_{FA}	Time of first tanker arrival	0 (clock time)
BerthingTime	Time required for the berthing operation	2 (hours)
DeberthingTime	Time required for the deberthing operation	1 (hours)
EmptyTravTime	Harbour to berth (and berth to harbour) travel time for tug when travelling without a tanker in tow	15 (minutes)
AvgArr	Part of the specification for DM. InterArrTime	8 (hours)
PerSml	Percentage of arriving tankers that are small	25 %
PerMed	Percentage of arriving tankers that are medium	25 %
PerLrg	Percentage of arriving tankers that are large	50 %
SmlMin	Part of the specification for DM.LoadTimeSml	8 (hours)
SmlMax	Part of the specification for DM.LoadTimeSml	10 (hours)
MedMin	Part of the specification for DM.LoadTimeMed	10 (hours)

Constants		
Name	**Role**	**Value**
MedMax	Part of the specification for DM.LoadTimeMed	14 (hours)
LrgMin	Part of the specification for DM.LoadTimeLrg	15 (hours)
LrgMax	Part of the specification for DM.LoadTimeLrg	21 (hours)
Parameters		
Name	**Role**	**Values**
MaxBerth	Number of operational berths at the port	3 and 4

The tankers represent the only consumer entity class within the Project Description. The attributes selected for this class are summarised in Table 4.19.

TABLE 4.19. The tanker consumer entity class.

Consumer Entity Class: Tanker	
This consumer entity class represents the tankers that arrive at the port for loading.	
Attributes	**Description**
Size	The size of the tanker (value is one of SMALL, MEDIUM, LARGE) as assigned via DM.TankerSize
StartWait	A timestamp used to determine waiting times
TotalWait	Accumulated waiting time

The tugboat that provides the berthing/deberthing service is the one resource entity in the conceptual model. Its selected attributes are summarised in the Resource template given in Table 4.20.

In the perspective we have chosen, the tankers being loaded in the berths are regarded as being members of a group aggregate. In addition we

TABLE 4.20. The tugboat Resource Entity.

Resource Entity: Tug	
This resource entity represents the tugboat that is needed to berth tankers that arrive in the harbour and to deberth tankers that have finished loading.	
Attributes	**Description**
Status	Indicates the status of the tug as specified by one of the following values BERTHING – berthing a tanker DEBERTHING – deberthing a tanker TOHARBOUR – going to the harbour with no tanker in tow TOBERTHS – going to berth area with no tanker in tow PauseH – in the harbour following the completion of a deberthing operation PauseB – in the berth area following the completion of a berthing operation
Tnkr	Attribute-tuple of the tanker being towed (when applicable)

introduce two queue aggregates, one to accommodate the virtual queue formed by tankers in the harbour waiting for berthing service from the tug and the other to represent the virtual queue of tankers that have been loaded and are waiting in a berth for a tug in order to begin a deberthing operation. The specifications for these aggregates are provided in Table 4.21.

TABLE 4.21. Templates for the various aggregate entities.

(a) The group aggregate representing the berths.

Aggregate Entity: BerthGrp	
This group aggregate represents the collection of tankers that are being loaded in one of the berths.	
Attributes	**Description**
List	A list of the attribute-tuples of the tankers that occupy a berth
N	The number of entries in List (maximum value is MaxBerth)

(b) The queue aggregate representing the (virtual) tanker queue in the harbour.

Aggregate Entity: HarbourQue	
This queue aggregate represents the (virtual) queue of tankers in the harbour harbour waiting to be assisted into a berth by the tug.	
Attributes	**Description**
List	A list of attribute-tuples of the tankers in the harbour waiting for the tug Discipline: FIFO
N	Number of entries in List

(c) The queue aggregate representing the (virtual) tanker queue at the berths.

Aggregate Entity: DeberthQue	
This queue aggregate represents the (virtual) queue of tankers that have completed loading and are waiting in a berth for deberthing assistance from the tug.	
Attributes	**Description**
List	A list of attribute-tuples of the tankers waiting to be deberthed Discipline: FIFO
N	Number of entries in List

Input–Output Components

The tankers that flow through the port represent an input entity stream. The variable $\tilde{u}_T(t)$ is used to represent this stream and its characterisation is provided in Table 4.22.

TABLE 4.22. Input for *ABCmod.Port.ver1*.

Input				
Input Variable	**Description**	**Data Models**		**Action Sequence**
		Domain Sequence	**Range Sequence**	
$\tilde{u}_T(t)$	$\tilde{u}_T$ represents the input entity stream corresponding to the Tanker consumer entity class	First arrival: $t = t_{FA}$ interarrival time: EXP(AvgArr)	All values equal to 1	TankerArrivals (C.Tanker)

One of the project requirements is to determine the average number of berths that are occupied over the duration of the observation interval. This corresponds to the average value of the time variable *A.BerthGrp.N* which is an attribute of the group aggregate called BerthGrp (see Table 4.21a). The Trajectory Set associated with *A.BerthGrp.N* is summarised in

Table 4.23. The specific output of interest is provided by the DSOV AvgOccBerths which is described in Table 4.25 below.

TABLE 4.23. Trajectory set for *ABCmodPort.ver1*.

Trajectory Sets	
Name	**Description**
TRJ[A.BerthGrp.N]	The time variable A.BerthGrp.N is an attribute of the group aggregate BerthGrp and provides the number of occupied berths at any moment *t*

A second project requirement is to determine the average time which tankers spend waiting for assistance from the tug. Recall that such waiting can occur in two separate contexts: the first is upon arrival in the harbour (waiting to begin the berthing operation) and the second is upon completion of loading (waiting to begin the deberthing operation). These two durations are accumulated in the sample variable *C.Tanker.TotalWait* which is an attribute of the Tanker consumer entity class (see Table 4.19). The associated sample set is summarised in Table 4.24. The specific output value of interest is provided by the DSOV AvgWaitTime which is described in Table 4.25.

TABLE 4.24. Sample set for *ABCmod.Port.ver1*.

Sample Set	
Name	**Description**
PHI[Tanker.TotalWait]	Each member of PHI[Tanker.TotalWait] is the final value of the attribute Tanker.TotalWait for some instance of the consumer entity class called Tanker. This attribute serves to accumulate the time spent, by that instance, waiting for the tug

TABLE 4.25. Derived scalar output variables for *ABCmod.Port.ver1*.

Derived Scalar Output Variables (DSOVs)			
Name	**Description**	**Output Set Name**	**Operator**
AvgOccBerths	Average number of occupied berths	TRJ[A.BerthGrp.N]	AVG
AvgWaitTime	Average time that the tankers spend waiting for the tug	PHI[Tanker.TotalWait]	AVG

Behaviour Components

Time units: hours
Observation interval: $t_0 = 0$, t_f: cannot be predetermined because a steady state study is required

The attributes that require initialisation are listed in Table 4.26 with their chosen initial values.

TABLE 4.26. Initialisation requirements.

Initialise
R.Tug.Status ← PauseB A.BerthGrp.N ← 0 A.HarbourQue.N ← 0 A.DeberthQue.N ← 0

Several data modules are required in the formulation of ABCmod.Port.ver1. These are summarised in Table 4.27.

TABLE 4.27. Data modules for *ABCmod.Port.ver1*.

Data Modules		
Name	**Description**	**Data Model**
TankerSize	Returns a value for the Size attribute of an arriving tanker	Percent of small, medium, and large tankers is given by the constants PerSml, PerMed and PerLrg repectively
LoadTimeSml	Returns the loading time for a small tanker	UNIFORM(SmlMin, SmlMax)
LoadTimeMed	Returns the loading time for a medium tankers	UNIFORM(MedMin, MedMax)
LoadTimeLrg	Returns the loading time for a large tankers	UNIFORM(LrgMin, LrgMax)

One user-defined module has been identified to facilitate the ABCmod conceptual model formulation. It is summarised in Table 4.28.

TABLE 4.28. User-defined module for *ABCmod.Port.ver1*.

User-Defined Modules	
Name	**Description**
LoadingTime(Size)	The returned value is the loading time for a tanker assigned according to tanker size. This module accesses one of the data modules LoadTimeSml, LoadTimeMed, or LoadTimeLrg according to the value of the Size argument

The most important step in characterising the behaviour aspect of any ABCmod conceptual model is the identification of the relevant 'units of behaviour' that are embedded within SUI. Each of these is associated with an Activity construct. Table 4.29 summarises these constructs for the port project under consideration.

TABLE 4.29. Summary of Activity constructs for *ABCmod.Port.ver1*.

Summary of Activity Constructs	
Action Sequence	
TankerArrivals	The Input Entity Stream of arriving tankers
Activities	
Berthing	The tug moves an empty tanker from the harbour into an empty berth
Deberthing	The tug moves a loaded tanker from a berth out to the harbour
MoveToHarbour	The tug moves to the harbour without any tanker in tow, to get a tanker waiting to be berthed
MoveToBerths	The tug moves to the berth area from the harbour without any tanker in tow
Triggered Activities	
Loading	The loading of a tanker begins as soon as it is berthed. Thus the Loading Activity is triggered by the Berthing Activity

The details for each of the Activity constructs listed in Table 4.29 are provided in Table 4.30.

TABLE 4.30. Elaboration of the various Activity constructs.

(a) The Action Sequence called Tanker Arrivals.

Action Sequence: TankerArrivals(C.Tanker)	
Precondition	$t = M[\tilde{u}_T](t)$
Event	C.Tanker.Size ← DM.TankerSize C.Tanker.StartWait ← t C.Tanker.TotalWait ← 0 SM.InsertQue(A.HarbourQue, C.Tanker)

(b) The Action Sequence called Berthing.

Activity: Berthing[1]	
Precondition	(R.Tug.Status = PauseH)&(A.HarbourQue.N > 0)
Event	R.Tug.Status ← BERTHING SM.RemoveQue(A.HarbourQue, R.Tug.Tnkr) R.Tug.Tnkr.TotalWait +← (t – R.Tug.Tnkr.StartWait)
Duration	BerthingTime
Event	SM.InsertGrp(A.BerthGrp, R.Tug.Tnkr) TA.Loading(R.Tug.Tnkr) R.Tug.Status ← PauseB

(c) The Triggered Activity called Loading.

Triggered Activity: Loading(C.Tanker)	
Event	
Duration	UM.LoadingTime(C.Tanker.Size)
Event	C.Tanker.StartWait ← t SM.InsertQue(A.DeberthQue, C.Tanker)

(d) The Activity called Deberthing.

Activity: Deberthing	
Precondition	(R.Tug.Status = PauseB)&(A.DeberthQue.N ≠ 0)
Event	R.Tug.Status ← DEBERTHING SM.RemoveQue(A.DeberthQue, R.Tug.Tnkr) R.Tug.Tnkr.TotalWait +← (t – R.Tug.Tnkr.StartWait) SM.Put(PSI[Tanker.TotalWait], R.Tug.Tnkr.TotalWait) SM.RemoveGrp(A.BerthGrp, R.Tug.Tnkr)
Duration	DeberthingTime
Event	R.Tug.Status ← PauseH SM.Leave(R.Tug.Tnkr)

[1] The operator '+←' is an add and assign operator. For example, $x + \leftarrow 1$ should be interpreted as $x \leftarrow x + 1$.

(e) The Activity called MoveToHarbour.

Activity: MoveToHarbour	
Precondition	(R.Tug.Status = PauseB)& (A.DeberthQue.N = 0)& (A.HarbourQue.N > 0)& (A.BerthGrp.N < MaxBerth)
Event	R.Tug.Status ← TOHARBOUR
Duration	EmptyTravTime
Event	R.Tug.Status ← PauseH

(f) The Activity called MoveToBerths.

Activity: MoveToBerths	
Precondition	(R.Tug.Status = PauseH) & (A.HarbourQue.N = 0) & (A.BerGrp.N>0)
Event	R.Tug.Status ← TOBERTHS
Duration	EmptyTravTime
Event	R.Tug.Status ← PauseB

4.3.2 Example 2

The modelling and simulation project in this Example is an extension of the one considered in Example 1. The Project Description for Example 1 indicates that upon completion of a deberthing task, the tug returns to the berthing area without any tanker in tow when there is no tanker in the harbour queue and the berths are not empty. This could be regarded as unrealistically simplistic because it ignores the situation where a tanker arrives in the harbour 'shortly after' the tug embarks on its trip back to the berth area. In this example we introduce an alternate, and possibly more practical, strategy for dealing with this case. As becomes apparent in the discussion that follows, this change gives rise to the need for an interrupt.

Project Description

The harbour operation as outlined in the Project Description for Example 1 is changed in one respect which relates to the tug's behaviour on completion of a deberthing task. When there is no tanker in the tanker queue waiting to be berthed and the berths are not empty, the tug again embarks on a return trip to the berthing area without a tanker in tow. However, in this modified case, if a tanker arrives in the harbour and the tug has not yet completed 70% of its journey back to the berths or if there is no tanker waiting to be deberthed, then the tug returns to the harbour entrance to service the newly arrived tanker.

ABCmod.Port.ver2

Several extensions need to be incorporated into the conceptual model for Example 1 in order to accommodate the additional feature outlined in the revised project description of Example 2. These are summarised below.

- The time required for the tug to return to the harbour if such a requirement arises under the new policy, needs to be determined. We take this duration to be equal to the elapsed time between the tug's departure from the harbour and the arrival of a tanker in the harbour. This determination requires a new (timestamp) attribute for the tug which holds the value of time when the tug leaves the harbour. The additional attribute of the tug is shown in the revised Tug specification given in Table 4.31.

TABLE 4.31. Revised tug specification for *ABCmod.Port.ver2*.

Resource Entity: Tug	
This resource entity represents the tugboat that is needed to berth tankers that arrive in the harbour and deberth tankers that have finished loading.	
Attributes	**Description**
Status	Indicates the task that is being carried out by the tug as specified by one of the following values BERTHING – berthing a tanker DEBERTHING – deberthing a tanker TOHARBOUR – going to the harbour with no tanker in tow TOBERTHS – going to the berth area with no tanker in tow PauseH – in the harbour following the completion of a deberthing operation PauseB – in the berth area following the completion of a berthing operation
Tnkr	Attribute-tuple of the tanker being towed (when applicable)
StartTime	Timestamp indicating the time when the tug leaves the harbour to travel back to the berth area

- The newly introduced possibility of the tug returning to the harbour under certain conditions implies an intervention within the behaviour specified in the previous MoveToBerths Activity. More specifically, the possibility of an interrupt is introduced and hence this Activity needs to be replaced with an Extended Activity. Its details are given in Table 4.32.

TABLE 4.32. Revision of the MoveToBerths Activity required for *ABCmod. Port.ver2.*

Extended Activity: MoveToBerths	
Precondition	(R.Tug.Status = PauseH)&(A.HarbourQue.N = 0)& (A.BerGrp.N>0)
Event	R.Tug.Status ← TOBERTHS R.Tug.StartTime ← t
Duration	EmptyTravTime
Interrupt Precondition	(A.HarbourQue.N > 0)&(((t-R.Tug.StartTime) < 0.7 *EmptyTravTime)) | (A.DeberthQue.N = 0))
Event	TA.ReturnToHarbour Terminate
Event	R.Tug.Status ← PauseB

- The Event associated with the interrupt in Table 4.32 introduces a new unit of behaviour that we call ReturnToHarbour. This corresponds to a Triggered Activity whose details are provided in Table 4.33.

TABLE 4.33. The Triggered Activity called ReturnToHarbour.

Triggered Activity: ReturnToHarbour	
Event	R.Tug.Status ← TOHARBOUR
Duration	(t – R.Tug.StartTime)
Event	R.Tug.Status ← PauseH

4.3.3 Example 3

In this Example, we further modify the port's operating environment by introducing the possible occurrence of storms. The details surrounding such an occurrence are provided in the revised project description.

Project Description

The operation of the harbour, as outlined thus far, is now subjected to the occurrence of storms. The duration of storms is a random variable that is uniformly distributed; namely, UNIFORM(Short,Long) where Short = 2 hours and Long = 6 hours. Likewise the time between successive storms is a random variable which has an exponential distribution with a mean of AvgCalm = 144 hours. The first storm occurs at time t_{fst} = 7 hours. When a storm arrives the tug together with any tanker that is being towed, battens down hatches and drops anchor. When the storm is over, the tug resumes the Activity that was underway before the storm occurred.

ABCmod.Port.ver3

The input variable *SS*(t) is introduced to represent the storm phenomenon. More specifically, we assign a value of TRUE to this variable when a storm is raging and a value of FALSE when storm conditions are absent. Because the first storm occurs after the beginning of the observation interval, an initial value (of FALSE) needs to be explicitly assigned to *SS*.

Several additional constants have been introduced and these need to be reflected in an update of Table 4.18. This addition is shown in Table 4.34.

The updated Inputs table that incorporates the input variable *SS*(t) is given in Table 4.36. The updated Initialise table is given in Table 4.37.

TABLE 4.34. Addition to Table 4.18 required for *ABCmod.Port.ver3*.

Constants		
Name	**Role**	**Value**
Short	Part of the specification for the data model for input *SS*(t); see Table 4.36	2 (hours)
Long	Part of the specification for the data model for input *SS*(t)	6 (hours)
AvgCalm	Part of the specification for the data model for input *SS*(t)	144 (hours)
t_{fst}	Part of the specification for the data model for input *SS*(t)	7 (hours clock time)

Two additional attributes are introduced for the tug in order to deal with the modified behaviour of the SUI resulting from the storm. These are TravelTime and Anchored. The revised table for the tug is given in Table 4.35.

TABLE 4.35. Revised Tug Specification for *ABCmod.Port.ver3*.

Resource Entity: Tug	
This resource entity represents the tugboat that is needed to berth tankers that arrive in the harbour and deberth tankers that have finished loading.	
Attributes	**Description**
Status	Indicates the task that is being carried out by the tug as specified by one of the following values BERTHING – berthing a tanker DEBERTHING – deberthing a tanker TOHARBOUR – going to the harbour with no tanker in tow TOBERTHS – going to the berth area with no tanker in tow PauseH – in the harbour following the completion of a deberthing operation PauseB – in the berth area following the completion of a berthing operation
Tnkr	Attribute-tuple of the tanker being towed (when applicable)
StartTime	Timestamp indicating the time when the tug leaves the harbour to travel back to the berth area
TravelTime	The travelling time required to complete the current task being carried out by the tug
Anchored	Set to TRUE if tug has been forced to stop because of storm, set to FALSE otherwise

TABLE 4.36. Incorporation of *SS(t)* in the Inputs table for *ABCmod.Port.ver3*.

Inputs				
Input Variable	**Description**	**Data Models**		**Action Sequence**
		Domain Sequence	**Range Sequence**	
SS(*t*)	The input variable *SS*(*t*) represents the storm status	First storm $t \quad t_{fst}$ Duration NIF RM(Short Long) Interstorm time E P(AvgCalm)	alue of *SS* alternates between FALSE (calm prevails) and TR E (storm is raging)	Storm(*SS*)
$\tilde{u}_{Tk}(t)$	$\tilde{u}_{Tk}$ represents the input entit stream corresponding to the Tanker consumer entit class	First arrival $t \quad t_{FA}$ Interarrival time E P(AvgArr)	All values equal to 1	TankerArrivals(C.Tanker)

TABLE 4.37. The modified Initialise table for *ABCmod.Port.ver3*.

Initialise
R.Tug.Status ← PauseB
A.BerthGrp.N ← 0
A.HarbourQue.N ← 0
A.DeberthQue.N ← 0
SS ← FALSE

The Action Sequence Storm(*SS*) referenced in Table 4.36 is given in Table 4.38.

TABLE 4.38. The Action Sequence for *SS(t)* as required in *ABCmod.Port.ver3.*

Action Sequence: Storm(*SS*)	
Precondition	$t = M[SS](t)$
Event	SS = NOT(SS)

The various activity constructs are affected by the newly introduced stormy environment. Because the storm introduces the possibility of an interrupt, these Activities need to be replaced by Extended Activities. These are listed in Table 4.39(a) through Table 4.39(e). Note that often the Event associated with an Interrupt ends with a 'Terminate'. The implication here is that the behaviour in question is stopped, usually to be reinitiated when circumstances permit (in this case, when the storm is over).

TABLE 4.39. Extensions to Extended Activities as required in *ABCmod.Port.ver3.*

(a) The Extended Activity called Berthing.

Extended Activity: Berthing	
Precondition	(*SS* = FALSE) & (((R.Tug.Status = PauseH) & (A.HarbourQue.N > 0)) | ((R.Tug.Anchored = TRUE) & (R.Tug.Status = BERTHING)))

Event	If(R.Tug.Anchored = FALSE)
	R.Tug.Status ← BERTHING
	SM.RemoveQue(A.HarbourQue, Tug.Tnkr)
	R Tug.Tnkr.TotalWait +←
	(t – R.Tug.Tnkr.StartWait)
	R.Tug.TravelTime ← BerthingTime
	Else
	R.Tug.Anchored ← FALSE
	EndIf
	R.Tug.StartTime ← t

Duration	R.Tug.TravelTime
Interrupt Precondition	SS = TRUE
Event	R.Tug.TravelTime - ← t - R.Tug.StartTime R.Tug.Anchored ← TRUE Terminate
Event	SM.InsertGrp(A.BerthGrp, R.Tug.Tnkr) TA.Loading(R.Tug.Tnkr) R.Tug.Status ← PauseB

(b) The Extended Activity called Deberthing.

<table>
<tr><th colspan="2">Extended Activity: Deberthing</th></tr>
<tr><td>Precondition</td><td>(SS = FALSE) &
(((R.Tug.Status = PauseB) & (A.DeberthQue.N≠ 0))
|
((R.Tug.Anchored = TRUE) &
(R.Tug.Status = DEBERTHING)))</td></tr>
<tr><td>Event</td><td>If(R.Tug.Anchored = FALSE)
R.Tug.Status ← DEBERTHING
SM.RemoveQue(A.DeberthQue, R.Tug.Tnkr)
R.Tug.Tnkr.TotalWait +←(t – R.Tug.Tnkr.StartWait)
SM.Put(PSI[Tanker.TotalWait],
R.Tug.Tnkr.TotalWait)
SM.RemoveGrp(A.BerthGrp, R.Tug.Tnkr)
R.Tug.TravelTime ← DeberthingTime
Else
R.Tug.Anchored ← FALSE
EndIf
R.Tug.StartTime ← t</td></tr>
<tr><td>Duration</td><td>R.Tug.TravelTime</td></tr>
<tr><td>Interrupt
Precondition</td><td>SS = TRUE</td></tr>
<tr><td>Event</td><td>R.Tug.TravelTime - ← t - R.Tug.StartTime
R.Tug.Anchored ← TRUE
Terminate</td></tr>
<tr><td>Event</td><td>R.Tug.Status ← PauseH
SM.Leave(R.Tug.Tnkr)</td></tr>
</table>

(c) The Extended Activity called MoveToHarbour.

<table>
<tr><th colspan="2">Extended Activity: MoveToHarbour</th></tr>
<tr><td>Precondition</td><td>(SS = FALSE) &
(((R.Tug.Status = PauseA) &
(A.DeberthQue.N = 0) &
(A.HarbourQue.N > 0) &
(A.BerthGrp.N < MaxBerth))
|
((R.Tug.Anchored = TRUE) &
(R.Tug.Status = TOHARBOUR)))</td></tr>
<tr><td>Event</td><td>If(R.Tug.Anchored = FALSE)
R.Tug.Status ← TOHARBOUR
R.Tug.TravelTime ← EmptyTravTime
Else
R.Tug.Anchored ← FALSE
EndIf
R.Tug.StartTime ← t</td></tr>
<tr><td>Duration</td><td>R.Tug.TravelTime</td></tr>
<tr><td>Interrupt
Precondition</td><td>SS = TRUE</td></tr>
<tr><td>Event</td><td>R.Tug.TravelTime - ← t - R.Tug.StartTime
R.Tug.Anchored ← TRUE
Terminate</td></tr>
<tr><td>Event</td><td>R.Tug.Status ←PauseH</td></tr>
</table>

(d) The Extended Activity called MoveToBerths.

Extended Activity: MoveToBerths	
Precondition	(SS = FALSE)& (((R.Tug.Status = PauseH)& (A.HarbourQue.N = 0)&(A.BerthGrp.N>0)) | ((R.Tug.Anchored = TRUE)& (R.Tug.Status = TOBERTHS)))

Event	If(R.Tug.Anchored = FALSE)
	R.Tug.Status ← TOBERTHS
	R.Tug.TravelTime ← EmptyTravTime
	Else
	R.Tug.Anchored ← FALSE
	EndIf
	R.Tug.StartTime ← t

Duration	R.Tug.TravelTime
Interrupt 1 Precondition	 (A.HarbourQue.N > 0)&(((t - R.Tug.StartTime) < 0.7 *EmptyTravTime))| (A.DeberthQue.N = 0))
Event	TA.ReturnToHarbour Terminate
Interrupt 2 Precondition	 SS = TRUE
Event	R.Tug.TravelTime - ← t - R.Tug.StartTravel R.Tug.Anchored ← TRUE Terminate
Event	R.Tug.Status ← PauseB

(e) The Extended Triggered Activity called ReturnToHarbour.

Extended Triggered Activity: ReturnToHarbour	
Event	R.Tug.Status ← TOHARBOUR R.Tug.TravelTime ← t – R.Tug.StartTime R.Tug.StartTime ← t
Duration	R.Tug.TravelTime
Interrupt Precondition	SS = TRUE
Event	R.Tug.TravelTime - ← t - R.Tug.StartTime R.Tug.Anchored ← TRUE Terminate
Event	R.Tug.Status ← PauseH

4.4 Exercises and Projects

4.1 The dining philosophers problem is a classic vehicle for illustrating the occurrence of deadlock in an operating system and, as well, for exploring strategies to avoid its occurrence. The concern in this problem is to explore the dining philosophers problem from the perspective of a modelling and simulation project.

We imagine five philosophers seated around a circular table. Between each pair of philosophers there is a single fork and in the middle of the table is a large bowl of spaghetti. These philosophers have a very focused existence which consists of a continuous cycle of thinking and eating. There is, however, a basic prerequisite for the eating phase; namely, a philosopher must be in possession of both the fork on his[2] right and the fork on his left in order to access and eat the spaghetti at the center of the table. Inasmuch as the philosophers are an orderly group, they have a protocol for acquiring the forks. When any particular philosopher finishes his thinking phase, he must first acquire the fork on his right and only then can he seek to acquire the fork on his left (whose acquisition enables the initiation of the eating phase). When the eating phase is complete, the philosopher replaces both forks and begins his thinking phase.

The situation outlined above can, however, lead to deadlock. This is a situation where no philosopher is eating and no philosopher is thinking, but rather they are all holding their right fork and waiting to

[2] Although the presentation suggests a group of male philosophers, this should not be taken literally because the group is, in fact, gender balanced.

eat. Under these circumstances the left fork will never become available for any of them and hence none will ever eat!

Suppose that the eating time *ET* for each of the philosophers is an exponentially distributed random variable with the same mean of μ_E minutes. Likewise suppose that the thinking time *TT* for each of the philosophers is an exponentially distributed random variable with the same mean of μ_T minutes. It has been conjectured that there is an 'interesting' relationship between the ratio (μ_E/μ_T) and the time it takes for deadlock to occur (we denote this time interval by T_{dead}). A modelling and simulation study has been proposed to determine if a noteworthy relation can indeed be identified.The assessment is to be based on two graphs.The first is a graph of T_{dead} versus (μ_E/μ_T) with (μ_E/μ_T) in the range 1 to 10 and the second is a graph of T_{dead} versus (μ_E/μ_T) with (μ_E/μ_T) in the range 0.1 to 1.

Formulate an ABCmod conceptual model for the project as outlined above. By way of initialisation, assume that the five philosophers enter the SUI in a sequential manner at times: $t = 0$ (the left-hand boundary of the observation interval), $t = 0.1\ \mu_E$, $t = 0.2\ \mu_E$, $t = 0.3\ \mu_E$, and $t = 0.4\ \mu_E$. Upon entry, each philosopher begins a thinking phase.

4.2 The repeated occurrence of deadlock has greatly annoyed the dining philosophers described in Problem 4.1 After due consideration, they agreed to alter their fork acquisition protocol in one small (but significant) way. Instead of having to first acquire the fork on his right, the fifth philosopher will henceforth be required to first acquire the fork on his left, and only then can he seek to acquire the fork on his right. (It can be readily demonstrated that with this altered rule deadlock will indeed be avoided.) In this modified context the goal of the modelling and simulation project is to develop a graphical presentation of the average waiting time to eat as a function of (μ_E/μ_T) where waiting time is measured from the moment a philosopher stops thinking to the moment when he begins eating.

Formulate an ABCmod conceptual model for these modified circumstances of the dining philosophers. Use the same initialisation procedure that was outlined in Problem 4.1

4.3 A lock system in a waterway provides the means for diverting boat traffic around a section of turbulent water. One (or more) locks are placed in a manmade parallel water channel and each functions as an elevator, raising or lowering boats from one water level to another. In this way boat traffic is able to bypass the nonnavigatable portion of a river. A representation of a typical lock's operation is given in Figure 4.6a.

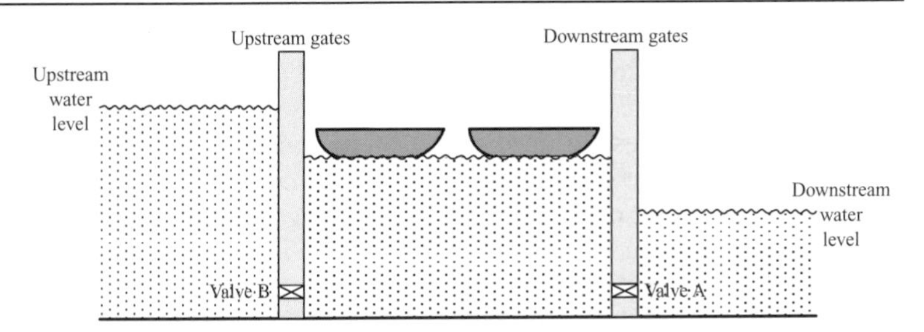

FIGURE 4.6a. Lock operation.

The lock's operation can be divided into two very similar cycles which we refer to as the up-cycle and the down-cycle. The up-cycle begins with the upstream gates closed, downstream gates open, and the water within the compartment at the downstream level. Boats waiting at the downstream end to move upstream, enter the compartment. When the compartment is suitably filled with boats, the downstream gates close, valve B opens (valve A is closed), and water fills the compartment to raise the boats to the upstream level. The upstream gates then open and the boats exit and continue on their journey. The stages of the down-cycle are the reverse of those described for the up-cycle.

The number of boats that can be accommodated within the lock compartment during either phase is naturally restricted by the physical size of the compartment. This is basically dependent on the length of the lock because boats must be moored along the edges of the compartment for safety reasons. Hence the linear length of the compartment is a major constraining parameter. This constraint cannot be directly translated into a specific number of boats because boats have varying individual lengths. Furthermore, there is a requirement for a one meter separation between boats and between the boats adjacent to the moving gates and the gates themselves. A typical configuration of boats in the lock during an up-phase is shown in Figure 4.6b.

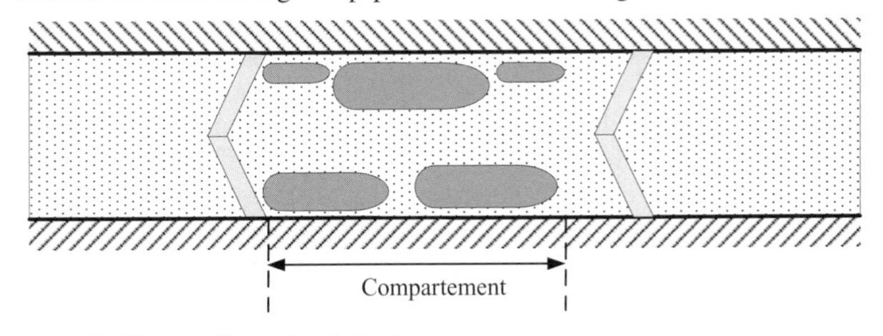

FIGURE 4.6b. Boat configuration in lock.

Both the up-cycle and the down-cycle have the same three phases which we call the loading phase, the transport phase, and the exit phase. The loading phase is the phase when boats from the waiting queue position themselves within the lock compartment. The duration of this phase is dependent on the number of boats that enter the compartment. A reasonable approximation for this loading duration is: $(d_1 + n\ d_2)$ where $d_1 = 4$ minutes, $d_2 = 2$ minutes, and n is the number of entering boats. The boats selected to enter the compartment are taken sequentially from the waiting queue in the order of their arrival until the 'next' boat cannot fit any available space. Then the next boat in the queue that can fit into the available space is selected and this continues until no further boats can be accommodated.

The transport phase includes the closing of the open gate of the lock compartment and the filling/emptying of the water from the compartment by appropriate opening/closing of valves A or B. The duration of this phase is 7 minutes. The exit phase, during which the boats leave the compartment, has a duration of 5 minutes. Thus the total cycle time is: $(d_1 + n\ d_2) + 7 + 5 = (16 + n\ d_2)$ minutes.

The management of the lock system (the particular lock in question is one of a series of locks on the waterway) has been receiving complaints about long delays during the peak traffic period of the day which extends from 11:00 AM to 5:00 PM. (The lock system operates from 8:00 AM to 8:00 PM.) Fortunately traffic rapidly diminishes after 5:00 PM and the queues that normally exist at the end of the busy period generally empty by the 8:00 PM closing time.

The usable length of the lock compartment is 40 meters. One option that is being considered by management to resolve the excessive delay problem is to increase the compartment length to 50 meters. There are significant costs involved in such a reconfiguration and there is uncertainty about what impact it would have on the delays experienced by the various boat categories. A modeling and simulation study has been proposed as a means for acquiring insight into the effectiveness of the plan. Furthermore, an increase in boat traffic is anticipated over the short term and it has been suggested that the proposed compartment extension would also be able to accommodate at least a 15% increase. This possibility is also to be investigated by the study.

The boats traveling in this waterway fall into three categories which we reference as 1, 2, and 3. These categories are intended to reflect a size attribute (i.e., small, medium, and large, respectively). The actual length of arriving boats in category k is uniformly distributed in the range $[L_k - \Delta_k, L_k + \Delta_k]$ meters. During the high-traffic portion of the day the interarrival time for boats in category k is exponentially distributed with mean μ_k.(minutes). The values of the various constants are given in Table 4.40.

TABLE 4.40. Size and arrival attributes of the three boat categories.

Size	μ_k (minutes	L_k (meters)	Δ_k (meters)
Small (k = 1)	5	6	1
Medium (k = 2)	15	9	1.5
Large (k = 3)	45	12	1.5

a) Formulate a set of performance measures that would likely be of value for assessing the effectiveness of the proposed lock extension within the context of a modeling and simulation study.
b) Develop an ABCmod conceptual model that captures the various relevant aspects of the problem.

4.4 HappyComputing Inc. is a personal computer service, sales, and rental shop. Customers who arrive at the shop fall into one of four categories depending on the nature of the 'work' which results from their visit. These are labeled as follows.

- C1: This customer wishes to purchase or rent a PC.
- C2: This customer is returning a rental PC.
- C3: This customer has brought in a PC that requires service of a relatively minor nature (e.g., upgrade of hard drive or installation of additional memory). The customer typically waits for the service to be completed or possibly returns later in the day to pick up the machine.
- C4: This customer's PC has a problem that needs to be diagnosed before repair can be undertaken. In this case the customer leaves the PC in the shop with the understanding that he or she will be telephoned when the problem has been corrected.

The shop has three employees: one is salesperson and the other two are technicians. One technician (the senior technician) has extensive training and considerable experience. The other (the junior technician) has limited training and skills. The salesperson is the initial point of contact for all arriving customers. The needs of both type C1 and type C2 customers are handled exclusively by the salesperson.

The shop is open Monday through Saturday inclusive from 9:00 AM to 6:00 PM. The salesperson (or a substitute) is always present. The senior technician has a day off on Mondays and the junior technician's day off is Thursday. Each employee has a one hour lunch break.

Customers of the type C3 category are handled by the junior technician on a first-in-first-out basis. However, in about 20% of the cases the junior technician is obliged to consult with the senior technician in order to deal with some aspect of the servicing requirement. This draws the senior technician away from his normal work activity which is the servicing of the PCs that are brought to the shop by category C4 customers. Note that on Thursdays the senior technician takes responsibility for the C3 work on a priority basis (i.e., he always interrupts his C4 task to accommodate the C3 customer).

It is the policy of the shop to carry out a comprehensive examination of all rental PCs when they are returned and to carry out any necessary refurbishing before they placed back into the rental pool. The refurbishing includes a range of possible tasks that usually involve hardware and software upgrades. This refurbishing activity is the responsibility of the junior technician but it is carried out only when there are no PCs from category C3 customers that require service.

TABLE 4.41. Interarrival times for customer categories.

Customer Category	Min (minutes)	Max (minutes)
C1	70	130
C2	110	170
C3	180	260
C4	120	210

The interarrival times of each customer category over the course of a business day are uniformly distributed; however, the parameters of the distributions vary according to customer type. The boundaries of the various uniform distributions are summarised in Table 4.41.

Each arriving customer, in effect, generates a service requirement which requires time to complete. The service time requirement for each customer category is a random variable. The assumed distribution for each of the categories is given in Table 4.42 (together with associated parameter values).

TABLE 4.42. Service time requirements for each customer category.

Customer Category	Distribution of Service Time Requirement	Distribution Parameters * (minutes).
C1	Normal	$\mu = 25$, $\Sigma^2 = 10$
C2	Uniform	min = 25, max = 35
C3	Triangular	a = 30, b = 75, c = 45
C4	Triangular	a = 45, b = 175, c = 140

* See Section A1.4.4 of Annex A.

The owner of the shop (who, in fact, is the senior technician) wants to decrease the turnaround time for the PCs brought in by C4 customers because that is the part of the business that he is especially interested in 'growing'. He has, furthermore, noticed that the current workload of the junior technician often leaves him with significant amounts of idle time. He is therefore considering asking the junior technician to take a number of courses in order to upgrade his technical skills. This will have two benefits. First, it will enable the junior technician to deal with the service requirements of C3 customers without having to request assistance and second, it will enable the (upgraded) junior technician to assist with the servicing of the PCs brought to the shop by the C4 customers when he is not otherwise occupied with his current responsibilities. The owner anticipates that the net impact will be a reduction of about 25% in the turnaround time for the C4 category of service.

The goal in this modelling and simulation project is to determine if the owner's expectation is correct.

a) The problem statement as given above omits several details that need to be provided before the project can be realistically undertaken. Identify these and suggest meaningful clarifications.
b) What is a realistic performance measure for this study? Do you regard this as a bounded horizon study or a steady-state study?
c) Develop an ABCmod conceptual model.

4.5 Balking occurs when a customer in a queue (or anything else that is enqueued) has waited too long for service and abandons the queue. The length of the wait time that triggers the balking event may be fixed or may be a random variable. There are various ways in which balking can be handled within the ABCmod framework and the purpose of this exercise is to formulate at least one approach.

In Section 5.3 of Chapter 5 we outline a simple modeling and simulation project formulated around a fast-food outlet called Kojo's Kitchen. An ABCmod conceptual model that evolves from the project description is also presented. The SUI, as presented, does not include customer balking. Your task is to introduce this feature and duly modify/extend the given ABCmod conceptual model so that balking is incorporated.

Suppose we use the variable *balk-time* to represent the length of time that a customer will wait in the queue before becoming disgruntled and leaving. For definiteness, assume that *balk-time* is a random variable and that it has a triangular distribution with parameters as shown in Figure 4.7.

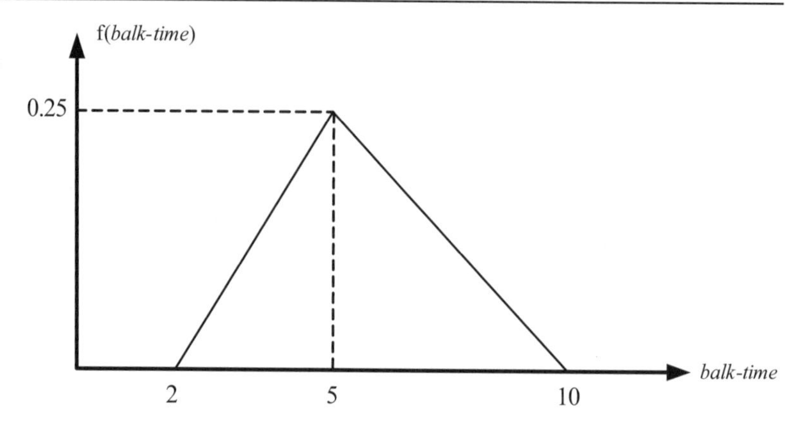

FIGURE 4.7. Distribution for the random variable *balk-time*.

Note finally that a necessary part of any balking specification is the clarification of what happens to a customer that balks. In the case of Kojo's Kitchen such a customer simply disappears from the SUI.

4.5 References

4.1. Gershwin, S.B., (1991) Hierarchical flow control: A framework for scheduling and planning discrete events in manufacturing, in Y.-C. Ho (Ed.), *Discrete Event Dynamic Systems*, IEEE Press, Piscataway, NJ.
4.2. Kreutzer, W., (1986), *System Simulation: Programming Styles and Languages*, Addison-Wesley, Sydney-Wokingham, UK.
4.3. Martinez, J.C., (2001), EZStrobe: General-purpose simulation system based on activity cycle diagrams, in *Proceedings of the 2001 Winter Simulation Conference*, pp. 1556–1564.
4.4. Peterson, J.T., (1981), *Petri Net Theory and the Modeling of Systems*, Prentice-Hall, Englewood Cliffs, NJ.
4.5. Shi, J.J., (2000), Object-oriented technology for enhancing activity-based modelling functionality, in J.A. Joines, R.R. Barton, K. Kang, and P.A. Fishwick (Eds.), *Proceedings of the 2000 Winter Simulation Conference*, Orlando, FL, December 10–13, 2000, pp. 1938–1944.
4.6. Tadao, M., (1989), Petri nets: Properties, analysis and applications, *Proceedings of IEEE*, **77**: 541–580.
4.7. Zeigler, B.P., (1976), *Theory of Modeling and Simulation*, Wiley Interscience, New York.
4.8. Zeigler, B.P., Praehofer, H., and Kim, T.G., (2000), *Theory of Modeling and Simulation: Integrating Discrete Event and Continuous Complex Dynamic Systems*, Academic Press, San Diego.

Chapter 5 DEDS Simulation Model Development

5.1 Constructing a Simulation Model

The simulation model associated with a modelling and simulation project is a computer program that captures the behavioural and structural details of the SUI as specified by the conceptual model. There are two important features of this computer program. The first is simply the fact that, like any computer program, its development must respect the rules of the programming language/environment chosen as the vehicle for implementation. The second feature, however, is distinctive to the modelling and simulation paradigm. It relates to the perspective the program writer takes with respect to the manner in which the model dynamics are 'packaged'. This perspective is often dictated by the programming language/environment being used.

For example, a reasonable choice might appear to be a direct implementation of the Activity constructs used in formulating the ABCmod conceptual model itself, as outlined in Chapter 4. This, however, is not a practical choice because it does not lend itself to an efficient time-advance mechanism which is an important constituent in the execution of any simulation model. The inherent difficulty relates to the initiation of the various Activity constructs. The 'trigger' in each case is the logical expression within the Activity's precondition. In most cases the precondition is formulated in terms of the model's state variables, hence from an implementation point of view, there is little choice but to move time forward in small increments until some meaningful event occurs (e.g., the end of a duration) and results in state variable changes that may trigger one or more other Activities. In principle, this approach is straightforward but in practice it is awkward and exceedingly inefficient.

The perspective outlined above is called 'Activity scanning' and is one of three world views that are commonly recognised in the formulation of discrete-event simulation models (See Chapter 3 of Banks et al. [5.1]). The other two are called 'event scheduling' and 'process oriented', respectively. Essentially all simulation programming languages/environments that are currently available have a bias, or even a total commitment, to one or the other of these world views. Each of these is briefly outlined below.

Event Scheduling: The essential constituent of this perspective is a set of future events that are scheduled to occur. A future event is a collection of actions that includes state variable changes and the possible scheduling of other future events. These actions all occur at the same value of (simulated) time. The simulation model's behaviour is formulated in terms of these future events which are maintained in a time-ordered list. A simulation run therefore unfolds in discontinuous jumps in time which correspond to the time stamps of the events in the future event list.

Process Oriented: Recall that our intent in developing our conceptual modelling framework was to identify atomic units of behaviour which emerged as Activities. In the process-oriented approach, various Activities are assembled into larger units that have a natural affinity in terms of capturing a higher level of behaviour. Each of these is called a process and the simulation model in the process-oriented world view is typically formulated as a collection of such processes which interact as they unfold over time. A process portrays the flow of entities from one Activity to another. In most cases, a process corresponds to the lifecycle of some entity within the ABCmod conceptual model. A commonly occurring example is a process that captures the entry of a consumer entity instance into the model, its participation in a number of Activities, and finally its departure from the model. Processes can also be defined for resources that participate in one Activity after another (possibly in a circular manner) without ever leaving the model.

5.2 Relationship Between the World Views

At its most fundamental level, DEDS model behaviour takes place as a result of discrete events at specific points in time which change the value of some of the model's state variables. As discussed in Section 4.2.4, there are two types of events: conditional events and scheduled events. A simulation model behaviour specification must provide the means to determine when these events occur and also to execute them, in other words, to carry out the status change specifications (SCS) associated with the event. Each of the three world views described in the Section 5.1 has its own approach for organising these event specifications for execution. Because all views are necessarily built from the same discrete events, it is natural to expect that relationships can be identified that permit transformations among the various 'views' to be carried out. These relationships provide the basis that enables the translation of an ABCmod conceptual model into either an event-scheduling simulation model or a process-oriented simulation model.

Recall that the basic Activity in our ABCmod conceptual modelling framework is composed of a conditional event (the starting event which is linked to a precondition) followed by a duration which is then followed by a scheduled event (the terminating event). The various specifications included in both the starting event and the terminating event are called SCSs (status change specifications) and these always include state variable changes. Because the development of an ABCmod conceptual model is a conceptual modelling exercise, the management of these SCSs and the advancement of time have no relevance.

As indicated earlier in Section 5.1, in the event-scheduling world view a DEDS simulation model is expressed in terms of a set of future events. An ABCmod conceptual model is transformed into an event-scheduling world view by reorganising the preconditions and the SCSs in the starting events and in the terminating events of the various Activities into members of this set. The basic concepts, data structures, and algorithms for our perspective of this world view are outlined in Section 5.4.1. The transformation of an ABCmod conceptual model into an event-scheduling simulation model is presented in Section 5.4. The Java programming language is used to illustrate the creation of this category of simulation model.

The event-scheduling approach breaks Activities down into constituent parts. The process-oriented approach on the other hand, interconnects Activities into units that correspond to entity lifecycles. Consider again Figure 4.2 that presents the lifecycle of three different shoppers moving from activity to activity. A process-oriented simulation program would execute these lifecycles for each instance of the shopper entity. During a simulation run, entity instances typically interact with other entities within the ABCmod Activities that constitute the lifecycle. A prerequisite for the transformation of an ABCmod conceptual model into a process-oriented simulation model is an intermediate step that identifies these lifecycles (or processes).

Section 5.5 introduces an approach for developing processes from an ABCmod conceptual model. These processes are applicable to any process-oriented programming environment, and can be viewed as an additional step added to the conceptual modelling phase. The remainder of Section 5.5 shows how an ABCmod conceptual model can be transformed into a GPSS (a process-oriented programming environment) simulation model.

5.3 Kojo's Kitchen

A simple project formulated around a fast-food outlet in the food court of a shopping mall is used to illustrate the main concepts required for the translation of an ABCmod conceptual model into an event-scheduling model.

We begin with a statement of the project followed by the development of an ABCmod conceptual model.

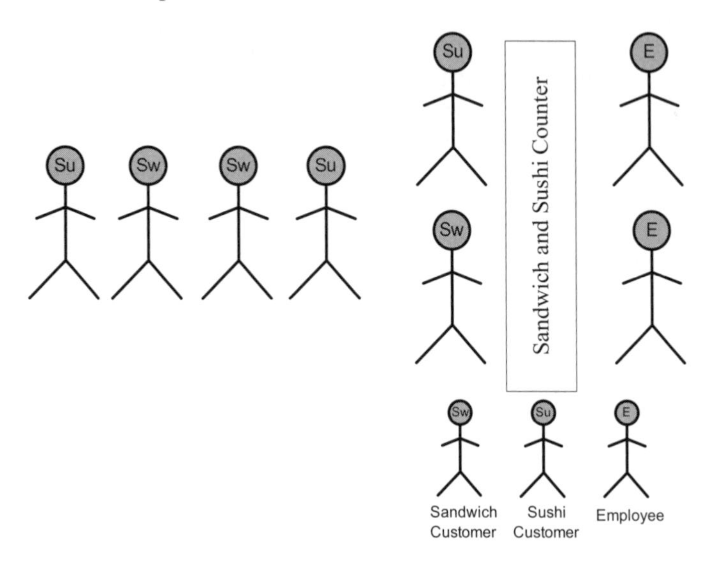

FIGURE 5.1. Kojo's Kitchen fast-food outlet.

Project Description

Kojo's Kitchen is one of the fast-food outlets in the food court of a shopping mall. The mall (and hence Kojo's) is open between 10:00 AM and 9:00 PM every day. Kojo's serves only two types of product; namely, sandwiches and sushi. We assume there are only two types of customer: one type purchases only sandwiches and the other type purchases only sushi products. Two rush-hour periods occur during the business day, one between 11:30 AM and 1:30 PM, and the other between 5:00 PM and 7:00 PM.

The stochastic process for customer interarrival times is nonhomogeneous but, for convenience, is taken to be piecewise homogeneous (see Section A1.8 of Annex 1). Interarrival times in each of the homogeneous segments are taken to be exponentially distributed with means as given in Table 5.1.

TABLE 5.1. Kojo's Kitchen customer arrival data model.

Customer Type	Period	Mean Inter-Arrival Time (min)
Sandwich Customer	10:00 AM–11:30 AM	15
	11:30 AM–1:30 PM	3
	1:30 PM–5:00 PM	12
	5:00 PM–7:00 PM	4
	7:00 PM–9:00 PM	10
Sushi Customer	10:00 AM–11:30 AM	22
	11:30 AM–1:30 PM	4
	1:30 PM–5:00 PM	15
	5:00 PM–7:00 PM	7
	7:00 PM–9:00 PM	12

Currently two employees work at the counter throughout the day preparing sandwiches and sushi products for the customers. Service times are product-dependent and they are both uniformly distributed, ranging from three to five minutes for sandwich preparation and from five to eight minutes for sushi preparation.

Goal: Kojo's manager is very happy with business, but has been receiving complaints from customers about long waiting times. He is interested in exploring staffing options to reduce these complaints. The specific interest is in comparing the current situation (base case) to an alternative where a third employee is added during the busy periods (between 11:30 AM and 1:30 PM and between 5:00 PM and 7:00 PM). The performance measure of interest is the percentage of customers that wait longer than five minutes for service over the course of a business day.

The various components of an ABCmod conceptual model for the flow of customers through Kojo's Kitchen are given in Tables 5.2 through 5.14.

Structural Components

TABLE 5.2. Constants and parameters for Kojo's Kitchen project.

Constants		
Name	**Role**	**Value**
WMean1	Mean interarrival time for sandwich customer between 10:00 AM and 11:30 AM	15
WMean2	Mean interarrival time for sandwich customer between 11:30 *AM* and 1:30 PM	3
WMean3	Mean interarrival time for sandwich customer between 1:30 PM and 5:00 PM	12
WMean4	Mean interarrival time for sandwich customer between 5:00 PM and 7:00 PM	4
WMean5	Mean interarrival time for sandwich customer between 7:00 PM and 9:00 PM	10
UMean1	Mean interarrival time for sushi customer between 10:00 AM and 11:30 AM	22
UMean2	Mean interarrival time for sushi customer between 11:30 AM and 1:30 PM	4
UMean3	Mean interarrival time for sushi customer between 1:30 PM and 5:00 PM	15
UMean4	Mean interarrival time for sushi customer between 5:00 PM and 7:00 PM	7
UMean5	Mean interarrival time for sushi customer between 7:00 PM and 9:00 PM	12
STWMin	Minimum service time for sandwich customer	3
STWMax	Maximum service time for sandwich customer	5
STUMin	Minimum service time for sushi customer	5
STUMax	Maximum service time for sushi customer	8
NumEmpReg	Number of employees at counter during non-busy times	2
NumEmpBusy	Number of employees at the counter during busy times	3
Case1	Identifier for the base case	1
Case2	Identifier for the alternative	2
Parameters		
Name	**Role**	**Values**
EmpSchedCase	Set to Case1 for the base case and Case2 for the alternative case (when a third employee is present at the counter during busy periods)	Case1, Case2

TABLE 5.3. Customer Consumer Entity Class.

Consumer Entity Class: *Customer*	
The customers that purchase items at the Kojo's Kitchen.	
Attributes	**Description**
Type	Set to the type of customer, either 'W' (sandwich) or 'U' (sushi)
TimeEnterQu	The time the customer enters the queue

TABLE 5.4. Counter group.

Aggregate Entity: *CounterGroup*	
This group contains the customers being served at Kojo's Kitchen.	
Attribute	**Description**
List	Set of attribute-tuples of the customers that are being served
N	Number of entries in List. This value is always less than or equal to the input variable EmpNum (number of employees at the counter)

TABLE 5.5. Customer queue.

Aggregate Entity: *CustQue*	
Queue of customers in front of the Kojo's Kitchen.	
Attribute	**Description**
List	List of the attribute-tuples of the customers that are waiting for service Discipline: FIFO
N	Number of entries in List

Input/Output Components

TABLE 5.6. Inputs for Kojo's Kitchen project.

Inputs				
Input Variable	**Description**	**Data Models**		**Action Sequence**
		Domain Sequence	**Range Sequence.**	
$U_W(t)$	The input variable U_{W} represents the input entity stream corresponding to the sandwich type customer consumer entity class	First arrival: $t = t_{\Gamma}$ (random using inter-arrival time data model) Interarrival time: Exponential(X) where X is: WMean1, $0 \leq t < 90$ WMean2, $90 \leq t < 210$ WMean3, $210 \leq t < 420$ WMean4, $420 \leq t < 540$ WMean5, $540 \leq t < 660$	All values equal to 1	Warrivals (C.Customer)
$U_U(t)$	The input variable U_{U} represents the input entity stream corresponding to the sushi type customer consumer entity class	First arrival: $t = t_{\Gamma}$ (random using inter-arrival time data model) Interarrival time: Exponential(X) where X is: UMean1, $0 \leq t < 90$ UMean2, $90 \leq t < 210$ UMean3, $210 \leq t < 420$ UMean4, $420 \leq t < 540$ UMean5, $540 \leq t < 660$	All values equal to 1	UArrivals (C.Customer)
EmpNum(t)	This input variable represents the number of employees at the counter serving customers.	EmpNumReg, when Case is equal to Case1, otherwise when EmpShedCase equal to Case2: EmpNumReg, $0 \leq t < 90$, $210 \leq t < 420$, $540 \leq t < 660$ EmpNumBusy, $90 \leq t < 210$, $420 \leq t < 540$	EmpNumReg or EmpNumBusy	SchedEmp

TABLE 5.7. Output for Kojo's Kitchen project.

Sample Sets	
Name	**Description**
PHI[WaitTime]	Each value in the sample set PHI[WaitTime] is the time spent spent waiting in the queue for service by some instance of the consumer entity class called Customer

(a) Sample Set for Kojo's Kitchen project.

Derived Scalar Output Variable (DSOV)			
Name	**Description**	**Output Set Name**	**Operator**
PropLongWait	Proportion of customers that wait longer than 5 minutes in the queue	PHI[WaitTime]	PropGT(5,PHI [WaitTime])

(b) DSOV for Kojo's Kitchen project.

Behaviour Components

Time units: minutes

Observation interval: $t_0 = 0$, t_f: = 660 minutes (11 hour business day).

TABLE 5.8. Initialisation for Kojo's Kitchen project.

Initialise
EmpNum(t_0) = 2 A.CounterGroup.N ← 0 A.CustQue.N ← 0

TABLE 5.9. Data modules for Kojo's Kitchen project.

Data Modules		
Name	**Description**	**Data Model**
WSrvTm	Returns a value for the service time of a sandwich customer	UNIFORM(STWMin, STWMax)
USrvTm	Returns a value for the service time of a sushi customer	UNIFORM(STUMin, STUMax)

TABLE 5.10. User-defined modules for Kojo's Kitchen project.

User-Defined Modules	
PropGT(Val,SampleSet)	This procedure analyses the set SampleSet and returns the proportion of entries in the set that exceed the value Val

TABLE 5.11. Summary of Activity constructs for Kojo's Kitchen project.

Summary of Activity Constructs	
Action Sequences	
WArrivals	The Input Entity Stream of arriving sandwich customer
UArrivals	The Input Entity Stream of arriving sushi customer
SchedEmp	Extra employee scheduling
Activities	
ServingW	Service for a sandwich customer
ServingU	Service for a sushi customer

TABLE 5.12. Sandwich customer Arrival Action sequence.

Action Sequence: WArrivals(C.Customer)	
Precondition	$t = M[U_w](t)$
Event	C.Customer.Type ← W C.Customer.TimeEnterQu ←t SM.InsertQue(A.CustQue, C.Customer)

TABLE 5.13. Sushi customer Arrival Action sequence.

Action Sequence: UArrivals(C.Customer)	
Precondition	$t = M[U_U](t)$
Event	C.Customer.Type ← *U* C.Customer.TimeEnterQu ← *t* SM.InsertQue(A.CustQue, C.Customer)

TABLE 5.14. Employee scheduling Action sequence.

Action Sequence: SchedEmp	
Precondition	(*t* = *M*[EmpNum](*t*)) &(EmpShedCase = Case2)
Event	IF(*t* = 90) EmpNum = 3 ELSE IF(*t* = 420) EmpNum = 3 ELSE EmpNum = 2

TABLE 5.15. Serving sandwich customer Activity.

	Activity: ServingW
Precondition	(A.CounterGroup.N < EmpNum) & (A.CustQue.N ≠ 0) & (A.CustQue[0].Type = W)
Event	SM.RemoveQue(A.CustQue, C.Customer) SM.InsertGroup(A.CounterGroup,C.Customer) SM.Put(PHI[WaitTime], (*t*-C.Customer.TimeEnterQu))
Duration	DM.WSrvTm()
Event	SM.RemoveGroup(A.CounterGroup, C.Customer) SM.Leave(C.Customer)

TABLE 5.16. Serving sushi customer Activity.

	Activity: ServingU
Precondition	(A.CounterGroup.N < EmpNum) & (A.CustQue.N ≠ 0) & (A.CustQue[0].Type = S)
Event	SM.RemoveQue(A.CustQue, C.Customer) SM.InsertGroup(A.CounterGroup,C.Customer) SM.Put(PHI[WaitTime], (*t*-C.Customer.TimeEnterQu))
Duration	DM.USrvTm()
Event	SM.RemoveGroup(A.CounterGroup,C.Customer) SM.Leave(C.Customer)

5.4 Transforming an ABCmod Conceptual Model into an Event-Scheduling Simulation Model

5.4.1 Event-Scheduling Simulation Models

The execution of an event-scheduling simulation model is concerned with the processing of future events. The model's behaviour over the course of the simulation run can be represented as a sequence of snapshots of the model taken at those the discrete points in time when the events in the future events list occur.

These snapshots contain:

- The state of the model
- The list of scheduled future events
- The value of the simulation clock

A future event is more complex than the notions of a scheduled event or a conditional event that have been previously introduced (see section 4.2.4). In some respects a future event can be regarded as a composite of these two notions. A future event is composed of a sequence of actions whose impact on the model is captured by a status change specification (SCS). This sequence begins with a scheduled event (e.g., the Activity terminating event in an ABCmod Activity) whose changes to the model might enable the preconditions for one or more conditional events which then cause further changes to the model (recall that the starting event of an Activity is a conditional event). The changes to the model that result may enable more conditional events, and this cascade may continue. This can lead to additional future events being scheduled (i.e., placed on the future event list) as becomes apparent in the discussion that follows.

The view outlined above is fundamental to the translation of an ABCmod conceptual model to an event-scheduling simulation model. The rest of this section describes the programming mechanisms that are required to implement the event-scheduling approach.

In our particular view of event scheduling, a future event list (FEL) is used in the scheduling of future events. It is composed of a list of records called *event notices.* Each contains a *future event name* that identifies a future event and a *time attribute* that defines the time when it must occur. Event notices are ordered on the FEL according to their time attribute. A notice contains, as a minimum, the future event name and a value for the time attribute. The future event name serves primarily to reference a future event routine (FER) that carries out the actions specified in the future event.[1] The notice may also contain a third element that references model entities or other values required by the FER.

Essential to executing a simulation model is a time advance routine. An algorithm for such a routine is provided in the flowchart of Figure 5.2. The routine contains a loop that processes the event notices on the FEL until a stop event notice or stop condition is encountered. The processing of an event notice has two steps: the clock is advanced to the value of the time attribute, and the referenced FER is called to change the model's state and possibly add notices to the FEL. The notice at the head of the FEL is the next one to be processed and is called the *imminent event notice*.

The handling of an input function (exogenous events) is generally achieved by creating event notices based on the times prescribed by some input domain sequence (the timing map M[]). Consider the arrival of sandwich customers at Kojo's Kitchen. An event notice for the first customer arrival is placed on the FEL with a time attribute value that is greater than or equal to the left boundary of the observation interval (see Table 5.12).

[1] It is sometimes convenient to use the future event name as a reference to the event notice itself. Such usage is always clearly apparent.

When this notice becomes the imminent event notice the referenced FER will:

- Generate an interarrival time, say a^*, using a data model associated the input variable $U_W(t)$ (see Table 5.6).
- Establish the time of the next arrival as $t^* = t + a^*$ (t^* corresponds to the next time in the domain sequence $CS_D[U_w]$).
- Insert into the FEL a new arrival event notice having t^* as the value of the time attribute.

This procedure (which we call bootstrapping) results in a stream of customer arrivals for the simulation model.

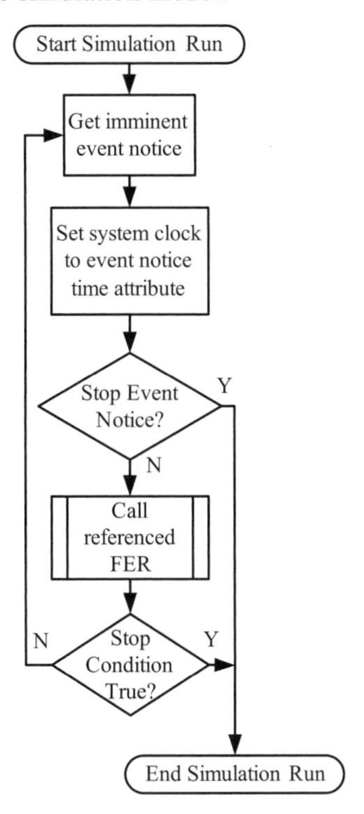

FIGURE 5.2. Time advance algorithm for event scheduling.

The same procedure is used for generating most inputs for a simulation model. Consider a manufacturing problem with machines that have a 'down-time' (e.g., they become unserviceable and require repair). An initial breakdown event notice is created and placed on the FEL. When the breakdown event notice is processed, an end-of-repair event notice is scheduled to bring the machine back into service. When the end-of-repair event notice is processed, then another breakdown event notice is placed on the FEL which indicates the end of the 'up-time', and so on.

Stopping the simulation run can be achieved using one of two methods. The first is to run the simulation until a predetermined time. In this case, a stop event notice is placed on the FEL. Its time attribute value is the prescribed termination time for the simulation run. In our considerations below of the Kojo's Kitchen project, a stop event notice with the event name StopEvent and time attribute value of 660 minutes (11 hours) is placed on the FEL.

The second method for stopping a simulation run is to specify some condition for termination, such as the production of the 100th component, the occurrence of a catastrophic system failure (suitably defined), or the last carton shipped by a distribution centre. The actual time of termination is not known at the start of the simulation run, and the state of the model must be checked after the processing of the imminent event to determine if the conditions for termination have been established.

FERs contain the specifications for the required changes to the model's status. There are two main steps in the FER's execution:

1. Carry out the changes to the model's status associated with a scheduled event.[2] Typically, this would correspond to carrying out the SCS of some Activity's terminating event.
2. Check the various preconditions to determine if any conditional events (i.e., starting events) can be activated. This involves testing the preconditions for all Activities (more efficient approaches are possible). For each precondition that is found to be TRUE, the FER
    a) Carries out state changes associated with that conditional event (i.e., the SCS of the corresponding Activity's starting event).
    b) Schedules a future event derived from the SCS of the corresponding Activity's terminating event.

   Generally, Step 2 needs to be repeated until there are no preconditions that are TRUE.

As an illustration of the various possible interactions between the FEL and the FER we examine the processing that is associated with the handling of a customer at Kojo's Kitchen. Consider the situation where the imminent event notice on the FEL is an EndServing event notice. We assume that such a notice is associated with the scheduled event that corresponds to the completion of some customer's service at the counter. As many as two (or three) event notices called EndServing can simultaneously exist on the FEL because there can be two (or three)

[2] Data output operations may also be carried out.

customers at the counter.[3] Each such notice would have been placed on the FEL when the customer arrived at the counter.

The following processing is carried out by the FER.

1. The customer just served leaves the model. Note that the event notice must necessarily contain a reference to the specific customer entity instance to be removed from the counter because there can be two (or three) customers simultaneously there.
2. In the case of the EndServing event, space becomes available at the counter. This could enable either a ServingW Activity or a ServingU Activity (see Table 5.15 and Table 5.16). If a customer entity instance is waiting in the queue to be served:
   a) Remove the customer entity instance from the queue and place it at the counter i.e., in the CounterGroup.
   b) Insert that customer's computed wait time into the PHI[WaitTime] sample set.
   c) Schedule another EndServing event notice (based on the terminating terminating event of the enabled Activity) by:
      i. Determine the serving time s (using the appropriate data model) for the customer entity instance newly placed at the counter and establish the end of service time as $t_{{}^{B}end^{B}} = t + s$.
      ii. Insert a new event notice in the FEL with time attribute set to $t_{{}^{B}end^{B}}$, with name EndServing, and with a reference to the customer entity instance placed at the counter.

The next subsection outlines how the event-scheduling routine can be implemented using Java.

5.4.2 Java-based Tools for Event Scheduling

Event-scheduling simulation models and simulation programs are typically created using general programming languages. Often libraries are available for supporting this development. They provide functions for random number generators, random variate functions, list processing, and so on.

[3] In this simple project, the counterGroup List attribute is not necessary. Only the attribute N is required to give the number of customers being serviced at the counter. When an endServing event notice is processed, then the corresponding FER would simply decrement the attribute N; that is, the customer attribute-tuple serves no purpose once it is removed from the customer queue. Any time a customer is brought to the counter the attribute N is incremented. In more complex models, the original attribute-tuple could be required for future processing once it left the counter (e.g., could be interested in the operation of the whole fast-food court).

The indent of this section is to demonstrate how event scheduling programming tools can be developed using the Java programming language. These tools will be used in the next section in the presentation of the translation process from an ABCmod conceptual model to an event scheduling simulation model. Java is an object-oriented language that offers many predefined classes. Most if not all conceptual model entities can be mapped directly into Java classes (the following section provides some suggested mappings). This section presents an overall approach to using Java for implementing event-scheduling simulation models and the time advance algorithm (for creating simulation programs). It is expected that the reader is familiar with Java.[4]

Java supports Abstract classes.[5] Such classes cannot be instantiated as executing objects, but provide the mechanism for supplying an abstract definition for extension to create classes that can be instantiated. An *EvSched* Abstract class (see the UML class diagram in Figure 5.3) is presented for creating Java event-scheduling simulation models. The class contains the variables, abstract methods, and methods for implementing event scheduling. The class is intended for use with other classes (EventNotice, ESAttributeList, and ESOutputSet) used to define future event notices, attribute-tuples, and collecting output. Abstract methods must be defined when extending the class. These components and other classes used by *EvSched* are summarised in Table 5.17 and include the following.

a) The *fel* variable references a *PriorityQueue* object[6] to implement the FEL. This class provides the functionality to rank event notices in a list (see *EventNotice* class below).
b) The *clock* variable is the implementation of the simulation clock.
c) The variables *time0* and *timef* correspond to the boundaries of the observation interval. The *timef* variable is set when the simulation execution ends. Its value will only be known at that point when the right-hand boundary of the observation interval is implicit. These values are required for calculating DSOV values for trajectory sets.

---

[4] Many books are available on the subject (e.g., [5.2], [5.3]) and complete reference material for Java version 5.0 is offered by Sun Microsystems [5.5].

[5] A Java class provides the specification for creating instances of objects. Thus an object corresponds to memory being allocated in a computer program that contains data and can be executed. The class is much like a data type whereas the object is much like the variable declared using the data type.

[6] The *PriorityQueue* class is provided by the Java Collections Framework.

TABLE 5.17. The EvSched Abstract class.

	Name	Description
Variables	fel	The Future Event List implemented as a PriorityQueue object
	clock	The simulation clock
	time0, timef	These two variables define the boundaries of the observation interval
	StopEvent	A numeric identifier representing the future event name in the event notice for implementing the explicit right-hand boundary of the observation interval
Methods	initEvSched	Used to initialise the simulation clock and timing. There are two versions – one with the time for the end of the observation interval and one without
	setTimef	Used to change the right- hand boundary of the observation interval
	runSimulation	This method implements the event-scheduling time advance algorithm and controls the execution of the simulation
	addEventNotice	The method provides the means to add an event notice to *fel*
	removeEventNotice	Used to remove a notice from *fel*. This method allows implementation of interrupts and pre-emption
Abstract methods	processEvent	This method is used to execute the FERs associated to the future event names
	implicitStopCondition	This method can be used to implement the implicit right-hand boundary of the observation interval
Other classes	EventNotice	Class used to instantiate event notice objects added to *fel*
	ESAttributeList	Class used to instantiate attribute-tuples for implementing consumer and resource entities
	ESOutputSet	Class to instantiate an output set (either trajectory set or sample set). The class provides methods to compute DSOV values (i.e., values for a scalar output variable)

d) The constant *StopEvent* (integer constant set to –1) serves as the event name in an event notice to schedule termination of the simulation at a predefined time (explicit right-hand boundary of the observation interval).
e) The simulation model execution is centered on the *runSimulation* method that implements the time advance algorithm (see Figure 5.2) for processing event notices on *fel*.
f) The method *initEvSched* initialises the *fel* variable, the simulation clock, *time0* variable, and optionally sets up a StopEvent notice. The two forms of the method *initEvSched* initialise *clock* and *time0* to its *startTime* argument. The second form provides *endTime* that is used to create an event notice on *fel* with the name *StopEvent* and *timeAttr* set to *endTime*.
g) The method *setTimef* provides the means to change the right-hand boundary of the observation interval. Its effect is to add an event notice containing the StopEvent constant. This method can be used to handle warm-up periods or to increase the run length as described in Chapter 6.
h) The method *addEventNotice* provides the means to instantiate an event notice object and add it to *fel*. Two forms of the method are available, one where only a future event name and time attribute are provided and the other where in addition to these two parameters, a reference to an object is provided. In the first case, the *obj* variable of the event notice object is simply set to *null*.
i) The *removeEventNotice* method supports the implementation of interrupts and pre-emption. Recall from the previous Chapter that such interventions terminate ongoing Activities. Terminating an Activity is equivalent to removing the event notice that corresponds to the Activity's terminating event from *fel* (see Section 5.4.4 for details).
j) The *processEvent* is an abstract method (to be created when extending the *EvSched* Class) for calling future event routines that carry out the SCS associated with the events. The *runSimulation* method calls *processEvent* with arguments *evName* (event name) and *obj* (reference to objects associated with the event).
k) The *implicitStopCondition* abstract method provides the mechanism for implementing an implicit right-hand boundary of the observation interval. This method should evaluate the model variables to determine when the condition for terminating the simulation exists. For example, the method could return TRUE when the 100th widget was completed in a manufacturing model.

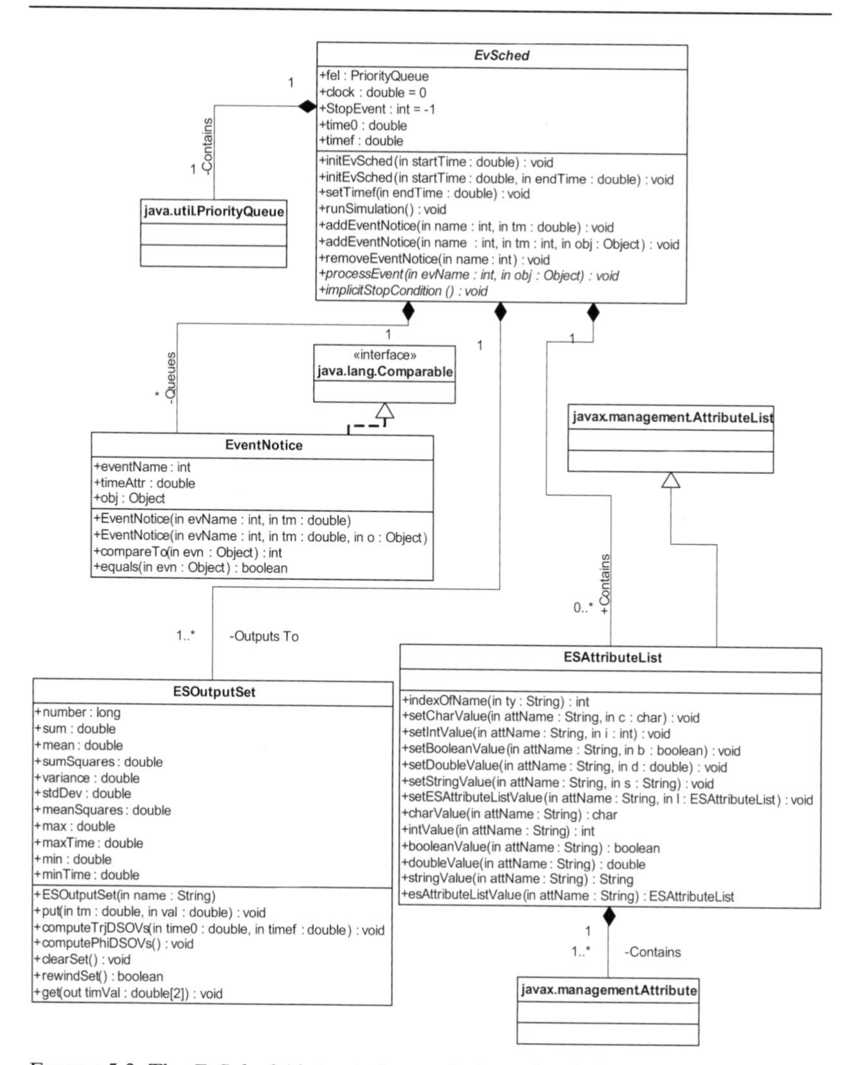

FIGURE 5.3. The *EvSched* Abstract class and other related classes.

l) The *EventNotice* class supports instantiation of future event notices. An *EvSched* object queues one or more of the *EventNotice* objects on *fel.* Three variables are defined in the class: *eventName* (future event name[7]), *timeAttr* (event notice time attribute), and *obj* (reference to any object associated to the event). The *EventNotice* class implements the Comparable Interface which is a requirement for use with the *PriorityQueue* class.

[7] The equivalent numeric identifier of the future event name is stored in the EventNotice object. For examples, see Table 5.20.

The method *compareTo* is used by the *PriorityQueue* class methods to order the event notices according to their *timeAttr* attribute.

m) The *ESAttributeList* class supports instantiation of attribute-tuples that can represent consumer and resource entities. This class extends the *AttributeList* class[8] The *AttributeList* object manipulates a list of *Attribute* objects that relate a name to any Java object. Extensions to the *AttributeList* and *Attribute* classes are necessary to find the index of an *Attribute* with a specific name (the *indexOf* method provided by *AttributeList* matches both the name and value during a search) and for setting/getting the values of attributes based on their names. The *ESAttributeList* provides the method *indexOfName* and a set of methods for setting attribute values (e.g., *setIntValue(attName,i)*) and getting attributes values (e.g., *intValue(attName)*).

n) The *ESOutputSet* class provides the means to collect output data during the simulation run (with the method *put*[9]) and then to compute derived scalar values (using methods *computeTrjDSOVs* or *computePhiDSOVs*). The PSOV values are recorded in a file with the name provided in the constructor's argument (*ESOutputSet(File-Name)*). The class also provides a number of other methods to facilitate collection and analysis of the output data:

   i. *clearSet*: This method removes all data currently recorded. This method can be used to accommodate warm-up periods (see Chapter 6) or to re-initialise the simulation program.
   ii. *rewindSet*: This method provides the means to point to the beginning of the recorded data (i.e. beginning of the file that contains the data). This method is required before making calls to the *get* method.
   iii. *get*: This method allows access to output data collected for a PSOV so that project specific operators can be applied when the project has distintive DSOV requirements. The class offers access to the variable *number* that contains the number of elements in the output data set. Table 5.18 shows the variables that contain DSOV values after *computeTrjDSOVs* or *computePhiDSOVs* is called.

---

[8] The *AttributeList* and *Attribute* classes are provided by Java Management Extensions (JMX).

[9] Note that collecting data for either a trajectory set or a sample set uses the same method; both a time value and data value are stored in both cases.

The complete listings for the classes EvSched.java, ESAttribute.java, ESOutput.java, and EventNotice.java can be downloaded from the textbook Web site. The method *runSimulation* is shown in Figure 5.4. The execution of this method relates to the various elements in the algorithm as follows.

a) The *fel.poll()* method removes the event notice (*EventNotice* object) at the head of *fel.* A cast is necessary to cast the *Object* reference returned by the method to an *EventNotice* object reference.
b) The *clock* is updated with the value of the event notice attribute *nxtev.timeAttr*. This implements the time advance mechanism of the algorithm.
c) A check is then made to see if the event id corresponds to a stop event. Calling either *initEvSched(double startTime, double endTime)* or *setTime(double endTime)* places on *fel* an event notice with the name *StopEvent* (-1) and with its *timeAttr* set to the value of *endTime*. When such an event notice is found at the head of *fel*, the *runSimulation* method breaks out of its processing loop (while loop). This logic accommodates the case where the observation interval has an explicit right-hand boundary, that is, the simulation runs must stop at a predetermined time.
d) If no stop event is detected, the event name *nxtev.eventName* and object*nxtev.obj* are passed as an argument to the *processEvent* method. This method will be specific to the simulation model (see the next section for an example of how this abstract method is used). The function of this method is to call the FER associated with the event name.
e) Finally the *implicitStopCondition* abstract method is called to test for an implicit stop condition, that is, some state of the model that indicates the end of the simulation. When this method returns TRUE,*runSimulation* will break out of its processing loop. As with the *processEvent* method, this method is specific to the model being implemented.
f) The *timef* variable is set to the current value of the clock after breaking out of the main processing loop. At the end of a simulation run, the variables *time0* and *timef* represent the boundaries of the observation interval.

The stage has been set to create an event-scheduling simulation model from a conceptual model developed using the ABCmod framework described in Chapter 4. The next Section describes this translation, illustrating it using the conceptual model of the Kojo's Kitchen project from Section 5.3 and the Abstract class *EvSched* from this Section.

TABLE 5.18. Computing DSOV values in the *ESOutputSet* class.[10]

ESOutputSet Variable	***computeTrjDSOVs***	***computePhiDSOVs***
sum	$y_K(t_f - t_K) + \sum_{i=1}^{K} y_{i-1}(t_i - t_{i-1})$	$\sum_{i=0}^{K} y_i$
sumSquares	$y_K^2(t_f - t_K) + \sum_{i=1}^{K} y_{i-1}^2(t_i - t_{i-1})$	$\sum_{i=0}^{K} y_i^2$
mean	$\frac{sum}{t_f - t_0}$	$\frac{sum}{K}$
max, maxTime	Maximum value in the trajectory set and the time it was recorded	Maximum value in the sample set (with time it was recorded)
min, minTime	Minimum value in the trajectory set and the time it was recorded	Minimum value in the sample set (with time it was recorded)
meanSquares	$\frac{sumSquares}{t_f - t_0}$	$\frac{sumSquares}{K}$
variance	$meanSquares - mean^2$	$meanSquares - mean^2$
stdDev	$\sqrt{variance}$	$\sqrt{variance}$

[10] The pair (t_i, y_i) represents the recorded time/value pairs, K equals the total number of recorded pair values (i.e., equals *number*), and t_0, t_f, are the two *computeTrjDSOVs* arguments *time0* and *timef*.

```
// Run simuation
public void runSimulation()
{
  while(true) // set up loop
  {
    EventNotice nxtev = (EventNotice) fel.poll();
    if(nxtev == null)
    { // This is a safety check, in case system is not properly specified
      System.out.println("FEL is empty - terminating");
      break;
    }

    clock = nxtev.timeAttr; // update the clock

    if(nxtev.eventName == StopEvent) // Check for a timed stop event
    {
       System.out.println("Encountered stop event - terminating");
       break;
    }

    processEvent(nxtev.eventName, nxtev.obj);  // Call referenced FER

    if(implicitStopCondition())
    {
       System.out.println("Implicit stop condition is true - terminating");
       break;
    }
  }
    timef = clock;
}
```

FIGURE 5.4. Code for the runSimulation method.

5.4.3 Translating to an Event-Scheduling Simulation Model

An ABCmod conceptual model provides a specification of the behaviour of a DEDS model. Figure 5.5 shows the various components in an ABCmod conceptual model and how they relate to those of the event-scheduling simulation model. The ABCmod conceptual model structural components consisting of the entities, constants, parameters, input variables, and output sets are represented by data structures in the simulation model. Components representing behaviour will be translated to routines.

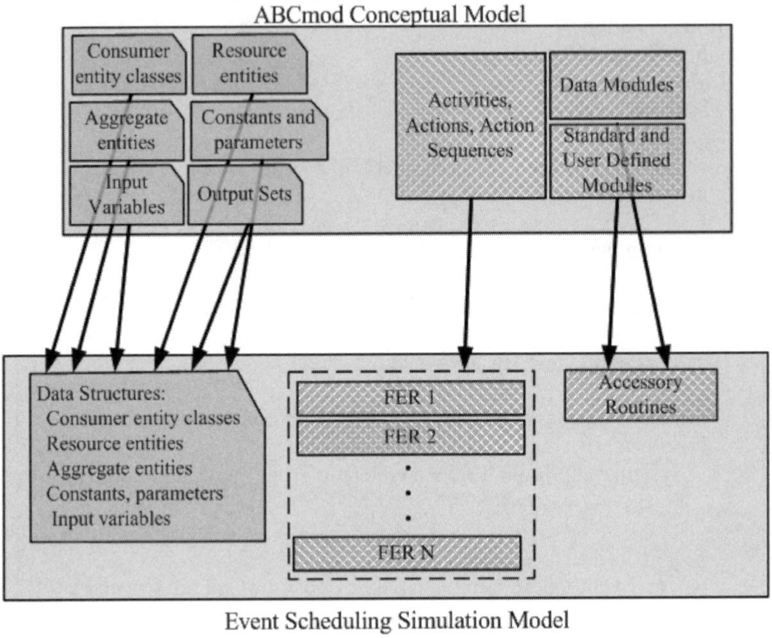

FIGURE 5.5. Relating ABCmod components to event-scheduling model components.

Translating the structural components is relatively straightforward. Typically there is a one-to-one correspondence between data structures and the conceptual model component. Table 5.19 lists a set of possible Java classes that can be used and extended to represent the various ABCmod entities. It is by no means complete and the reader is encouraged to review available Java classes for representing entities.

Translating the behavioural components consisting of the Action Sequences, Activities, data modules, and user-defined modules, is not as straightforward. Dealing with data modules and user-defined modules is relatively easy because they are typically coded as routines (e.g., Java methods). Often libraries provide a number of routines for supporting simulation functionality such as random variate generation and list processing. For example, a Java package, cern.colt,[11] provides a number of

[11] CERN (European Organisation for Nuclear Research) makes available a number of Java packages as Open Source Libraries for High Performance Scientific and Technical Computing in Java [5.4]. It provides a number of classes that implement stochastic data models. Version 1.2.0 was used during the writing of this textbook.

objects for implementing random variates. The challenge in translating behaviour from an ABCmod conceptual model to the simulation model is capturing the Activity specifications in a set of FERs.

TABLE 5.19. Representing ABCMod entities with Java objects.

ABCmod Entities	Java Object Class
Consumers/ Resources	*ESAttributeList* – This class, developed in Section 5.4.2, is a subclass of the *AttributeList* class that creates and manipulates a list of *Attribute* objects. The *Attribute* class associates a name with any type of object. *Java Class* – The Java class provides the means to define attributes (as class variables) and methods to manipulate these attributes
Queues	*Queue Interface* with a number of Queue classes: *ArrayBlockingQueue* (bounded FIFO queue) *ConcurrentLinkQueue* (Thread-safe FIFO queue using linked nodes) *DelayQueue* (unbounded queue of delayed elements – when elements delay is expired, they are presented at the head of the queue in FIFO fashion) *LinkedBlockingQueue*(bounded queue implemented with linked nodes) *ProrityBlockingQueue*(similar to PriorityQueue, but offers also blocking retrieval functions when queue is empty) *PriorityQueue*(unbounded queue that orders elements according to priority) SychronousQueue (queue that provides a rendezvous mechanism – does not collect but implements single put and take)
Groups	*Set Interface* with a number of Set classes: *HashSet* – This class is an extension of the AbstractSet and supports the set methods using a hash table. This class provides constant performance for basic operations such as add, remove, contains, and size. But the time for construction of the HashSet object depends on the size of the hash table *LinkedHashSet* – Similar to the HashSet class but stores objects in the set using a doubly linked list. This provides an order to the added objects; that is, it is possible to read the objects in the same order they were stored

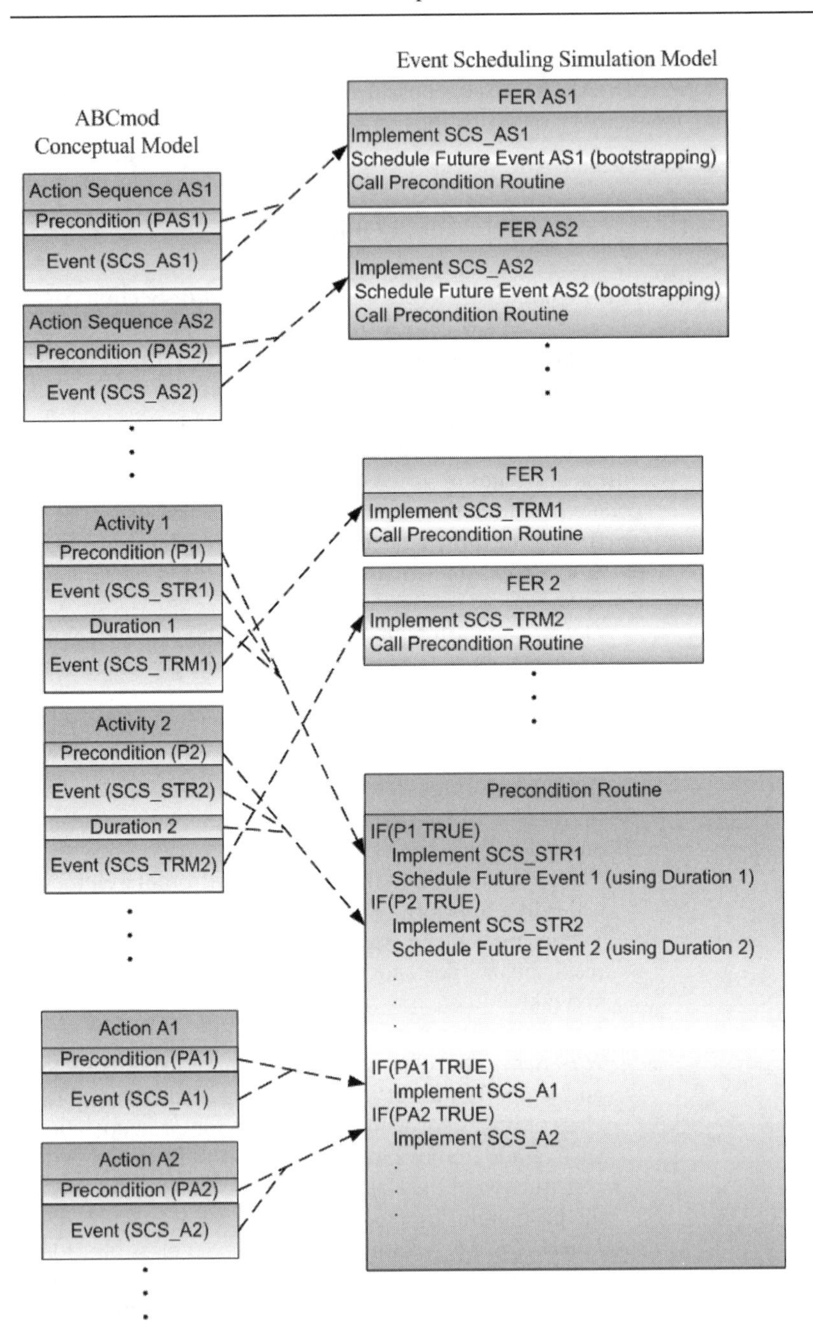

FIGURE 5.6. Creating FERs.

Each Activity in the ABCmod conceptual model is separated into its constituents; namely, the precondition, the SCS of the starting event, the SCS of the terminating event and the duration. These are re-organised into a precondition routine and a collection of FER's. Fortunately, there exists a pattern in this re-organisation as shown in Fig. 5.6.

A FER is created for each Activity in the ABCmod conceptual model. The program code of a FER has two parts, the first is an implementation of the SCS that is associated with the terminating event of the Activity and the second is an invocation of the precondition routine.

A simple modular approach can be taken in organizing the program code for the precondition routine as illustrated in Figure 5.6. Each segment of the routine is associated with some particular Activity and encompasses its precondition, the SCS of its starting event and its duration. The segment is entered only if the precondition of the Activity is TRUE. The program code for the segment has two parts. The first is an implementation of the SCS that is associated with the starting event of the Activity and the second is code that schedules a future event at time (t+D) where t is the current value of (simulated) time and D is the duration of the Activity. The future event is, in fact, the invocation of the FER that is associated with the Activity in question.

A FER is likewise created for each action sequence in the ABCmod conceptual model. However the structure of each such FER is different from the FER that evolves from an Activity. In part this is because an action sequence has no terminating event. The FER in this case has three parts. The first is an implementation of the SCS of the Action Sequence's starting event. The second is program code that schedules a future event that is the FER's own invocation at some a (future) time determined by a bootstrapping approach. The bootstrapping references the timing map that is associated with the action sequence. The third part is an invocation of the precondition routine. There is no segment needed in the precondition routine for an action sequence.

The transformation of an Action is straightforward. It does not give rise to a FER but does give rise to a segment in the precondition routine. This segment is entered only when the precondition for the Action is TRUE. The segment has only one part which is the program code that implements the SCS of the Action's starting event.

In Figure 5.6 the precondition routine is called from all FERs. This modular approach is adequate for our introductory discussion. In general, it is not efficient because checking all preconditions in each FER may not be necessary inasmuch as any particular SCS will typically affect only a limited number of preconditions. As well, the precondition routine can become quite long when the model is complex.

Recall the example of the EndServing future event of the Kojo's Kitchen project. The corresponding FER carry's out the SCS of the terminating event of the ServingW Activity (remove the C.Customer from the A.CounterGroup aggregate) and then tests for the presence of a C.Customer in the customer queue (the precondition for

ServingW and precondition for ServingU). If a precondition is TRUE, the C.Customer is moved to the counter (starting event for ServingW or ServingU), and another EndServing event is scheduled (i.e., the terminating event for the ServingW or ServingU Activity).

The steps for creating an event-scheduling simulation model from an ABCmod conceptual model can be summarised as follows.

- Step 1 – Define appropriate data structures to accommodate the entities, constants, parameters, and input variables within the ABCmod conceptual model. The status of the model will be identified with these data structures.
- Step 2 – Develop the required data modules and user-defined modules specified in the conceptual model together with an initialisation routine. In Java, data models usually correspond to available objects. For example the classes Exponential and Uniform implement, respectively, exponential and uniform distributions. User-defined modules are implemented as Java methods. The Java class constructor provides the ideal place for initialising the simulation model object and including initialisation code specified in the 'Initialise table' of the ABCmod conceptual model.
- Step 3 – Develop a FER for each Activity and each Action Sequence within the ABCmod conceptual model. In Java, each FER is implemented as a Java method.
- Step 4 – Develop a precondition routine to start Activities or Actions when their precondition is TRUE. In Java the precondition routine is implemented as a method.
- Step 5 – Develop appropriate program code to generate the required output data. For generating sample set output, the SCSs in the ABCmod conceptual model explicitly state what is required (i.e., SM.Put( )). However, for trajectory set output, the requirement is implicit and program code must be formulated to provide time/value pairs each time the variable associated with the trajectory set changes in value.

Let's examine how these steps can be applied to the ABCmod conceptual model for Kojo's Kitchen (see Tables 5.2 through 5.16) to create a Java simulation model. Table 5.19 was used to select Java classes for representing the consumer entity instances (C.Customer), the group of customers being served at the counter (A.CounterGroup) and the queue at the counter (A.CustQue). The C.Customer entity instances are represented in Java as *ESAttributeList* objects with two attributes; namely Type (value is a *Character* object with possible values W or S) and TimeEnterQu (double value corresponding to the time the customer enters the queue). The *counterGroup* object is defined as a variable of the *KojoKitchen* class and implemented as a *HashSet* object (see Figure 5.7). The *HashSet* class provides the methods to add an object

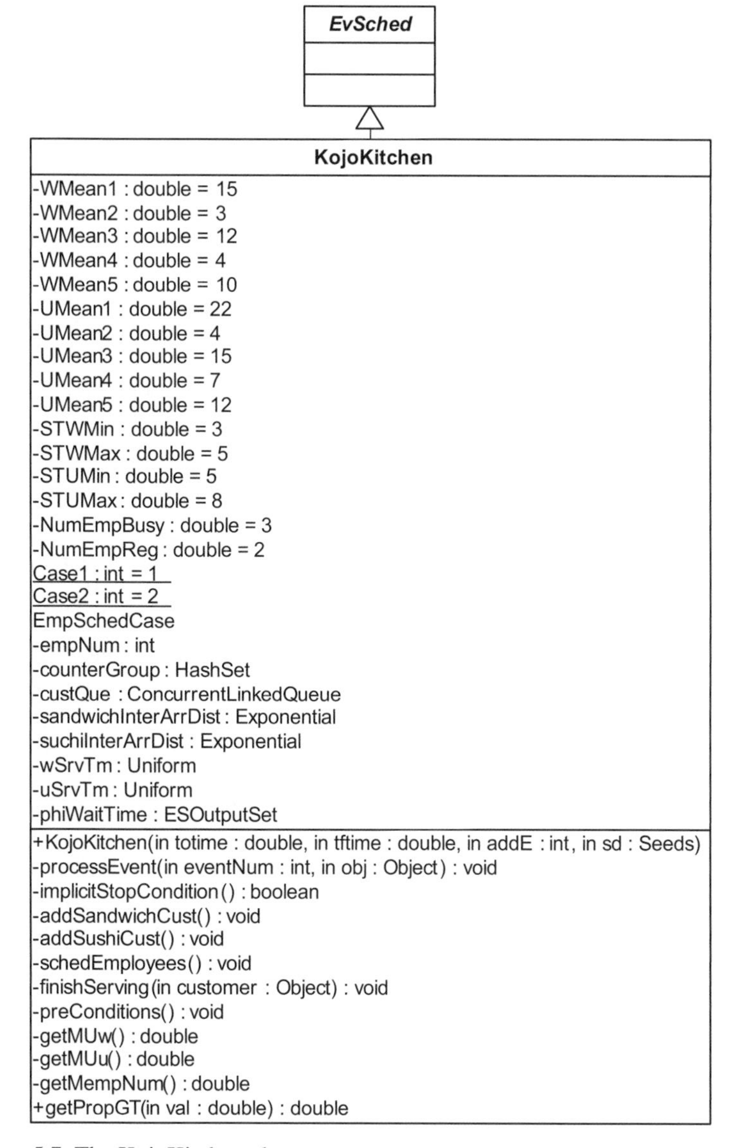

FIGURE 5.7. The KojoKitchen class.

to the set (*add*), remove an object from a set (*remove*), and get the number of elements in the set (*size*). Finally the queue A.CustQue is defined as a *ConcurrentLinkedQueue* object called *custQue* that implements an unbounded FIFO queue.

Step 1 is completed by encoding the constants, parameter, and input variable found in the conceptual model into the *KojoKitchen* class:

- WMean1, WMean2, WMean3, WMean4, WMean5
- UMean1, UMean2, UMean3, UMean3, UMean5
- STWMin, STWMax, STMin, STMax
- NumEmpReg (2 – regular number of employees), NumEmpBusy (3 – number of employees during the busy period)
- Case1 (1 indicating base case for executing the model), Case2 (2 indicating alternate case for executing the model)
- EmpSchedCase – (parameter) An int variable set to AECase1 for the base case (only two employees serving at the counter) and AECase2 when an additional employee is added during busy periods
- EmpNum – (input variable) An integer variable that represents the number of employees at Kojo's Kitchen

Four objects serve as data modules: two for arrivals (i.e., the two *Exponential* objects *sandwichInterArrDist* and *sushiInterArrDist*) and two others for service times (i.e., the two *Uniform* objects *coldCutSrvTm* and *sushiSrvTm*). Three user-defined modules, *getMUw*, *getMUu*, and *getMEmpNum*, implement, respectively, the M[U_W] (t), M[U_U] (t), and M[EmpNum](t) functions, that is, the timing maps for the three Action Sequences. The class constructor *KojoKitchen* contains all necessary code to set up the various data model objects as well as the required initialisation. Step 2 is now complete.

The class variables provide the set of data structures for representing the model structure. Methods are created to operate on these data structures in Step 3 and 4 to capture model behaviour specified in the ABCmod conceptual model.

In Step 3, future events and corresponding FERs are defined. Table 5.20 lists the four future events for the Kojo's Kitchen model and the corresponding FERs implemented as Java methods. The *processEvent* method, implemented with a simple *switch* statement as shown in Figure 5.8, is responsible for calling these methods. Recall from Section 5.4.2 that this method is called by *runSimulation*.

TABLE 5.20. Future events for the Kojo's Kitchen simulation model.

Future Event Name	Identifier	Java Method	ABCmod Activity Constructs
SandwichArrival	1	addSandwichCust	WArrivals
SushiArrival	2	addSushiCust	UArrivals
SchedEmp	3	schedEmployees	SchedEmp
EndServing	4	finishServing	ServingW ServingU

The *addSandwichCust* and *addSushiCust* methods are almost identical as can be expected when examining the corresponding conceptual model Action Sequences, WArrivals and UArrivals. Figure 5.9 shows the code for the *addSandwichCust* and *getMUw* methods.

```
public void processEvent(int eventNum, Object obj)
{
        switch(eventNum)
        {
         case SandwichArrival: addSandwichCust(); break;
         case SushiArrival: addSushiCust(); break;
         case SchedEmp: schedEmployees() ; break;
         case EndServing: finishServing(obj) ; break;
        default:
             System.out.println("Error: Bad event identifier" + eventNum); break;
        }
}
```

FIGURE 5.8. Implementation the *processEvent* method for Kojo's Kitchen.

In *addSandwichCust* a C.Customer entity is instantiated as an *ESAttributeList* object and the Type and TimeEnterQu attributes are initialised respectively, to a *Character* object (with value W) and a *Double* object (whose value is the current time). The C.Customer object is then added to the customer queue using the *custQue.add* method. Bootstrapping is used in *addSandwichCust* to create a stream of arriving customers using the *getMUw* method and scheduling the next arrival with the *addEventNotice* method. Arrival event notices are placed on the FEL as a *SandwichArrival* event. Finally the method invokes the *preConditions* method to test Activity preconditions (more on this later).

Fig. 5.10 show the *schedEmployees* method (that uses the *getMempNum* method) to implements the Action Sequence SchedEmp. The input variable *EmpNum* is updated according to the current time per the conceptual model SCS. Again, bootstrapping is used to schedule the next update of the input variable. Note that in this case however, *getMempNum* generates the timing map deterministically. Note also that *preConditions* is not called at the end of the method because this Action Sequence will not cause any Activity to start.

Figure 5.11 shows the code for the *finishServing* method (the *endServing* FER). Note that in this simple project, both terminating events for the Activities ServingW and ServingU are identical and thus implemented using a single FER. Normally two separate future events would be used. This method illustrates how a reference to an object in the *EventNotice* is used. As shown in Figure 5.12, when an event notice for the *finishServing* event is created, a reference to the customer added to the *counterGroup* is included in the notice. When *processEvent* (see Figure 5.8) is called by *runSimulation,* this reference is passed on until it reaches the *finishServing* method that will use the reference to remove the appropriate *customer* object from *counterGroup.* Finally, the method *preConditions* is called to test Activity preconditions. This completes Step 3.

```
private void addSandwichCust()
{
        // WArrival Action Event SCS
   ESAttributeList customer = new ESAttributeList();
   customer.add(new Attribute(Type,new Character('W')));
   customer.add(new Attribute(TimeEnterQu,new Double(clock)));
   custQue.add(customer);
   addEventNotice(SandwichArrival,getUw());// Schedule next arrival
   preConditions(); // check preconditions
}

private double getMUw()  // for getting next value of Uw(t)
{
   double nxtInterArr;
   double mean;

   if(clock < 90) mean = WMean1;
   else if (clock < 210) mean = WMean2;
   else if (clock < 420) mean = WMean3;
   else if (clock < 540) mean = WMean4;
   else mean = WMean5;
   nxtInterArr = sandwichInterArrDist.nextDouble(1.0/mean);
   return(nxtInterArr+clock);
}
```

FIGURE 5.9. Implementation of the WArrivals Action Sequence.

Step 4 includes adding code for testing the preconditions of starting events. Because of the simplicity of Kojo's Kitchen project, all testing of preconditions has been collected in a single method *preConditions* that is called by the methods *addSandwichCust, addSushiCust,* and *finishServing* (these methods correspond to FERs as discussed in Step 3). As shown in Fig. 5.12, the *custQue.peek* method is used to examine the head of the queue first to test the various Activity preconditions. The code implementing the starting events includes a statement calling the *poll* method to remove the head of *custQue* (note it is possible to use the *poll* method in place of the *peek* method making the code more efficient but this would deviate from the ABCmod conceptual model specifications). The starting event includes the logic to generate the output and is implemented as the method *phiWaitTime.put* method (more on this in the discussion of Step 5). The *addEventNotice* method implements the Activity durations, that is, schedules the future event that corresponds to the Activity's terminating event.

```
private void schedEmployees()                    // SchedEmp event
{
   double mEmpNum; //Timing map value

   switch(empSchedCase)
   {
    case Case1: return; // no changes when addEmpCase is 1
    case Case2:
          if((clock==90) || (clock==420)) empNum = 3;
          else empNum=2;
            // Schedule next change
          mEmpNum = getMempNum();
           if(mEmpNum != -1) addEventNotice(SchedEmp,mEmpNum);
           break;
     default:
           System.out.println("Invalid empSchedCase:"+empSchedCase);
   }
}

private double getMempNum() // for getting next value of EmpNum(t)
{
  if(clock == 90.0) return(210.0);
  else if(clock == 210.0) return(420.0);
  else if(clock == 420.0) return(540.0);
  return(-1); // no more
}
```

FIGURE 5.10. Implementation of the SchedEmp Action Sequence.

Step 5 deals with the collection of sample set and trajectory set data. However, there is no requirement for trajectory set output in the Kojo's Kitchen project (a relevant illustration is provided in the following Section). The acquisition of sample set data indicated by the SM.Put in the starting event SCS's of ServingW and ServingU Activities (Tables 5.15 and 5.16) is achieved using the *phiWaitTime.put* method as shown in Figure 5.12. The *phiWaitTime.put* method records the customer waiting time calculated by subtracting the time a C.Customer exits the queue from the time the C.Customer entered the queue as recorded in its *timeEnterQu* attribute. The *put* method records the output data in the file created when the *phiWaitTime* object was instantiated. A new method *getPropGT* is added to the *KojoKitchen* object to compute, from this recorded data, the proportion of customers that waited longer than five minutes.

```
private void finishServing(Object customer)        // finishServing event
{// ServingW and ServingSU terminating event SCS
   if(counterGroup.remove(customer)==false)
        System.out.println("Error: Customer not in counterGroup");
   preConditions(); // start Activities
}
```

FIGURE 5.11. Implementation of the *ServingW and ServingU* Terminating Events.

```
// Check for starting events of ServingW and ServingU
private void preConditions()
{
            char typeCh;
            ESAttributeList customer;
            if(counterGroup.size() < empNum) // Space at counter
            {
              customer = (ESAttributeList)custQue.peek();   // check head of queue
              if(customer != null)  // not null when customer is present
              { // Get the Type attribute object
                typeCh = customer.charValue(Type);
                switch(typeCh)
                {
                 case 'W': // ServingW starting event
                        customer = (ESAttributeList)custQue.poll();
                        counterGroup.add(customer);
                        phiWaitTime.put(clock,
                                    clock-customer.doubleValue(TimeEnterQu));
                        addEventNotice(EndServing,
                                      clock+sandwichSrvTm.nextDouble(),
                                      (Object)customer);
                        break;
                 case 'U': // ServingU starting event
                        customer = (ESAttributeList)custQue.poll();
                        counterGroup.add(customer);
                        phiWaitTime.put(clock,
                                    clock-customer.doubleValue(TimeEnterQu));
                        addEventNotice(EndServing,
                                      clock+sushiSrvTm.nextDouble(),
                                      (Object)customer);
                        break;
                 default:
                        System.out.println("Invalid customer type found "
                                    + typeCh + " ignored\n");
                        break;
                } //switch
              }//if
            }//if
}
```

FIGURE 5.12. Implementation of Activity preconditions and starting events.

5.4.4 Implementing Other Functionalities

This Section focuses on some event scheduling functionality requirements not illustrated in the discussion of the Kojo Kitchen project: e.g., handling triggered activities and interrupts. For this, we return to our discussions of the Port project in Chapter 4. In particular, we examine some aspects of the ABCmod conceptual model called ABCmod.Port.ver2 that was formulated in Example 2 in Section 4.1.2[12].

Table 5.21 show the future events defined for the event scheduling simulation model corresponding to the activity constructs of the conceptual model given by ABCmod.Port.ver2.

TABLE 5.21. Future events for the port simulation model.

Future Event Name	Identifier	Java Method	ABCmod Activity Construct
TankerArrival	1	addTanker	TankerArrival
BerthingDone	2	addToBerth	Berthing
DeberthingDone	3	tankerExits	Deberthing
ReachedHarbour	4	atHarbour	MoveToHarbour ReturnToHarbour
ReachedBerths	5	atBerths	MoveToBerths
LoadingDone	6	addToDeberthQue	Loading

We first consider the implementation of Triggered Activities. Typically, a method for implementing the starting event's SCS and scheduling the future event associated with the terminating event needs to be created. Then, a reference to TA.Name becomes a call to that defined method.

In the Port project the reference to TA.Loading in the Berthing Activity is implemented as a call to the *addEventNotice* method that schedules the *LoadingDone* future event as shown in Fig. 5.14. Because there is no starting event in the Loading Activity, no future action is needed.

The method *loadingTime* shown in Fig. 5.15 illustrates an implemention of a user defined module. It generates a loading time for a given tanker.

There is a requirement to observe the berth group size and this gives rise to the need for trajectory set output. Appropriate code for this task is shown in Figure 5.16. The class variable *lastBerthGrpN* is used to represent the last recorded value for the berth group size. Whenever a change in the size of the *berthGrp* aggregate is detected, it is recorded using the *put* method.

[12] Note that an event scheduling Java based program for ABCmod.port.ver2 is available at the Web site for this textbook. The file is called PortVer2.java. The reader is encouraged to examine this program and identify how the five translation steps outlined in the previous section have been carried out.

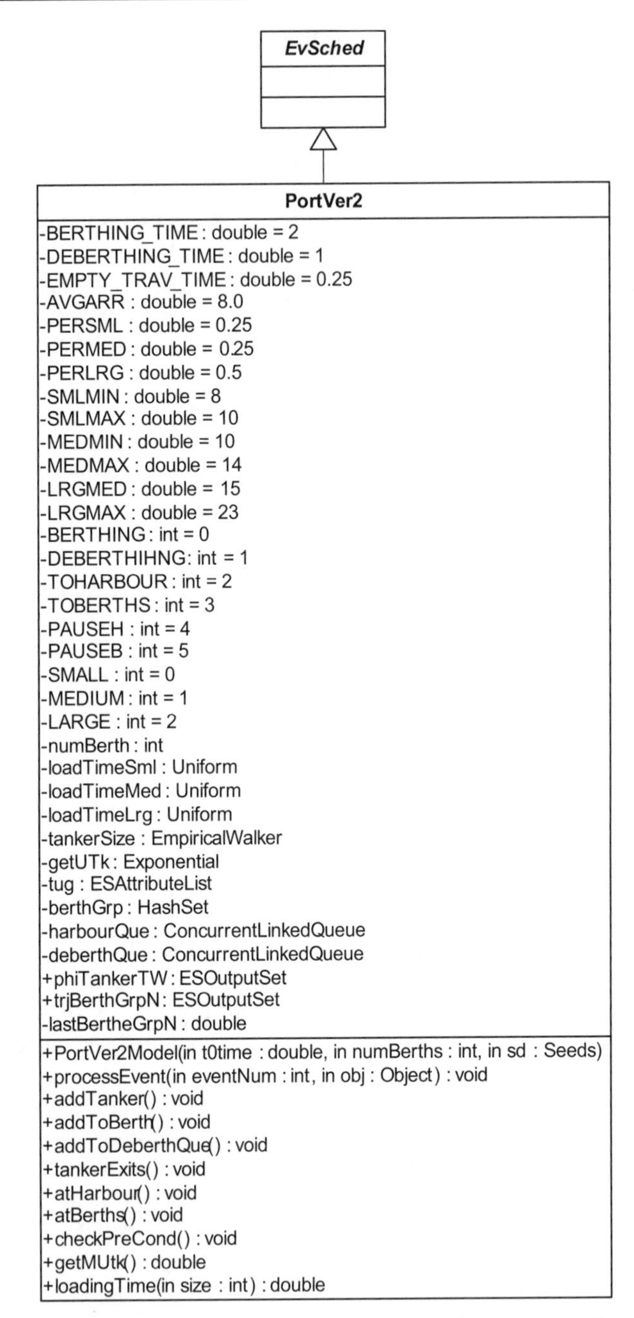

FIGURE 5.13. The PortVer2 class (corresponding to ABCmod.Port.ver2).

```
private void addToBerth()      // BerthingDone event
{
   // Berthing Activity terminating event SCS
   ESAttributeList tker = (ESAttributeList) tug.esAttributeListValue("Ship");
   berthGrp.add(tker);
   // Start Loading Activity (TA.Loading(T.Tug.Tnkr))
   addEventNotice(LoadingDone,clock+loadingTime(tker.intValue("Size")),tker);
   // R.Tug.Status <- PAUSEB
    tug.setIntValue("Status", new Integer(PAUSEB));
   checkPreCond(); // check for other preconditions to start next Activities
}
```

FIGURE 5.14. Implementation of the Triggered Activity Loading(C.Tanker).

```
private double loadingTime(int size)
{
          switch(size)
          {
            case SMALL: return(loadTimeSml.nextDouble());
            case MEDIUM: return(loadTimeMed.nextDouble());
            case LARGE: return(loadTimeLrg.nextDouble());
            default: break;
          }
          return(0);  // bad value
}
```

FIGURE 5.15. Implementation of user-defined module *LoadingTime*.

```
public void processEvent(int eventNum, Object obj)
{
          switch(eventNum)
          {
           . . . . .
          }
          // Update the trjBerthGrpN Trajectory set
          double n = (double) berthGrp.size();
          if(lastBerthGrpN != n)
          {
            trjBerthGrpN.put(clock,n);
            lastBerthGrpN = n;
          }
}
```

FIGURE 5.16. Implementation of the *processEvent* method.

The implementation of interrupts and pre-emption requires the termination of some Activity's duration and carrying out an event SCS to update the model's status. Recall from Section 4.2.4 that the difference between interruption and pre-emption is that the former is invoked when a precondition becomes TRUE whereas the latter is explicitly invoked in an SCS of some other Activity.

In both cases, the challenge is to correctly terminate the duration of an Activity. In any event-scheduling simulation model, the duration of an Activity is implemented by placing an event notice on the FEL to indicate the point in time when the duration ends. Thus to terminate the duration before its "natural ending", this event notice must be removed from the FEL.[13] The method *removeEventNotice* presented in Section 5.4.2 is used to implement this functionality.

```
// Check for Activity preconditions
private void checkPreCond()
{
        int tugSt;  // tug state
        double dbl1;

        // get state of tug
        tugSt = tug.intValue("Status");
        . . .
        . . .
        . . .

        // precondition for interrupting the MoveToBerths Activity
        if( (tugSt == TOBERTHS) && (harbourQue.size() > 0) )
        {
          // find out how close we are to the Berths
          dbl1 = tug.doubleValue("TimeLeftHarbour"); // get TimeLeftHarbour
          if( (clock-dbl1) < (0.7 * EMPTY_TRAV_TIME) )
          {
            // Terminates MoveToBerths Activity
            removeEventNotice(ReachedBerths);
            // Start the ReturnToHarbour Activity
            tug.setIntValue("Status",new Integer(TOHARBOUR));
           // Schedule end of ReturnToHarbour - same as MoveToHarbour
           addEventNotice(ReachedHarbour, clock+(clock-dbl1) );
          }
        }
}
```

FIGURE 5.17. Implementing the interrupt in the MoveToBerths Extended Activity of Table 4.32.

[13] It may be possible that the Activity is not terminated, but the duration is modified. For example, when a tug is towing a tanker into a port and a storm sets in, the tug could simply slow down instead of dropping anchor. In such a case, the event notice is removed from the FEL and replaced with a new event notice that takes into account the slowing down of the tug.

Recall that in Example 2 the MoveToBerths Extended Activity can be interrupted when the tug is returning to the berths with no tanker in tow and a tanker arrives in the port. If no tanker is waiting to be deberthed or the tug has not travelled more than 70% of the way back to the berths, the tug will return to the harbour to berth the tanker that has just entered the port. The event associated with the interrupt triggers the ReturnToHarbour Activity (a Triggered Activity) and explicitly terminates the MoveToBerths Activity (the Terminate 'instruction').

The interrupt precondition is checked in the same fashion as are preconditions for Activities. In the port simulation model, the method *checkPreCond* contains the code for testing both Activity preconditions and interrupt precondition (see Figure 5.17).

5.5 Transforming an ABCmod Conceptual Model into a Process-Oriented Simulation Model

5.5.1 Process-Oriented Simulation Models

The process view for a simulation model begins with a collection of process specifications. Each of these is formulated as an interconnection of some of the Activities within an ABCmod conceptual model. In one of the most common circumstances, the interconnections in a process specification are organised to reflect the lifecycle of one of the consumer entity classes that have been identified. The implication here is that there generally is a process specification for each consumer entity class. A basic (and natural) requirement is that every Activity in the ABCmod conceptual model needs to be included in at least one process specification. Sometimes this gives rise to a situation where one of the resources in the ABCmod conceptual model exhibits a lifecycle that needs to be captured in a process specification; that is, the resource participates in an Activity that is independent of consumer entities. Inputs not involving consumer entity classes are also typically captured in a process specification. Each process specification has a graphical representation which we illustrate below.

It is important to realise that specifying a process for a consumer entity class means that each consumer entity instance in the model will 'live' its own lifecycle, in other words, its own process instance. Each of these individual process instances can interact with each other, often by sharing (usually competing for) the various resources in the model. Consider again the example of the shoppers as shown in Figure 4.2. The process for the shopper entity class has several distinct phases: arrival in the store, followed by movement between browsing activities and payment activities, and finally departure from the store.

An entity's flow from one Activity to another within a process depends on the status of the 'downstream' Activity's precondition. If it is FALSE then flow is interrupted and a delay occurs. This status can, however, change from FALSE to TRUE when a change in the model occurs. But what changes the model? In the ABCmod framework changes to the model result from the occurrence of events. When some on-going Activity terminates, the SCS of its terminating event changes the status of the model and this could enable many pending Activities within the various process instances. This is how processes interact with each other. This situation is illustrated by the shopper that must wait in a queue before acquiring the service desk resource. The desk becomes available to a specific shopper only when that shopper is at the head of the queue and a server becomes available.

As an illustration of how processes can also be defined for resource entities, consider the Port project discussed in Chapter 4. The Activities MoveToHarbour and MoveToBerths involve only the tug resource entity. Because they do not involve the tankers, these Activities are not part of the tanker lifecycle. Thus it is necessary to define a process for the tug resource such that these two Activities become part of a tug process specification. Furthermore, the tanker process instances will interact with this tug process. For example, the arrival of the tanker in the port changes the model's status and this could result in a TRUE value for the precondition of the MoveToHarbour Activity, thereby initiating it. Similarly when the tug completes the Deberthing Activity (which always involves a tanker) a Berthing Activity could be initiated to move a waiting tanker in the harbour to the berths. Activities such as Berthing and Deberthing that involve both the tug and a tanker become part of both the tanker process and tug process. This also illustrates how interaction between processes can occur.

The formulation of a process-oriented simulation model from an ABCmod conceptual model is best carried out by first developing a group of process diagrams. These provide the means for organising the Activities in the ABCmod conceptual model into a collection of process specifications. Each of these can then be easily transformed into the process construct of the programming environment being used. The following subsection outlines the construction procedure for these process diagrams.

5.5.1.1 Formulating Process Diagrams from an ABCmod Conceptual Model

A process diagram organises a subset of the Activities in the ABCmod conceptual model into a directed graph which then serves as a process specification. The purpose of the directed graph is to represent the flow, over time, of an arbitrary instance of some entity type within the model.

These specification set the stage for the development of a process-oriented simulation model (in whatever programming environment that is to be used). Any particular Activity within a ABCmod conceptual model is typically included in multiple process diagrams and this reflects the inherent interaction among entity types. All Activities in an ABCmod conceptual model must be found in at least one process diagram.

The directed graph has two node types: namely, labelled rectangles and unlabelled circles. A labelled rectangle corresponds to an Activity in an ABCmod conceptual model and the label is the name of that Activity. In some special cases, a directed arc may connect two rectangles signifying that the downstream Activity is immediately initiated upon completion of the upstream Activity (as in the case where a Triggered Activity is invoked). However, in most cases, a path between two Activities is constructed from two directed arcs that are interconnected through a circle. The circle, which is called an interaction point, indicates a potential interruption in the flow of time between the completion of the upstream Activity and the initiation of the downstream Activity.

Recall that the unit of behaviour that is embedded in an Activity can begin only when its precondition is TRUE. Generally the precondition is a Boolean expression that incorporates attributes from a number of different entity types within the ABCmod conceptual model. When an entity instance completes an Activity within the process diagram there is no assurance that the conditions required to initiate a sequel Activity are necessarily in place; in other words, the precondition requirements for none of its potential sequel Activities are satisfied. Consider, for example, a shopper who has completed his or her Browsing activity. He or she is not necessarily able to immediately begin a payment activity because the server may be busy with another shopper and/or other shoppers may be waiting in a queue to participate in their own payment activities. The interaction point in a process diagram provides the means for representing such a potential delay. It can be viewed as a point where the flow of the entity to the next Activity is interrupted until the precondition for some possible sequel Activity is satisfied.

In Section 5.3 we introduced the Kojo's Kitchen project and an ABCmod conceptual model was formulated (see Tables 5.1 to Table 5.16). The transition to a process-oriented view requires the specification of three processes, hence three process diagrams. These are shown in Figure 5.18. Notice that all five Activity constructs summarised in Table 5.10 are included.

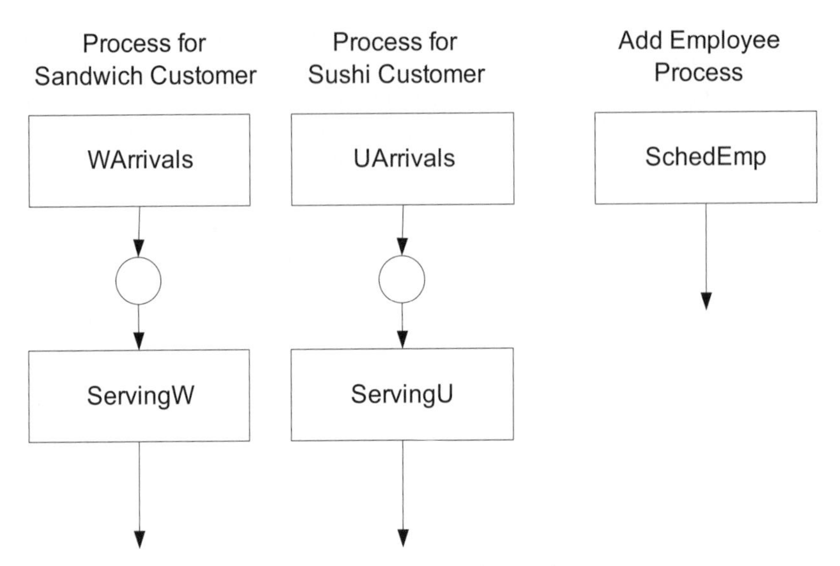

FIGURE 5.18. Process diagrams for Kojo's Kitchen project.

The Sandwich Customer process and the Sushi Customer process are essentially identical. A Customer arrives (either WArrivals or UArrivals), waits for service (the circle), receives service (ServingW or ServingU), and then leaves the model. The Activity SchedEmp (an Action Sequence) whose purpose is to appropriately modify the input variable EmpNum, gives rise to its own distinct process,

Our notion of process diagrams is further illustrated in the following subsection within the context of the Port project. Process diagrams are derived from the ABCmod conceptual model developed for this project in Chapter 4. These process diagrams are then used in Section 5.5.2 as the basis for developing a GPSS simulation model.

5.5.1.2 Process Diagrams for the Port Project

We begin with Version 1 of the Port project (see Section 4.3.1). Two process diagrams are sufficient to capture the complete behaviour of the model: one for the tanker consumer entity class and another for the tug resource entity. As we indicated earlier, the need for the latter specification arises because the MoveToHarbour and MoveToBerths Activities involve only the tug and make no reference to a tanker.

We begin by observing that a tanker which has arrived in the harbour must participate in a sequence of three Activities before leaving the port (and hence leaving the simulation model): namely, berthing, loading, and deberthing. The Berthing Activity can begin only when the tug is available and the specific tanker in question is eligible for its service (i.e., it is at the head of the harbour queue). An interaction point is thus required between the Arrival Activity and the Berthing Activity. The Loading Activity is a Triggered Activity and can be immediately initiated upon the completion of the Berthing Activity. The Deberthing Activity parallels the Berthing Activity in terms of the existence of a precondition, hence an interaction point is needed; that is, the path from Loading to Deberthing requires an interaction point. The resulting process diagram for the Tanker process is shown in Figure 5.19.

The lifecycle associated with the tug is circular; that is, the tug moves in an endless loop between the harbour and the berth area over the course of the observation interval. If the tug is at the berths, it will move to the harbour area in one of two possible ways (i.e., with or without a tanker in tow) depending on conditions that exist with respect to the tanker population. Likewise, when the tug is in the harbour area, it will move to the berths in one of two possible ways (again, with or without a tanker in tow). This circular flow is apparent in the process diagram shown in Figure 5.20. Note the various interaction points that are necessary to reflect the conditions superimposed by the tanker population.

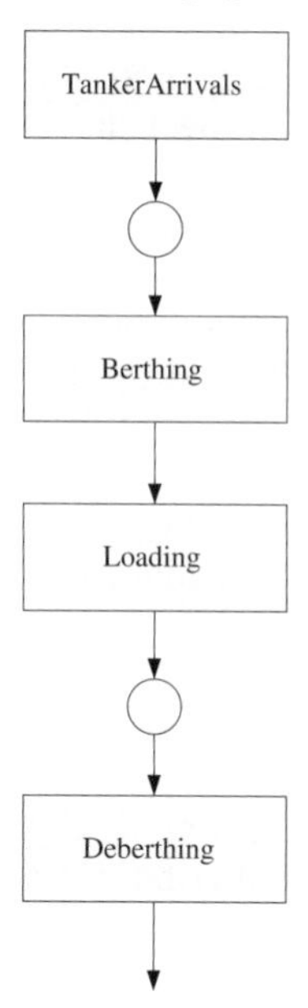

FIGURE 5.19. Tanker process diagram.

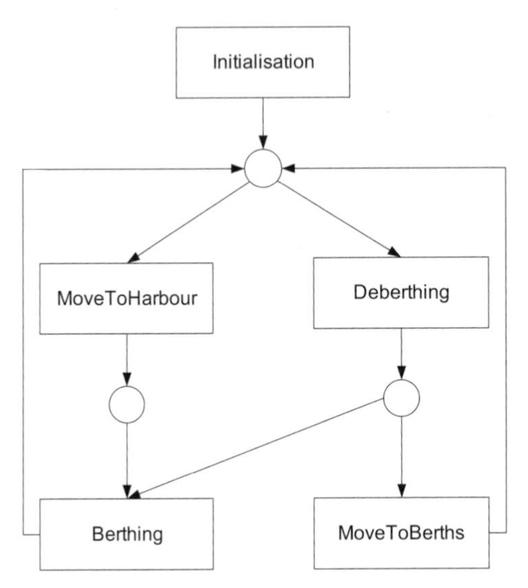

FIGURE 5.20. Tug process diagram.

Possible interventions in the flow of an Activity need to be reflected in process diagrams. This is achieved with the use of dashed arrows. Recall that in Version 2 of the Port project the MoveToBerths Activity can be interrupted.[14] Figure 5.21 shows the additional ReturnToHarbour Activity that is invoked (as a triggered Activity) when the MoveToBerths is interrupted. A dashed exit path from an Activity is used to identify the reaction to an interrupt (or to a pre-emption).

Consider now Version 3 of the Port project, where a storm can interrupt the Activities that involve the tug's movement. Figure 5.22 shows the tug process diagram for this version of the Port project. When an Activity is interrupted by a storm, its duration is terminated. In the context of the Port project under consideration, the state of the tug also changes inasmuch as it drops anchor interrupted. Activity will restart when the storm is over. Points of interaction have been added to illustrate this behaviour. Consider the Deberthing Activity. When interrupted, this Activity is terminated and the tug is placed in a state (R. Tug.Anchored = TRUE) during which it waits for the storm to be over. When the storm is over, the Deberthing Activity is started again (see the precondition of the Deberthing Activity).

---

[14] The tug's movement to the berths with no tanker in tow can be interrupted if a tanker arrives in the harbour and, if there is no tanker ready for deberthing or the tug has not travelled more that 70% of the way back to the berths.

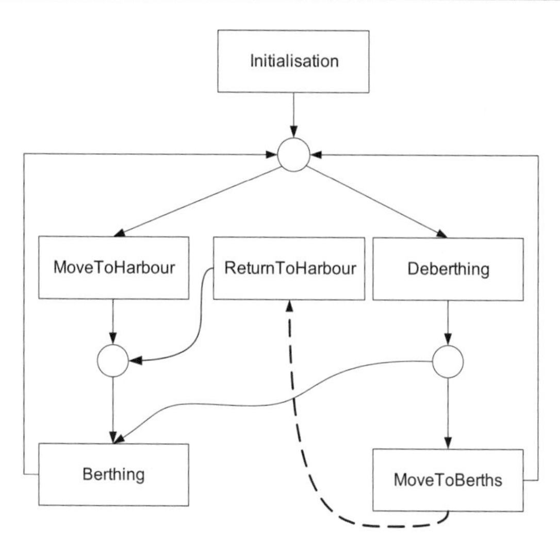

FIGURE 5.21. Tug process diagram with possible interrupt of the MoveToBerths Activity.

Similar interaction points representing the delay caused by the storm have been associated with the MoveToHarbour and MoveToBerths Activities in Figure 5.22. It is not necessary to add such a point for the Berthing Activity because a suitable interaction point already exists. When the ReturnToHarbour is interrupted, the MoveToHarbour Activity is started after the storm is over. It is interesting to note that the MoveToHarbour Activity can be interrupted in either of two ways: by the storm or by a tanker arriving in the harbour.

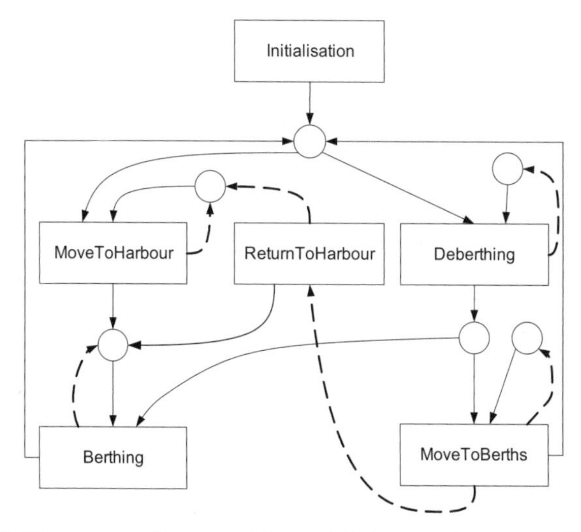

FIGURE 5.22. Tug process diagram with possible interruptions caused by storms.

Process diagrams are not linked to any simulation language or environment. Their purpose is to simplify the transition from an ABCmod conceptual model to any process-oriented language or environment. Section 5.5.2 shows how the process diagrams developed in this section are used in formulating a simulation model in GPSS.

5.5.2 Overview of GPSS

We provide a brief introduction to the GPSS simulation environment. More extensive background material is provided in the GPSS primer given in Annex 2. Readers who are not familiar with GPSS should take the time to review Annex 2.

GPSS provides a process-oriented environment for developing simulation models. These models are formulated in terms of processes for GPSS *Transactions.* A Transaction is composed of a collection of parameters (this corresponds to an attribute-tuple in an ABCmod conceptual model). Each Transaction includes several standard parameters that support GPSS processing, for example, a time parameter for scheduling and two references to GPSS Blocks (one that references the current Block in which the Transaction resides and the other the next Block the Transaction wishes to enter).

GPSS manages Transactions on a number of lists using list-processing techniques. Two lists are especially important in GPSS: namely, the future event chain (FEC) and the current event chain (CEC). The FEC contains a list of Transactions ordered according to a standard time parameter. To advance time the GPSS Scheduler will move the Transaction at the head of the FEC to the CEC and update the simulation clock to the Transaction's standard time parameter (see Annex 2 for details). Transactions are scheduled by placing them on the FEC with a future time value stored in its time parameter. The Scheduler processes the Transactions on the CEC by invoking the functions of the Block referenced in the Transaction Next Block parameter. When the CEC becomes empty, the Scheduler returns to the FEC for another Transaction.

GPSS Blocks are associated with specific functions and they provide the basic processing elements for executing a simulation model. Conceptually, Transactions trigger these functions as they traverse the Blocks. For example, when a Transaction enters an ADVANCE Block, it will be delayed for some defined time before exiting the Block. This Block provides a natural mechanism for implementing an Activity's duration.[15]

[15] The ADVANCE Block in fact schedules the entering Transaction on the FEC.

The GPSS Block functions act on structural entities (representing internal data structures). It is important to be aware of such structural entities when creating a GPSS simulation model. For example, when a Transaction traverses the ENTER Block, the Block inserts the Transaction in a Storage structural entity. Table 5.22 shows possible mapping from ABCmod consumer entity classes and service entities to the most common GPSS structural entities (see Annex 2 for a complete list of GPSS structural entities).

TABLE 5.22. Mapping ABCmod entities to GPSS structural entities.

ABCmod Entity	GPSS Structural Entity Options
Consumer entity instance	Transaction
Resource	Facility or Transaction
Group aggregate	Storage or Transaction Group
Queue aggregate	User Chain

The mapping is not perfect. For example, consider the case where a resource is represented by a Transaction and it is necessary to attach a consumer entity instance, also represented by a Transaction, to the resource. In the ABCmod framework, the cei is assigned to an attribute of the resource. In GPSS, Transaction parameters are simple numeric values and consequently it is not possible to assign the Transaction representing the cei to a parameter in the resource Transaction. An alternate mechanism to accommodate this 'attachment' needs to be identified and possible approach is outlined in the following section.

Development of a structure diagram as presented in Annex 2 is a recommended first step in creating a simulation model in GPSS. GPSS Blocks operate on the structural entities that appear in this diagram.

Elements of behaviour in a GPSS simulation model are expressed in terms of sequences of GPSS Blocks, each of which is called a GPSS Block segment (or simply a segment). A process in GPSS is formulated in terms of one or more such segments and has a graphical representation (see Figure 5.23). In most cases a process has a single segment. Segments start with a GENERATE Block through which Transactions enter the simulation model (this provides the means for handling input entity streams) and end with the TERMINATE Block when Transactions leave the simulation model. An example is shown in Figure 5.23. It is shown later that additional segments are needed to implement interventions. The collection of segments that completely implement a process diagram (as outlined in Section 5.5.1.1) can be regarded as a GPSS process.

A GPSS simulation model is a collection of GPSS processes. Each Block within the graphical view of a GPSS process corresponds to a GPSS statement thereby facilitating the construction of the corresponding GPSS program code. As an illustration, Figure 5.24 shows the GPSS program fragment that corresponds to the GPSS process shown in Figure 5.23.

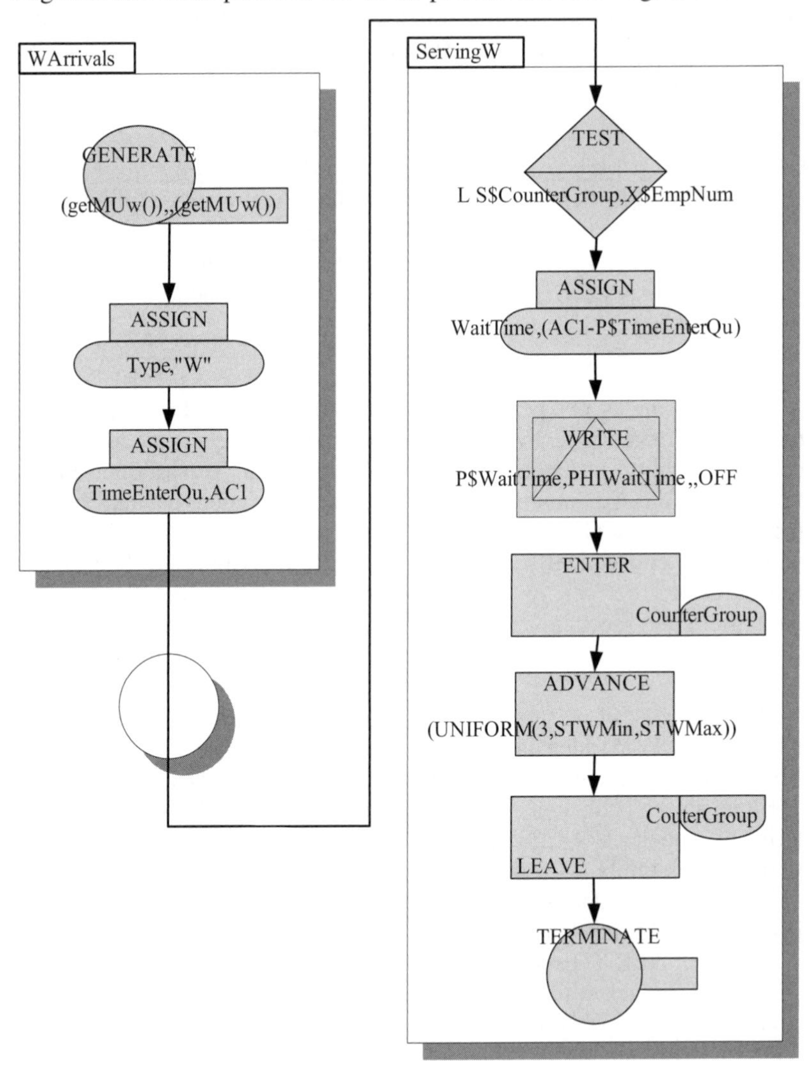

FIGURE 5.23. GPSS process for a sandwich customer at Kojo's Kitchen.

There are many situations where a TERMINATE Block is not part of a process because the entity never leaves the simulation model. This is illustrated in the tug process for the Port project that is examined later (see Figure 5.29).

```
*********************************************************
*  Sandwich Customer Process
*********************************************************

***WArrivals Activity
WCust GENERATE (getMUw()),,(getMUw()) ; Bootstrap. Block
      ; Action Sequence SCS
      ASSIGN Type,"W"              ; Update type
      ASSIGN TimeEnterQu,AC1       ; Mark with current time
***ServingW Activity
      ; Precondition
      TEST L S$CounterGroup,X$EmpNum  ; Precondition
      ; Starting Event SCS
      ASSIGN WaitTime,(AC1-P$WaitTime)
      WRITE P$WaitTime,PHIWaitTime,,OFF  ; To Sample Set
      ENTER CounterGroup          ; Enters the group
      ; Duration
      ADVANCE (UNIFORM(3,STWMin,STWMax))
      ; Terminating Event SCS
      LEAVE CounterGroup          ; Leaves the counter
      TERMINATE                   ; Leave
```

FIGURE 5.24. GPSS code for the GPSS process shown in Figure 5.23.

5.5.2 Developing a GPSS Simulation Model from an ABCmod Conceptual Model

In this section we illustrate the procedure for transforming an ABCmod conceptual model (augmented with process diagrams) into a process-priented simulation model in GPSS. Each process diagram becomes a GPSS process. There are two basic steps:

1. The consumer entity classes and service entities specified in the ABCmod conceptual model are mapped to GPSS structural entities (the result is a GPSS Structure Diagram).
2. Each process diagram and the ABCmod behaviour constructs that they reference are transformed into a GPSS process.

Blocks are used to implement the preconditions, SCSs, and durations that appear in the ABCmod Activities. The GPSS simulation modeller must carefully consider how the ABCmod constructs affect the GPSS structural entities and then implement the action using GPSS Blocks. When an Activity appears in multiple process diagrams, its various components need to be separated and translated into Blocks within different segments. These segments appear in the GPSS processes that are the counterparts to the several processes where the Activity is located. These various notions are illustrated in the discussion that follows which is based on the Port project introduced in Section 4.3.

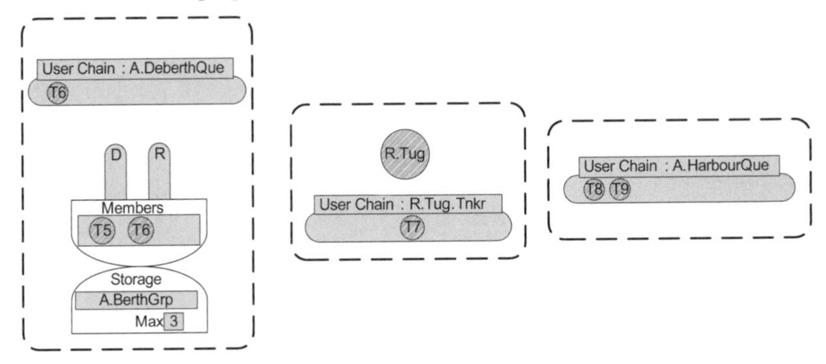

FIGURE 5.25. Components of the GPSS structure diagram for the Port project.

Figure 5.25 shows a GPSS structure diagram for the Port project. It provides a GPSS-oriented representation of the ABCmod entities: R.Tug, A.HarbourQue, A.BerthGrp, and A.DeberthQue and, as well, representative tanker Transactions called (T1, T2, T3, . . .). Note that R.Tug is represented using a Transaction (R.Tug) and a User Chain to handle the R.Tug.Tnkr attribute (this is an approach for dealing with the fact that GPSS cannot attach a tanker Transaction to the tug Transaction parameter).

Let's consider first the ABCmod Action Sequence TankerArrivals (C.Tanker) that defines arrivals of the consumer entity class called C.Tanker. It can be directly translated into the first section of the GPSS process as shown in Figure 5.26. The task of this section is to generate the arrival of tanker Transactions into the GPSS simulation model. The section begins with the GENERATE Block that provides the necessary function to create an input stream of tanker Transactions. The SCS of the event of the Action Sequence called TankerArrivals() serves to initialise the three tanker attributes: it is transformed into three ASSIGN Blocks. The reference to the SM.InsertQue procedure becomes an LINK Block which adds the tanker Transaction to the HarbourQue User Chain (see Figure 5.25).

Notice how the DM.TankerSize data model is implemented using the GPSS Function Entity with the same name (FN$TankerSize) and is used as an argument in the first ASSIGN Block. The rectangle backdrop in Figure 5.26 represents the Action sequence rectangle from the tanker process diagram of Figure 5.19. The main rectangle is divided into two parts that correspond to the standard components of an Action Sequence; namely, the precondition (PR) and the Action Sequence Event (E).

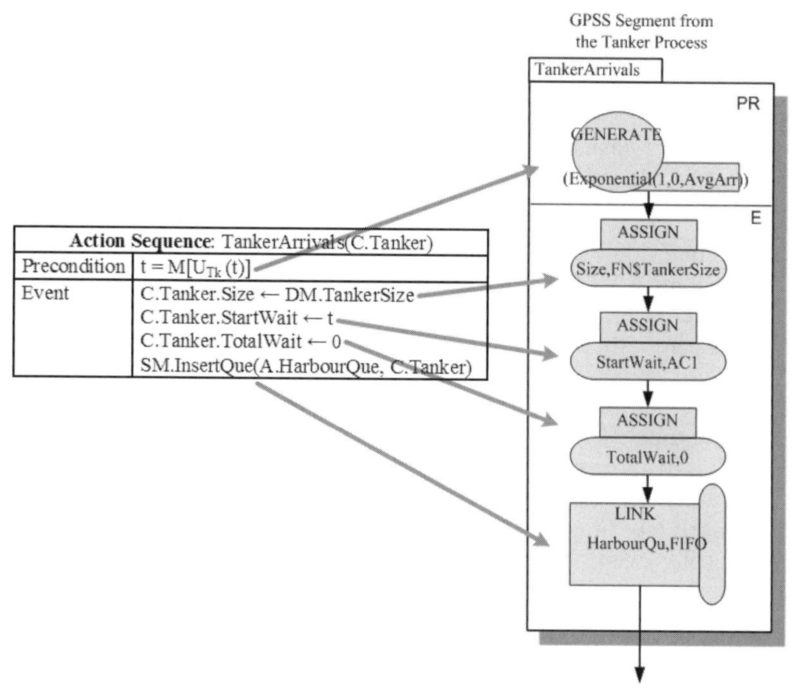

FIGURE 5.26. Translating the TankerArrival Action Sequence.

In subsequent figures that illustrate the GPSS processes, a backdrop of rectangles and circles is added to help illustrate Step (b) (as in Figure 5.26). The rectangles and circles are organised to reflect process diagrams where a rectangle corresponds to an Activity. These rectangles are broken down into the components of the ABCmod constructs and the following labels are used to identify these components.

- PR – **Pr**econdition
- E – **E**vent (for Action Sequences)
- SE – **S**tarting **E**vent
- DU – **DU**ration
- TE – **T**erminating **E**vent
- IPR – **I**nterrupt **PR**econdition
- IE – **I**nterrupt **E**vent
- PMPR- **P**ree**M**pt **Pr**econdition
- PME – **P**ree**M**pt **E**vent

Now we consider the case where a particular Activity appears in more than one process diagram. The Deberthing Activity is such a case because it is found in both the tanker and tug process diagrams (see Figures 5.19 and 5.20). Figure 5.27 shows how this ABCmod Activity is translated into GPSS Blocks located within the GPSS tanker and tug processes. The following comments elaborate on this translation.

a) Precondition: The tanker Transactions are placed in the DeberthQue User Chain waiting for the availability of the tug (see Figure 5.25). Consequently the tug Transaction is responsible for initiating the Deberthing Activity, that is, implements the precondition using a TEST Block in the tug process. The BV$DeberthingCnd is the GPSS Variable entity that represents the expression (CH$DeberthQu 'NE' 0). Note that checking the tug Status parameter is not required in the expression because it tries to enter the TEST Block only when its Status parameter is set to PauseB (see Figure 5.29).

b) Starting event: Parts of the starting event's SCS apply to the tug whereas others apply to the tanker. The parts that apply to the tug are implemented in the tug process; for example, setting the Status parameter to DEBERTHING and correspondingly, those that apply to the tanker are implemented in the tanker process; for example, manipulation of the tanker attributes (i.e. Transaction parameters). The WRITE Block saves the value of the WaitTime parameter in a data stream labeled PHITotalWait that represents the output sample set, called PHI [Tanker.TotalWait]. Removal of the tanker from the A.BerthGrp, SM.RemoveGrp(A.BerthGrp, R.Tug.Tnkr), is implemented by having the tanker Transaction traverse a LEAVE Block that references the BerthGrp Storage entity.

   The SM.RemoveQue in the starting event's SCS is more complex because it carries out two actions. The first removes the tanker from A.DeberthQue and the second attaches the tanker to the tug. This corresponds to removing a tanker Transaction from the DeberthQue User Chain and placing the Transaction in the Tug_Tnkr User Chain (see Figure 5.25). Two different Blocks are required; the UNLINK Block, traversed by the tug Transaction carries out the first action, and a LINK Block, traversed by the tanker Transaction, places the tug Transaction in the Tug_Tnkr User Chain.

c) Duration: Because tanker Transaction is in the Tug_Tnkr User Chain, the ADVANCE Block required for implementing the duration is placed in the tug process.

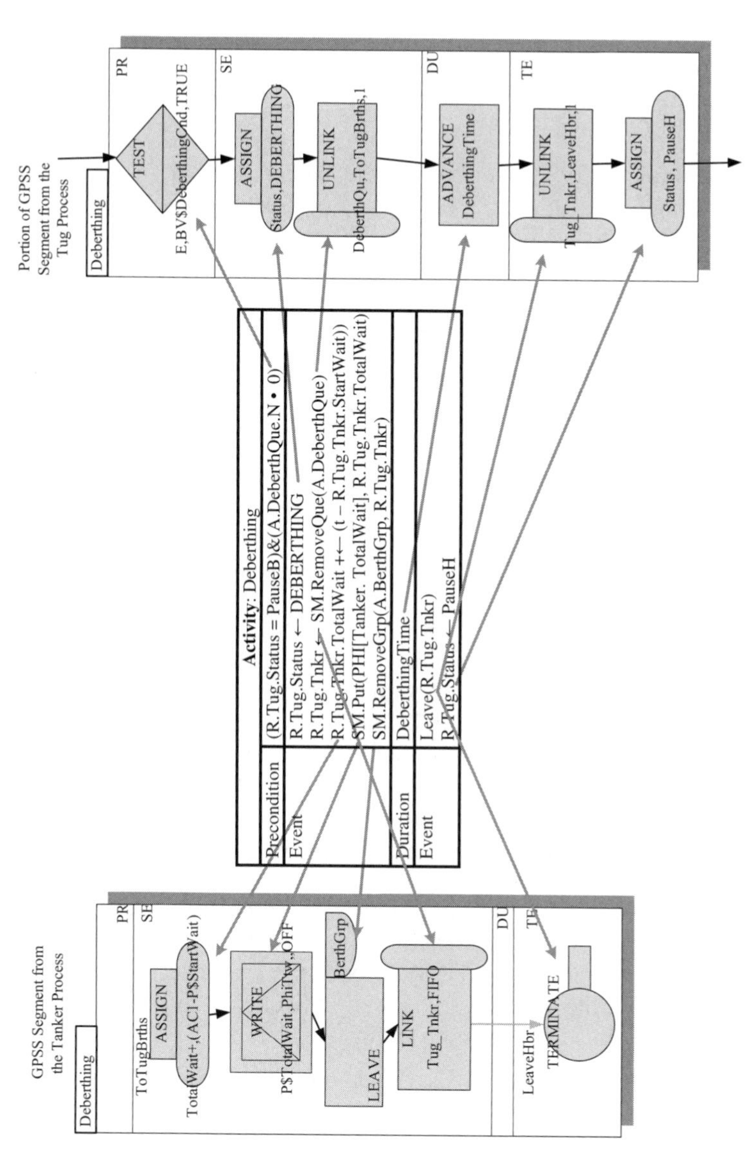

FIGURE 5.27. Translating the Deberthing Activity.

d) Terminating Event: The part of the terminating event's SCS that updates the Status attribute becomes an ASSIGN Block in the tug process. Leave(R.Tug.Tnkr) specifies that the tanker leaves the ABCmod conceptual model. The UNLINK Block in the tug process is used to remove the tanker Transaction from the Tug_Tnkr User Chain and send it to the TERMINATE Block (label TnkrLeave) that removes it from the GPSS simulation model.

The complete GPSS process for the tanker derived from the tanker Process Diagram of Figure 5.19 (and associated ABCmod constructs) is given in Figure 5.28. Similarly the GPSS process for the tug is given in Figure 5.29 (based on the Process Diagram of Figure 5.20). The translation is based on the principles described in the previous paragraphs. By way of additional clarification, we note the following

a) From the process diagrams presented earlier, it's apparent that the Loading Activity and the TankerArrivals Action Sequence are specific to the tanker process and the MoveToHarbour and MoveToBerths Activities are specific to the tug process. These Activities are consequently implemented in only one GPSS process.
b) The Berthing Activity, like the Deberthing Activity examined in detail earlier, is distributed into both the GPSS tanker and GPSS tug processes.
c) Movement of a Transaction from a LINK Block to a subsequent Block requires a number of actions (as represented by the grey arrows in Figure 5.28); A LINK Block moves the tanker Transaction into one of the User Chains shown in Figure 5.25. The Transaction is moved out of a User Chain by another Transaction (in this case the tug Transaction) when it traverses an UNLINK block. The tanker Transaction is then "sent" (see Annex 2 for details) to some GPSS Block. The grey arrows in Figure 5.28 represent both the action of the LINK Block that moves the tanker Transaction into the referenced User Chain and the action of the UNLINK Block traversed by the tug Transaction to move the tanker Transaction out of the referenced User Chain to the referenced GPSS Block. For example, the grey arrow leading from the TankerArrival Action Sequence to the Berthing Activity represents the following actions:
    i. Upon entering the LINK Block, the tanker Transaction is moved into the HarbourQue User Chain.
    ii. When the tug Transaction enters the appropriate UNLINK Block, the tanker Transaction is moved from the HarbourQue User Chain to the ASSIGN Block labelled: To TugHrb (see Figure 5.29).

d) The Loading Activity is implicitly triggered because the tanker Transaction moves automatically from the ENTER Block to the ADVANCE Block that implements the duration of the Loading Activity. No explicit action is required (this is the commonly used GPSS counterpart for a Triggered Activity).

e) The UM.LoadingTime User Module in the ABCmod conceptual model is implemented using a GPSS Plus Procedure with the same name.

f) The GPSS tug process has only one GPSS Block segment in which the tug Transaction circulates without ever leaving the simulation model. Testing preconditions of the ABCmod Activities is implemented using TEST Blocks as shown in Figure 5.29. Each precondition is implemented with Boolean Variable Entities that are referenced by the GPSS TEST Blocks. The definitions for these Variable Entities are as follows (as previously noted, testing the Status parameter is not required).

   i. MvTHarbPreCnd BVARIABLE ((CH$HarbourQu 'NE' 0) 'AND' (S$BerthGrp 'L' MaxBerth) 'AND' (CH$DeberthQu 'E' 0))
   ii. BerthPreCnd BVARIABLE (CH$HarbourQu 'G' 0))
   iii. DeberthPreCnd BVARIABLE (CH$DeberthQu 'NE' 0)
   iv. MvTBerthsPreCnd BVARIABLE (CH$HarbourQu 'E' 0) 'AND' (R$BerthGrp 'G' 0))

g) Note the use of the BUFFER Block in the TE area of the Berthing segment of the tug process. When the tug Transaction enters this Block, it allows the tanker Transaction to be processed first in order to ensure that the tanker Transaction traverses the ENTER Block (and hence becomes member of the BerthGrp Storage entity), before the tug Transaction moves on. This is important because the tug Transaction eventually tries to enter the TEST Block that implements the MoveToHarbour Activity's precondition. This precondition includes evaluating the number of tankers in the BerthGrp Storage entity.

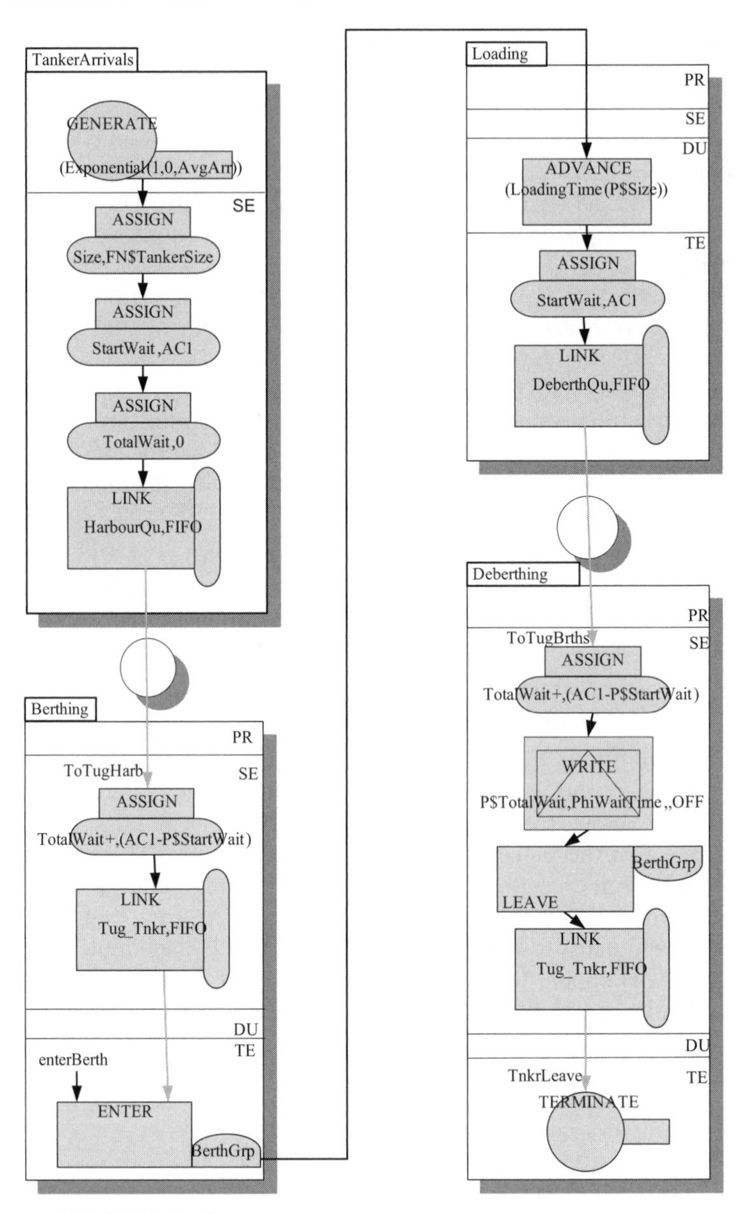

FIGURE 5.28. GPSS Tanker process.

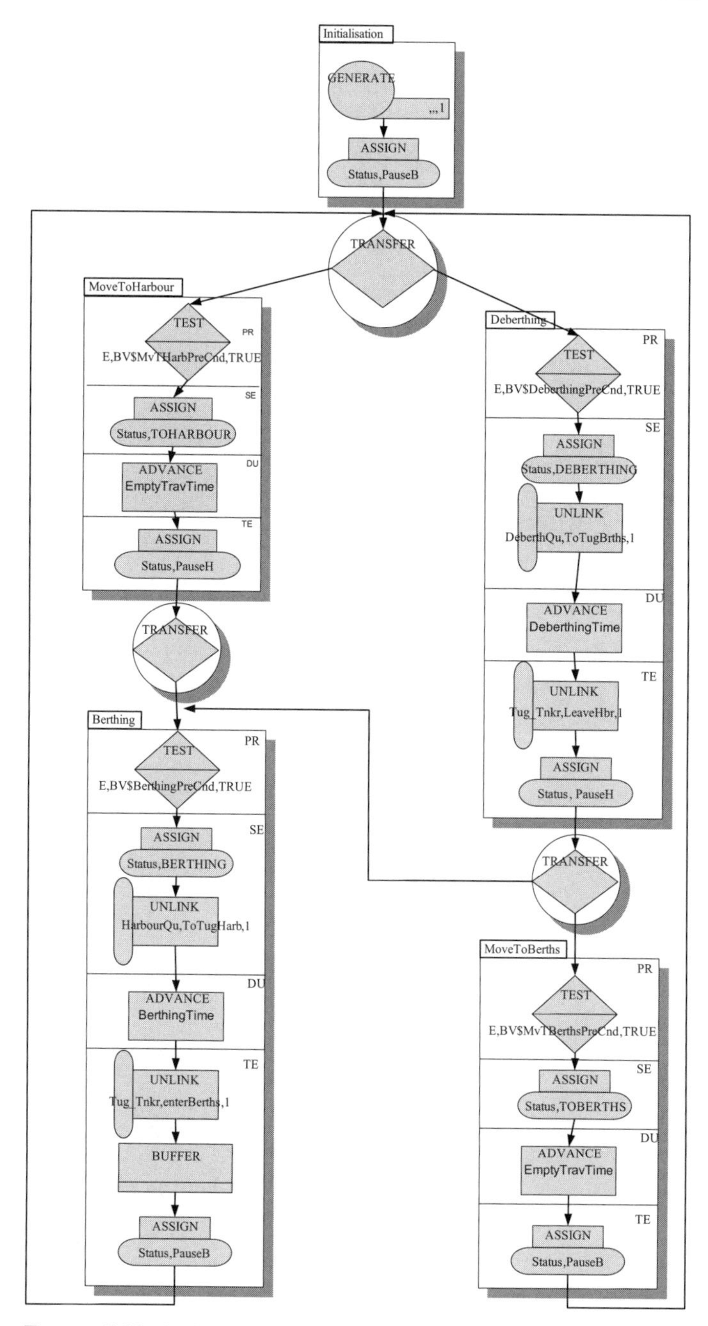

FIGURE 5.29. GPSS Tug process.

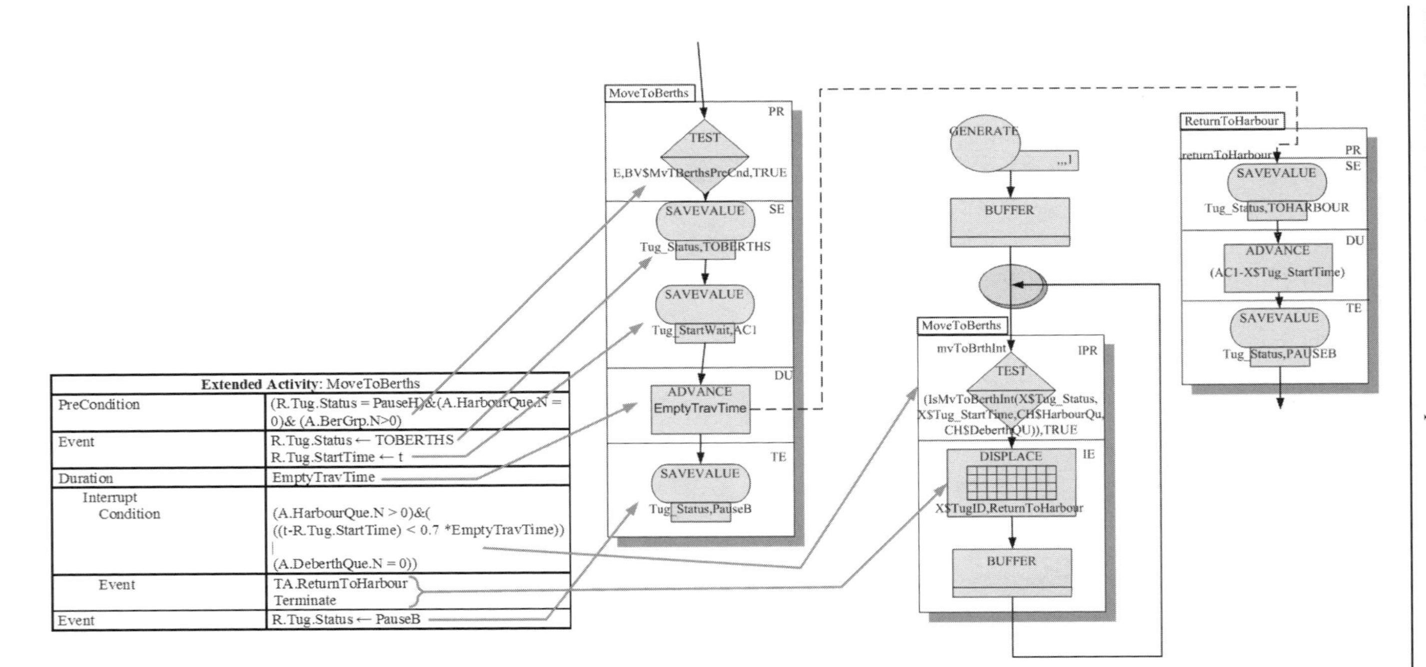

FIGURE 5.30. Translating the MoveToBerths Extended Activity.

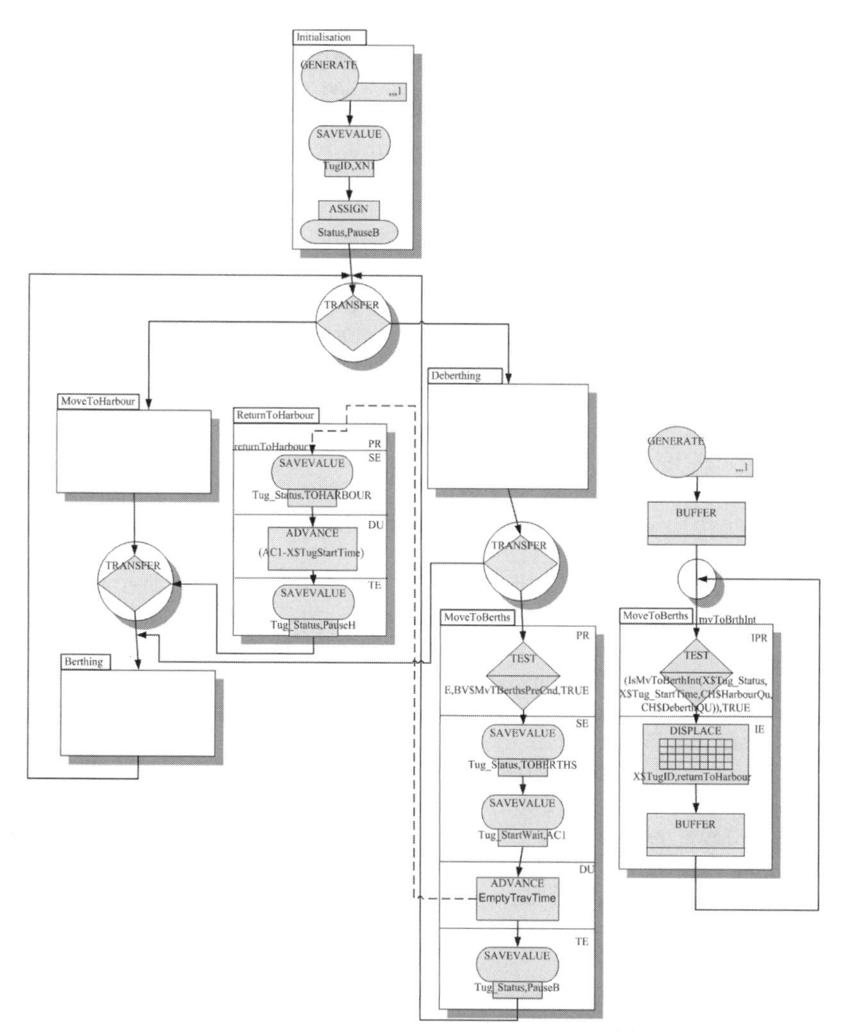

FIGURE 5.31. GPSS implementation of an ABCmod interrupt (Port Version 2).

A number of Blocks in GPSS are especially relevant for implementing interventions that may occur within an ABCmod Extended Activity. In particular, PREEMPT and RETURN Blocks can be used to implement pre-emption, whereas the DISPLACE Block can be used to implement an interrupt. Both the PREEMPT and DISPLACE Blocks remove Transactions that are on the FEC as a result of having entered an ADVANCE Block, that is, Transactions whose duration is currently 'in progress'.

Additional details about the operation of these Blocks can be found in Annex 2 or in the GPSS references that are provided there.

Version 2 of the port project introduces the possibility of an interrupt. In this case the tug, while carrying out its MoveToBerths Activity can (under certain conditions) become obliged to return to the Harbour to service a tanker that has arrived. Because of this interrupt possibility, the ABCmod Extended Activity construct is required and Figure 5.30 shows how it is translated into two GPSS segments that make up the GPSS tug process. An additional segment is required for a special Transaction, called the interrupt Transaction, to monitor the interrupt condition because the tug Transaction cannot monitor itself when scheduled on the FEC during the Extended Activity's duration. The interrupt precondition and interrupt event are implemented within this additional segment.

The MoveToBerths interrupt precondition is implemented with the TEST Block that the interrupt Transaction traverses only when the interrupt condition becomes TRUE[16]. This can occur when the tug is involved in the MoveToBeths Activity (has entered the corresponding ADVANCE Block and is scheduled on the FEC).

The DISPLACE Block implements MoveToBerths Interrupt Event's SCS (i.e., *TA.ReturnToHabour* and *Terminate)*. The DISPLACE Block uses the *X$TugId* argument to identify the tug (it contains the tug Transaction identifier) to be displaced from the FEC and the *ReturnToHarbour* argument as the label of the destination Block for the displaced Transaction. Thus the Block sends the tug Transaction to the first Block of the realisation of the ReturnToHabour Activity and by doing so halts the MoveToBerths Activity. The BUFFER Block that follows the DISPLACE Block allows the tug Transaction to move so that the interrupt condition becomes FALSE (the tug Status is changed) before the interrupt Transaction can test the state of the model. Figure 5.31 shows how the above changes fit into the overall GPSS implementation of the tug process.

---

[16] A PLUS procedure, IsMvToBerthInt, is called to evaluate the status of the model. Using a procedure provides a clearer means of expressing the precondition than Boolean Variable entity. All SNAs needed for testing are passed as arguments to the Procedure. Note that it is necessary to represent the tug attributes as SaveValue entities instead of Transaction parameters to support the testing of these attribute values. GPSS places the interrupt Transaction on the Retry chains of the Tug_Status SaveValue entity, Tug_StartTime SaveValue entity, HarbourQu User Chain entity, and DeberthQu User Chain entity. Whenever any of these entities changes, GPSS moves all transactions from the corresponding retry chain to the CEC, including the interrupt Transaction which will try again to traverse the TEST Block, that is, re-evaluate the TEST.

Version 3 of the Port project introduces the occurrence of storms which represent inputs to the SUI. The ABCmod conceptual model that is formulated in Chapter 4 handles the situation by the use of interrupts in the various Activities that are affected by storms. Implementation of these various interrupts in the GPSS simulation model can be accomplished using the same approach that is outlined above. Details are left as an exercise for the reader.[17]

5.5.1.3 Generating Output

GPSS automatically provides data relating to the entities[18] found in a simulation model; this data can be accessed using GPSS SNA's (see Annex 2). Often these SNA's correspond directly to values for ABCmod derived scalar output variables (DSOV's) that are stipulated in the project goals.

To determine values for DSOV's that are not automatically provided by GPSS, a two step process is proposed:

1. Save the sample set or trajectory set output into a GPSS data stream.[19]
2. When the simulation run has finished, use a Plus Procedure to compute the DSOV value using the contents of the data stream and store the results in a SaveValue entity.

Figure 5.32 illustrates the above two step method for the case of Kojo's simulation model to collect customer wait times in the PHI[WaitTime] sample set and then compute a value for *PropLongWait* using the values in this set. The WRITE Block records customer wait time values in the PHIWaitTime data stream (i.e. in the PHI[WaitTime] sample set). The data stream is set up as a file, *PHIWaitTime.txt,* during initialisation by the OPEN Block (the first segment shown in the figure is traversed by a single Transaction to initialise the simulation model).

The Plus procedure *propGT* (shown in Figure 5.32) executes at the end of the simulation run and its returned value is stored in the SaveValue entity called *PropLongWait*. This is accomplished by the second segment shown in Figure 5.32. A Transaction is generated at the end of the observation interval, that is, at time 660 minutes (11 hours) after the start of the simulation run. The Transaction traverses a WRITE Block to add a –1

[17] A GPSS simulation program for version 3 can be downloaded from the textbook Web site.

[18] The Queue Entities are dedicated to collecting statistics and can be used, for example, to collect statistics on the delay chains of the various structural entities, on the time Transactions spend in the model, and so on. See Annex 2 for details.

[19] A GPSS data stream can either be an internal data stream stored in internal memory or a file data stream store within a system file.

to the end of the PHIWaitTime data stream and then the SAVEVALUE Block to invoke *propGT* and store the results in *PropLongWait*. A similar approach is used for computing the *AvgWaitTime* DSOV in the Port simulation model (the Port Project GPSS code available on the textbook Web site provides details).

```
****************************************************
*  Initialisation
****************************************************
      GENERATE 0,,,1
      OPEN ("PHIWaitTime.txt"),PHIWaitTime
      SAVEVALUE EmpNum,2 ; Initialise for experiments
      TERMINATE

***************************************************
*  Stop simulation after 660 minutes (11 hours)
****************************************************
      GENERATE 660
      WRITE "-1",PHIWaitTime,,OFF
      SAVEVALUE PropLongWait,(propGT(5))
      CLOSE ,PHIWaitTime
      TERMINATE 1

PROCEDURE propGT(5) BEGIN
    TEMPORARY totalCount, countGT;
    SEEK(PHIWaitTime,1);   /* go to start of stream */
    totalCount = 0;
    countGT = 0;
    sample=READ(PHIWaitTime);
    WHILE(StringCompare(sample,"-1") 'NE' 0) DO BEGIN
      wtm=VALUE(sample);
      IF(wtm > val) THEN countGT=countGT+1;
      totalCount=totalCount+1;
      sample=READ(PHIWaitTime);
    END;
    return(countGT/totalCount);
END;
```

FIGURE 5.32. Output collection for the Kojo's Kitchen project.

5.6 Exercises and Projects

5.1 Develop an event-scheduling simulation program (and/or process-oriented simulation program) based on the ABCmod conceptual model formulated in Problem 4.1 of Chapter 4.

5.2 Develop an event-scheduling simulation program (and/or process-oriented simulation program) based on the ABCmod conceptual model formulated in Problem 4.2 of Chapter 4.

5.3 Develop an event-scheduling simulation program based on the ABCmod conceptual model formulated in Problem 4.3 of Chapter 4.

5.4 Assume that the development of the conceptual model of Problem 4.3 and the development of the simulation program of Problem 5.3 have been carried out by two teams where Team A has the primary responsibility for the ABCmod conceptual model and Team B has the primary responsibility for the simulation program. Carry out a verification exercise by:

   a) Giving Team B the task of reviewing the conceptual model before developing the event-scheduling simulation program.
   b) Giving Team A the task of reviewing the simulation program once it has been completed.

5.5 Develop a process-oriented simulation program based on the ABCmod conceptual model formulated in Problem 4.3 of Chapter 4.

5.6 Repeat the verification exercise of Problem 5.4 in the context of Problem 5.5.

5.7 Develop an event-scheduling simulation program (and/or process-oriented simulation program) based on the ABCmod conceptual model formulated in Problem 4.4 of Chapter 4.

5.8 Develop an event-scheduling simulation program (and/or process-oriented simulation program) based on the modified ABCmod conceptual model formulated in Problem 4.5 of Chapter 4.

5.7 References

5.1. Banks, J., Carson II, J.S., Nelson, B.L., and Nicol, D.M., (2005), *Discrete-Event System Simulation*, 4th edn., Prentice-Hall, Upper Saddle River, NJ.

5.2. Horatmann, C., (2003), *Computing Concepts with Java Essentials*, 3rd edn., John Wiley and Sons, New York.

5.3. Liang, Y.D., (2007), *Introduction to Java Programming: Fundamentals First*, 6th edn., Prentice-Hall, Upper Saddle River, NJ.

5.4. CERN, The Colt Project, (2004), Version 1.2.0, http://dsd.lbl.gov/~hoschek/-colt/.

5.5. Sun Microsystems, Java 2 Platform Standard Edition 5.0, TUhttp://java.sun.com/j2se/1.5.0/docs/index.htmlUT.

Chapter 6 Experimentation and Output Analysis

6.1 Overview of the Issue

In this chapter we explore the activities of experimentation and output analysis, which are both central to the success of any modelling and simulation project. In other words, we examine the process of correctly formulating and carrying out goal-directed experiments with the simulation program and then extracting meaningful information from the data acquired via its output variables. The underlying complexity here arises from the uncertainty that is superimposed on all variables in any DEDS model by the random nature of input variables and by the random behaviour of 'internal' processes (e.g., message service time at the nodes of a communications network or failure characteristics of machines in a manufacturing plant). As we have previously noted, these random phenomena represent one of the essential differences between models arising from the DEDS context and those arising from the realm of continuous-time dynamic systems.

A simulation program provides an observation window onto a variety of random phenomena that unfold as a result of the model's execution. Each can be linked to a random variable and some of these random variables are of special interest from the perspective of the project goals.

The notion of output variables was explored in the discussions of both Chapters 2 and 4 where it was stressed that any model necessarily has one or more such variables associated with it. This follows simply because they serve as the conduits for the data that are essential for the resolution of the project's goals. In these earlier discussions we introduced two categories of output variable called point-set output variables (PSOVs) and derived scalar output variables (DSOVs).

There are two types of variable in the PSOV category; namely, time variables and sample variables. These share a common means for delivering data from any particular execution of the simulation program, namely, through the accumulation of a finite set of (possibly) time-indexed values. However, the specific values in such a set are rarely of interest. Instead what is of interest is typically some property of these accumulated data, for example, minimum, maximum, average, or number (a count of the number of values in the set). Such a value is computed and assigned to a designated scalar variable. Such variables are necessarily random variables and they fall into the category of DSOVs. Our interest throughout this chapter is primarily with DSOVs and for convenience we refer to these simply as output variables.

Let's consider some examples of DSOVs that might arise at the level of the ABCmod conceptual modelling framework as discussed in Chapter 4. The list below demonstrates the most fundamental feature of any such variable; namely that it always has a 'definition', that is, a meaning in terms of the behaviours that are represented within the conceptual model. Although this may appear obvious, it is a feature that must be unambiguously documented in the statement of project goals.

- An output variable Y_A which represents the proportion of customers that waited for more than five minutes for service at Kojo's Kitchen in the food court
- An output variable Y_B, which represents the average time spent waiting for tugboat service by the tankers that pass through an ocean port model
- An output variable Y_C, which represents the maximum number of messages in the input buffer of a particular node P of a communications network, over a 24 hour period
- An output variable Y_D, which represents the portion of time that all the attendants in a full-service gas station are busy, over the course of a business day

Some details for these four variables are presented in Table 6.1 in terms of the notions in our ABCmod framework as discussed in Chapter 4. In particular the table shows how a value might be established for each of these variables by carrying out an operation on some underlying output set of data values.

TABLE 6.1. Elaboration of representative output variables.

Output Variable (DSOV)	SUI Context	Underlying Time Variable/Sample Variable (PSOV)	Output Set (Trajectory or Sample Set)	Operator on Output Set
Y_A	Kojo's Kitchen	Attribute Customer.WaitTime of the Customer consumer entity class that represents the customers flowing through Kojo's Kitchen	PHI[C.Customer.WaitTime]	PropGT
Y_B	Ocean port	Attribute Tanker.TotalWait of the Tanker consumer entity class is used to accumulate the time spent by each tanker instance waiting for tugboat service	PHI[C.Tanker.TotalWait]	AVG
Y_C	Communication network	Attribute Pnode.N of the queue representing the input buffer of the particular node of interest	TRJ[A.Pnode.N]	MAX
Y_D	Full-service gas station	Attribute Attend.AllBusy of the resource entity associated with the attendants whose value is set to 1 when all attendants are busy and is zero otherwise	TRJ[R.Attend.AllBusy]	AVG

Any particular output variable listed in Table 6.1 acquires a value as a consequence of the execution of its respective simulation program (i.e., as a consequence of a 'simulation run' or simply a 'run'). However, this value is not a direct outcome of the experiment but rather is obtained by carrying out an operation on a set of data values as illustrated in Figure 5.1. In the case of Y_A the data set is the sample set PHI[C.Customer.WaitTime] which is populated by values of the sample variable Customer.WaitTime (an attribute of the consumer entity class called Customer). Each Customer instance that passes through Kojo's Kitchen contributes a value to PHI[C.Customer.WaitTime], and for any particular simulation run, the value acquired by Y_A is obtained as PropGT(5,PHI(C.Customer.WaitTime)). Here *PropGT(Val,SampleSet)* is a user-defined module specified in the ABCmod conceptual model for the Kojo's Kitchen project in Chapter 5 (see Table 5.10).

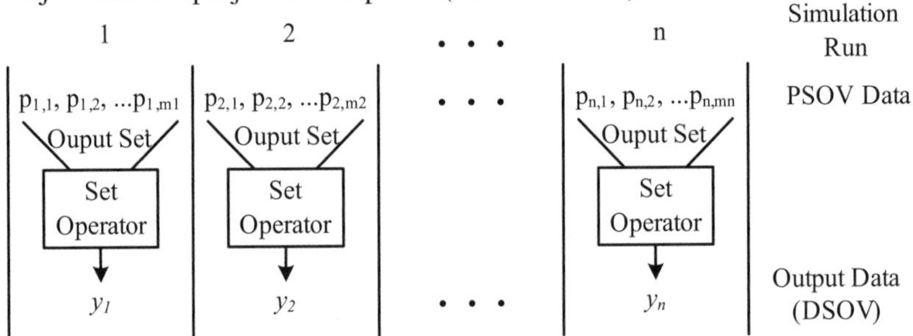

FIGURE 6.1. Generation of data from multiple simulation runs.

As previously observed, any DSOV is a random variable. There are certainly circumstances where interest in a random variable can focus simply on a particular value (e.g., the sum of the dots showing on a pair of dice when the dice are thrown during a game of chance). However, the value of a DSOV acquired from a single simulation run generally falls far short of providing useful information from the perspective of the requirements of project goals. The information that is needed typically relates to the values of the parameters of the distribution of the DSOV (e.g. mean, variance) and meaningful estimates of such parameters can only be formulated from results obtained from multiple runs that have been organised to yield independent observations.[1]

A frequent misunderstanding occurs when the project goals require a mean value estimate that appears to coincide directly with some DSOV

[1] Strictly speaking, this is not entirely correct. In the context of a steady-state study, there does exist an approach called the method of batch means where all required data are generated from a single long simulation run. A brief discussion can be found in Section 6.3.2.

(for definiteness, let's call it Y) that is defined as an average. Consider, for example, our port project where the mean waiting time of tankers is required. A particular simulation run will yield a sample set whose members are the waiting times of the tankers that passed through the port during that run. The average of the values in this sample set (which we denote by $\hat{y}$) would represent a single observation of the DSOV, Y.

One might be tempted here to use $\hat{y}$ as an estimate of the mean value that we seek (namely, the mean waiting time of tankers that pass through the port). However there generally is a correlation among the values because, for example, a long wait by some tanker will likely result in long waits by succeeding tankers thereby introducing a bias in the collected data. This circumstance precludes the use of the standard methods of statistics which depend on the assumption of independence, for example. for the determination of the confidence interval that we discuss below. It is for this reason that suitably replicated simulation runs (or other equivalent approaches) are required which will generate a collection of independent observations of Y from which the desired mean value estimate and a confidence interval can be formulated. This is achieved by proper management of the seeds used in the random variate generation procedures that are embedded in the simulation model.

For the most part, our interest in experimentation focuses on the mean values of designated output variables within the simulation program. It needs to be recognised, however, that the determination of an exact value for these is rarely feasible. Experiments with the simulation program can, at best, deliver the data from which an estimate of the mean (called a *point estimate*) can be formulated together with an assessment of the quality of the estimate (i.e., a *confidence interval*). Guidance for determining what experiments need to be carried out and how the acquired data need to be handled in order to obtain credible estimates are provided by some of the fundamental results from probability theory. An overview of these can be found in the latter sections of Annex 1. The topic is explored in the discussions below.

6.2 Bounded Horizon Studies

We now consider the basic problem of analysing the values acquired by an output variable in the simulation program in the case of a bounded horizon study. From a collection of values acquired from n simulation runs, we determine a point estimate of the mean (i.e., a single number whose validity has some credible basis) and then formulate an interval in which the point estimate lies with a prescribed degree of confidence.

The considerations that follow rely heavily on the results presented in Annex A1, in particular the results in Sections A1.5 through A1.7.

6.2.1 Point Estimates

Suppose Y is an output variable (i.e. DSOV) of the simulation program **M** and we seek an estimate of the mean of the distribution of Y, namely an estimate of $\mu = E[Y]$. The fundamental result from probability theory upon which we rely is the strong law of large numbers (see Section A1.5 of Annex A1). The interpretation in our context is that $\overline{Y}(n)$ approaches μ as n becomes large where:

$$\overline{Y}(n) = \frac{1}{n} \sum_{k=1}^{n} Y_k$$

and we regard the Y_ks as surrogate random variables for Y that are associated with a sequence of n correctly replicated experiments with **M** (the variable Y_k is associated with experiment k). All Y_k's have the same distribution as Y because they reflect the same process (namely the simulation program **M**). Furthermore, because they are linked to a sequence of correctly replicated experiments, we can assume that the Y_k's are independent. Hence the Y_k can be taken to be a set of independent identically distributed (IID) random variables.

Correctly replicated simulation runs are a key requirement in formulating the estimate that we seek. The implication here is that there is appropriate management of the seeds used to initialise the various random number generators from run to run to create a meaningful set of independent and identically distributed observations (initial conditions, however, must remain invariant except when their values are part of the random envelope).

On the basis of the above, a point estimate of μ can be obtained in the following way.

1. Choose a suitable value for n, the number of replications (in principle, n needs to be large, but a value in the order of 30 is generally satisfactory).
2. Collect the n observed values $y_1, y_2, \ldots, y_n$ for the random variables Y_k, $k = 1, 2, \ldots, n$, that result from n replicated simulation runs of the simulation program **M.**
3. Compute:

$$\overline{y}(n) = \frac{1}{n} \sum_{k=1}^{n} y_k \,.$$

The numerical value that results for $\overline{y}(n)$ is then taken to be the point estimate for $\mu = E[Y]$ that we seek.

6.2.2 Interval Estimation

We now expand our task by undertaking to find a suitable value for the number of replications n which will ensure a particular 'quality' for the estimate $\bar{y}(n)$. We know from Section A1.7 that an interval (called the confidence interval) can be established within which μ falls with a prescribed level of confidence. This interval has the form $[y(\bar{n}) - \zeta(n), y(\bar{n}) + \zeta(n)]$ where $\zeta(n) = (t_{n-1,a}\ s(n)/\sqrt{n})$, and $s(n)$ is an estimate of the standard deviation. We call $\zeta(n)$ the confidence interval half length. Its value is clearly dependent on the Student t-distribution value $t_{n-1,a}$.

The quality criterion we introduce is the requirement that, with confidence 100C% ($0 < C < 1$), $|\bar{y}(n) - \mu| < \zeta^*$. In other words, we want to ensure that (with a prescribed level of confidence) the interval half length $\zeta(n)$ is less than a specific value denoted by ζ^*. A possible choice for ζ^* is $r\bar{y}(n)$ where r is a value chosen in the range (0, 1). With this choice, the maximum displacement of the estimate from μ is proportional to the value of the estimate itself. Note that our quality measure can be interpreted as

$$\frac{\zeta(n)}{\bar{y}(n)} < r$$

The procedure is outlined below and is based entirely on the discussion in Section A1.7 of Annex 1 (Equation (A1.36) has particular relevance).

1. Choose values for r and for the confidence level parameter C, and as well, an initial value for n that is not smaller than 20.
2. Collect the n observed values y_1, y_2, . . . , y_n for the random variables Y_k , k = 1, 2, . . . , n, that result from n replicated simulation runs of the simulation program, **M**.
3. From tabulated data for the Student t-distribution, determine $t_{n-1,a}$ where $a = (1 - C)/2$.
4. Compute:

$$\bar{y}(n) = \frac{1}{n}\sum_{k=1}^{n} y_k$$

$$s^2(n) = \frac{\sum_{k=1}^{n}(y_k - \bar{y}(n))^2}{n-1} \tag{6.1}$$

$$\zeta(n) = \frac{t_{n-1,\alpha}\, s(n)}{\sqrt{n}}$$

5. If $\zeta(n) < \zeta^* = r\bar{y}(n)$ (or $\zeta(n) / \bar{y}(n) < r$) then accept $\bar{y}(n)$ as the point estimate of μ and end the procedure, otherwise continue to Step 6.
6. Choose Δn no smaller than 3 and collect additional observations $y_{n+1}, y_{n+2}, \ldots, y_{n+\Delta_n}$ through a further Δn replications, replace n with $n + \Delta n$, and repeat from Step 3.

6.2.3 Output Analysis for Kojo's Kitchen Project

This section examines how the analysis techniques described in Section 6.2.2 can be applied to achieving the goal set out in the Kojo's Kitchen project. Recall that the goal set out in Chapter 5 was to investigate the impact on the output variable *PropLongWait* (i.e., the proportion of customers waiting longer than five minutes) of adding an additional employee. The Java event-scheduling simulation program presented in Section 5.4.2 is used to experiment with the simulation model and generate data for analysis. The collected data are analysed using a number of useful data analysis tools available in Microsoft Excel.

Figure 6.2 shows the Java method used to carry out multiple simulation runs with the Kojo's Kitchen simulation model (see Section 5.3). Note the following.

- The first part of the method generates the random seeds used in all the simulation runs. The CERN Java package offers a Class *RandomSeedGenerator* that provides the means to generate appropriate (uncorrelated) random seeds. This ensures that the different simulation runs provide independent values for the *PropLongWait* output variable. Note also that the seeds are stored in an array of *Seeds* objects. Also note that four seeds make up a *Seeds*object, one for each random number generator used in the simulation program. Thus they can be reused when executing the runs for the alternate case. This is important for comparing the two cases as discussed in Section 6.4.
- For each simulation run, a new *KojoKitchen* object is created using the Class constructor. The constructor provides the data necessary for the simulation run, that is, specifies the observation interval (the first two arguments specify the right- and left-hand boundaries of the interval), a value for the *empSchedCase* parameter (either Case1 or Case2), and finally a *Seeds* object to seed the random number generators. Recall that Case 1 is the base case and Case 2 is the situation where the third employee is hired during busy periods.
- After each run, the value generated for *PropLongWait* is displayed along with the run number. The output of the running program can be redirected into a file and subsequently loaded into an Excel worksheet for analysis.

```
class KojoExperiment1
{
  public static void main(String[] args)
  {
     final int NUMRUNS = 10000;
     int i;
     Seeds[] sds = new Seeds[NUMRUNS];
     KojoKitchen kojo;  // simulation program
     double propLongWait;

     // Get a set of uncorrelated seeds
     RandomSeedGenerator rsg = new RandomSeedGenerator();
     for(i=0 ; i<NUMRUNS ; i++)
         sds[i] = new Seeds(rsg.nextSeed(),rsg.nextSeed(),
                            rsg.nextSeed(),rsg.nextSeed());

     // Loop for NUMRUN simulation runs for each case
     // Case 1
     System.out.println("Case 1 - no additional employee");
     for(i=0 ; i < NUMRUNS ; i++)
     {
       kojo = new KojoKitchen(0.0,660.0,KojoKitchen.Case1,sds[i]);
       kojo.runSimulation();
       propLongWait = kojo.getPropGT(5);
       System.out.println((i+1)+", "+propLongWait);
     }
     // Case 2
     System.out.println("Case 2 - add employee during busy times");
     for(i=0 ; i < NUMRUNS ; i++)
     {
       kojo = new KojoKitchen(0.0,660.0, KojoKitchen.Case2,sds[i]);
       kojo.runSimulation();
       propLongWait = kojo.getPropGT(5);
       System.out.println((i+1)+", "+propLongWait);
     }
  }
}
```

FIGURE 6.2. Java method for experimentation with the Kojo's Kitchen simulation program.

Table 6.2 shows the values for *propLongWait* for the first 20 simulation runs for each of the two cases. The values for the point estimate ($\overline{y}(n)$), the standard deviation ($s(n)$), and the confidence interval half length ($\zeta(n)$) are shown in the table, with n = 20. These were computed by using Equation (6.1) with a 90% confidence level (i.e C=0.9). The left boundary $\overline{y}(n)-\zeta(n)$ and right boundary $\overline{y}(n)+\zeta(n)$ of the confidence interval are given by CI Min and CI Max, respectively.

TABLE 6.2. Analysis of generated data from the first 20 simulation runs.

Run	Case 1	Case 2
1	0.634	0.263
2	0.595	0.209
3	0.256	0.067
4	0.532	0.335
5	0.282	0.049
6	0.649	0.278
7	0.458	0.024
8	0.515	0.158
9	0.618	0.062
10	0.667	0.348
11	0.483	0.238
12	0.524	0.107
13	0.663	0.447
14	0.235	0.053
15	0.404	0.051
16	0.472	0.112
17	0.425	0.094
18	0.565	0.124
19	0.392	0.048
20	0.381	0.123
$\overline{y}(n)$	0.487	0.160
$s(n)$	0.134	0.121
$\zeta(n)$	0.052	0.047
CI Min	0.436	0.113
CI Max	0.539	0.206

Table 6.3 shows for each of the two cases the values of $\overline{y}(n)$, $s(n)$, and $\zeta(n)$ (computed using Equation (6.1)) as well as the boundaries of the confidence interval (CI Min and CI Max) and the ratio $\zeta(n)/\overline{y}(n)$ when n (the number of simulation runs) is increased. Note from the rightmost column how the ratio $\zeta(n)/\overline{y}(n)$ decreases as n increases. This is mainly a consequence of a decreasing value for the confidence interval half length $\zeta(n)$.

Observe that for Case 1, with 20 runs the half length of the confidence interval is essentially 10% of the point estimate (see rightmost column where the value is 0.106). However with 20 runs, the interval half length in Case 2 is almost 30% of the point estimate (value in rightmost column is 0.293). For Case 2, 100 runs are required to achieve a comparable confidence interval as Case 1.

TABLE 6.3. Impact of number of runs on the confidence interval.

			Case 1			
n	$\overline{y}(n)$	$s(n)$	$\zeta(n)$	CI Min	CI Max	$\zeta(n)/\overline{y}(n)$
20	0.487	0.134	0.052	0.436	0.539	0.106
30	0.503	0.125	0.039	0.464	0.542	0.077
40	0.502	0.119	0.032	0.471	0.534	0.063
60	0.504	0.116	0.025	0.479	0.529	0.049
80	0.499	0.129	0.024	0.475	0.523	0.048
100	0.503	0.132	0.022	0.481	0.524	0.044
1000	0.510	0.120	0.006	0.504	0.517	0.012
10000	0.508	0.126	0.002	0.506	0.510	0.004

			Case 2			
n	$\overline{y}(n)$	$s(n)$	$\zeta(n)$	CI Min	CI Max	$\zeta(n)/\overline{y}(n)$
20	0.160	0.121	0.047	0.113	0.206	0.293
30	0.192	0.124	0.039	0.153	0.230	0.201
40	0.193	0.119	0.032	0.161	0.225	0.165
60	0.187	0.115	0.025	0.162	0.211	0.133
80	0.185	0.121	0.023	0.162	0.207	0.122
100	0.187	0.123	0.020	0.167	0.207	0.109
1000	0.188	0.121	0.006	0.181	0.194	0.034
10000	0.184	0.120	0.002	0.182	0.186	0.011

6.3 Steady-State Studies

The fundamental requirement in a steady-state study is the postponement of data collection during a simulation run until it is apparent that the simulation model is operating under steady-state conditions; that is, the stochastic processes associated with the output variables of interest have become stationary. A necessary (but not sufficient) condition for steady-state behaviour of the simulation model is the requirement that the underlying random variables associated with autonomous stochastic processes, such as arrival rates and service rates, are themselves stationary. But even when this is the case, the model's initial conditions usually give rise to circumstances that cause dependent stochastic processes in the simulation model to pass through a transient phase at the start of a simulation run.

Recall that for steady-state studies, the right-hand boundary of the observation interval is not specified. This provides the flexibility to execute a simulation run for as long as necessary in order to first reach steady-state conditions and then acquire sufficient data to permit meaningful conclusions. Consequently the execution of experiments for steady-state studies must address two important issues:

- Determining a warm-up period: A transient period is always present at the beginning of any simulation run. Behaviour data from this interval are (by definition) incompatible with the steady-state requirements of the study. The implication here is that a warm-up period that precedes the collection of data needs to be recognised. The duration of this period cannot be predicted and hence a mechanism for determining the end of the warm-up period must be incorporated into the experimentation procedure. Data collection can begin only after this transient, or warm-up period, has come to an end.
- Establishing confidence in the conclusions. A single simulation run can be executed for an extended observation interval to yield data from which a point estimate of the mean of the output variable (or variables) of interest can be calculated. Provided the length of the run has generated a sample of sufficiently large size, the estimate can have reasonable credibility (e.g., on the basis of the law of large numbers) However, a confidence interval for any such estimate requires a collection of independent observations in order to apply the techniques described in Section 6.2.2.

6.3.1 Determining the Warm-up Period

Considerable research effort has addressed the problem of establishing a suitable warm-up period for a simulation run, that is, an interval which allows sufficient time for the dependent stochastic process of interest to reach a steady-state (see, e.g., [6.4], [6.6], [6.7], [6.8]). The Welch moving average method is one of the many available approaches. It is graphically oriented, relatively straightforward, and provides reasonable estimates. This section outlines the application of this method (a more extensive presentation can be found in Law and Kelton [6.5]).

The Welch moving average method relies on a relatively small number of simulation run replications (e.g., five to ten). The duration of each replication needs to be sufficiently long so that it extends beyond the transient period. A typical replication is shown in Figure 6.3 which illustrates a representative transient condition at the start of the simulation run (e.g., a case where the simulation model begins without any consumer entity instances being present). The horizontal axis in Figure 6.3 corresponds to (simulated) time which has been compartmentalised into m time cells. The vertical axis shows how the average value for some output variable might change if separate averages were computed within the time cells. The changing shape of a hypothesised distribution function for this output variable is superimposed. The Figure shows that starting at time cell D, changes in average value no longer occur and hence steady-state can be assumed.

Selecting the size of the time cells and the number of time cells (which is equivalent to establishing the length of the simulation run) depends on the underlying nature of the simulation model. The size of the time cell should be large enough to be provide reasonable results (i.e., enough data points to compute a credible average within the cell), and yet short enough to be able to detect the existence of the transient.

Replication j generates an output set of n_j values; for example, $\{y_{k,j}: k = 1, 2, \ldots, n_j\}$. The average of those values that fall into time cell i is computed to produce $\bar{y}_{i,j}$ which is the ith cell average for the jth replication. Thus n replications will produce the set of n averages $\{\bar{y}_{i,j} : j = 1, 2, \ldots, n\}$ where i is the time cell index. The following steps are carried out to obtain an estimate of the time cell index where the system transient terminates, in other words, the system reaches steady-state.

1. Obtain the value $\bar{a}_i$ as the average over the n replications of the ith cell averages ($\bar{y}_{i,j}$); that is,

$$\bar{a}_i = \frac{1}{n}\sum_{j=1}^{n} \bar{y}_{i,j} \,.$$

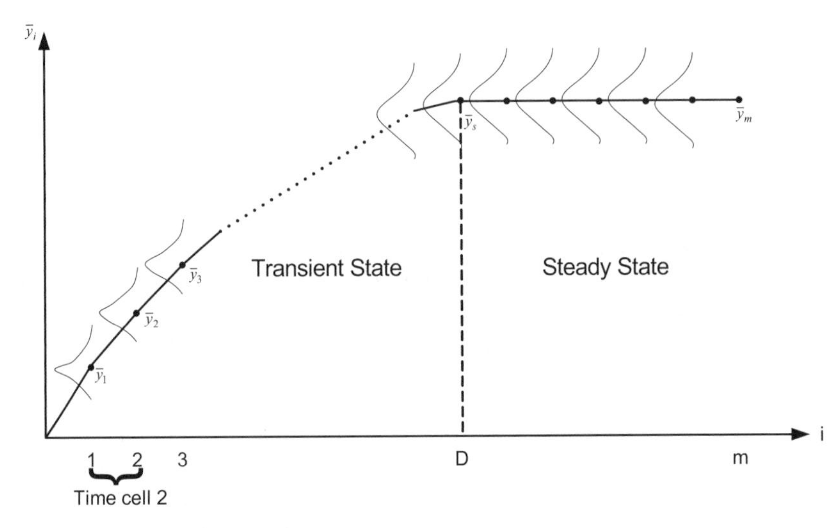

FIGURE 6.3. Reaching steady-state.

2. The values $\bar{a}_i$, i = 1, 2, . . . , m usually vary considerably. If plotted against index i the resulting graph is 'choppy' and difficult to interpret. A smoothing operation is required in order to smooth out rapid variations to obtain a smoother curve that captures the trend. For this purpose, the moving-average values $\bar{a}_i(w)$ are computed using Equation (6.2). The parameter w represents a window size that controls the smoothing operation. Its selection is by trial and error. Usually a number of values for w need to be tried. The objective is to find as small a value as possible that provides the desired smoothing effect.

$$\bar{a}_i(w) = \begin{cases} \dfrac{\sum_{l=-(i-1)}^{i-1} \bar{a}_{i+l}}{2i-1} & i = 1, \ldots, w \\ \\ \dfrac{\sum_{l=-w}^{w} \bar{a}_{i+l}}{2w+1} & i = w+1, \ldots, m-w \end{cases} . \tag{6.2}$$

3. Equation (6.2) is not as complex as it might appear. When $i > w$ there are w cell averages on either side of $\bar{a}_i$ that are averaged to produce the running average value $\bar{a}_i(w)$. When $i \leq w$ there are not enough values preceding time cell $i$to fill the window. In this

case w is replaced with $(i - 1)$. Table 6.4 shows how the running averages are computed for the case where $w = 3$.

The values $\overline{a}_i(w)$ are plotted against the cell index i and it should be apparent from this graph when steady-state has been achieved. A good practice is to extend the apparent length of the warm-up period (say by 30%). The idea here is to err on the safe side by making the warm-up period somewhat longer than necessary rather than inappropriately short.

TABLE 6.4. Welch running average with $w = 3$.

i	$\overline{a}_i(3)$ **Equation**	$\overline{a}_i(3)$ **Expansion**
1	$\frac{\sum_{l=0}^{0} \overline{a}_{i+l}}{1}$	$\frac{\overline{a}_1}{1}$
2	$\frac{\sum_{l=-1}^{1} \overline{a}_{i+l}}{3}$	$\frac{\overline{a}_1 + \overline{a}_2 + \overline{a}_3}{3}$
3	$\frac{\sum_{l=-2}^{2} \overline{a}_{i+l}}{5}$	$\frac{\overline{a}_1 + \overline{a}_2 + \overline{a}_3 + \overline{a}_4 + \overline{a}_5}{5}$
4	$\frac{\sum_{l=-3}^{3} \overline{a}_{i+l}}{7}$	$\frac{\overline{a}_1 + \overline{a}_2 + \overline{a}_3 + \overline{a}_4 + \overline{a}_5 + \overline{a}_6 + \overline{a}_7}{7}$
5	$\frac{\sum_{l=-3}^{3} \overline{a}_{i+l}}{7}$	$\frac{\overline{a}_2 + \overline{a}_3 + \overline{a}_4 + \overline{a}_5 + \overline{a}_6 + \overline{a}_7 + \overline{a}_8}{7}$
•	•	•
•	•	•
•	•	•
$m - 3$	$\frac{\sum_{l=-3}^{3} \overline{a}_{i+l}}{7}$	$\frac{\overline{a}_{m-6} + \overline{a}_{m-5} + \overline{a}_{m-4} + \overline{a}_{m-3} + \overline{a}_{m-2} + \overline{a}_{m-1} + \overline{a}_m}{7}$

We illustrate the use of the Welch moving average method using version 1 of our port project (no intervention and no storms). The output variables of interest are berth group size and the tanker total wait time. Figures 6.4 and 6.5 show the results of 10 simulation runs ($n = 10$) each of duration 15 weeks. The time cells have a width of 1 week which means that $m = 15$. The following observations are noteworthy.

- In the case of the berth group size, there is no apparent transient. Even without the use of running averages (see Figure 6.4a), the graph is relatively smooth. This result can be attributed to the small size of the group (namely, three) which results in the available berths being quickly filled by the first few arrivals of tankers.
- A transient is certainly apparent for the tanker total wait time as shown in Figure 6.5 and moving averages are required to smooth out the graph. A window size of five provides a suitable result and shows that the transient lasts for approximately three weeks. Either four or five weeks can be selected as a suitable warm-up period.
- The warm-up period has relevance for the elimination of the transient in the tanker total wait time output variable. However, this does not preclude the collection of berth group size data during the warm-up period.

6.3.2 Collection and Analysis of Results

Extending the right-hand boundary of the observation interval allows more data to be collected during a simulation run and this provides the basis for a number of methods for generating the necessary data for analysis (i.e., a set of IID values for the output variable). We examine two approaches. The replication–deletion method is described and illustrated using experimentation with the port simulation program as presented in Section 6.3.3. An overview of the method of batch means is also given. A more comprehensive presentation of the available options can be found in Law and Kelton [6.5].

Our problem continues to be the determination of an estimate of the mean of an output variable Y, i.e. $\mu = E[Y]$. However, as previously noted, in steady-state studies we must reduce the effect of transient data, and ideally eliminate it.

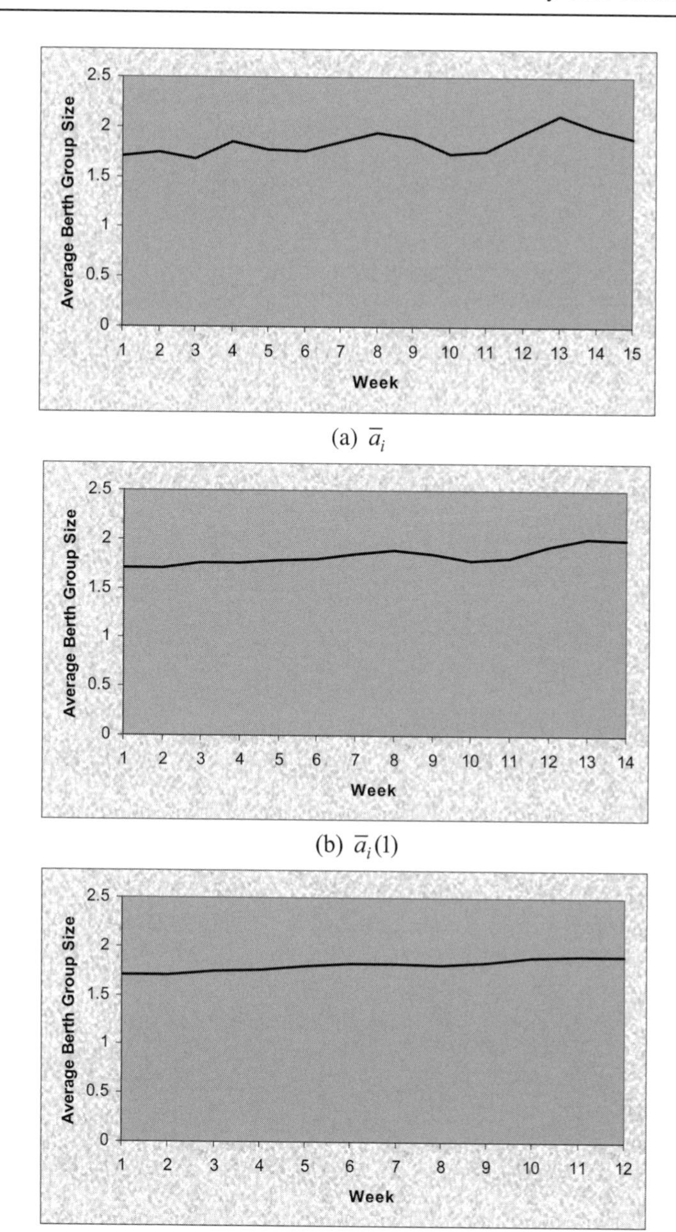

(a) $\overline{a}_i$

(b) $\overline{a}_i(1)$

(c) $\overline{a}_i(3)$

FIGURE 6.4. Welch method applied to berth group size.

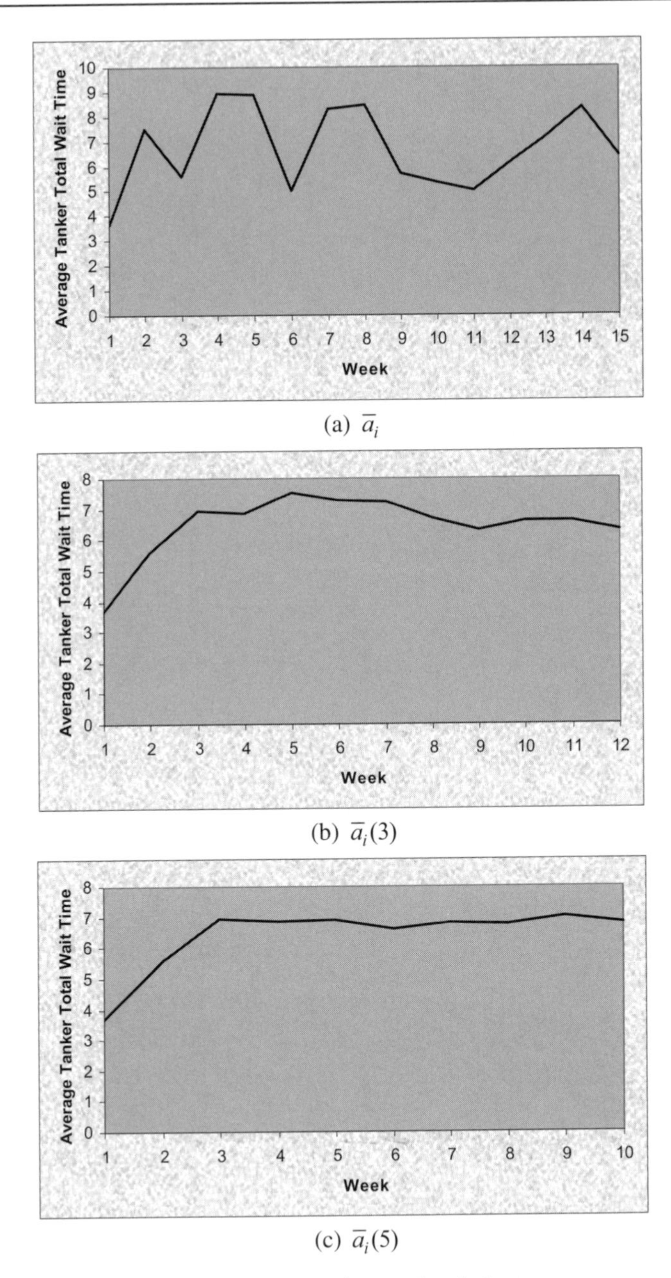

(a) $\bar{a}_i$

(b) $\bar{a}_i(3)$

(c) $\bar{a}_i(5)$

FIGURE 6.5. Welch method applied to tanker total wait time.

A practical approach in the replication–deletion method is to determine the right boundary of the warm-period t_w (using methods such as the one described in Section 6.3.1) and to delete any output set data generated prior to t_w. The right boundary of the observation interval (i.e., t_f) is simply taken to be the value of (simulated) time when a sufficient amount of data has been collected to generate a valid and meaningful collection of output observations. In fact a sequence of n simulation runs is executed to produce a set of n output values. An important feature of this method is that it naturally generates a set of IID values. It does, however, have the computing time overhead of repeating the warm-up period for each of the replications.

This approach resembles the experimentation and output analysis previously outlined for a bounded horizon study (see Section 6.2). From the output data, a point estimate and confidence interval can be obtained using Equation (6.1). In the discussion of Section 6.2.2 it was noted that increasing the number of simulation runs (i.e., replications) reduced the confidence interval half length $\zeta(n)$ and increased the quality of the point estimate. This equally applies in the replication–deletion approach for a steady-state study. However, in a steady-state study, $\zeta(n)$ can also be reduced by increasing the length of the simulation run, that is, by adjusting the right-hand boundary t_f of the observation interval. Based on these observations the procedure for the replication–deletion method can be formulated as a straightforward extension of the earlier procedure presented in Section 6.2.2. It is as follows.

1. Choose values for r, and for the confidence level parameter C, and as well an initial reasonable value for t_f, and an initial value for n that is not smaller than 20.
2. Collect the n observed values $y_1, y_2, \ldots, y_n$ for the random variables Y_k, $k = 1, 2, \ldots, n$, that result from n replicated simulation runs of the simulation program **M** that terminate at time t_f. (Note that the output set, from which y_i's are obtained include only data collected after the end of the warm-up period)
3. From tabulated data for the Student t-distribution, determine $t_{n-1,a}$ where $a = (1 - C)/2$.
4. Compute $\bar{y}(n)$ and $\zeta(n)$ using Equation (6.1).
5. If $\zeta(n) < r\,\bar{y}(n)$ (or $\zeta(n)/\bar{y}(n) < r$) then accept $\bar{y}(n)$ as the estimate of μ and end the procedure, otherwise continue to Step 6.
6. EITHER choose Δn no smaller than 3 and collect the additional observations $y_{n+1}, y_{n+2}, \ldots, y_{n+\Delta n}$ through a further Δn replications, replace n with $n + \Delta n$ and repeat from Step 3

   OR increase the value of t_f by at least 50% and repeat from Step 2.[2]

[2] In some environments (e.g., Java), it may be possible to save the state of the simulation program for each replication so that simulations runs can be continued from the previously specified t_f.

The batch means method is an entirely different approach that requires only a single (but potentially 'long') simulation run. An advantage of this approach is economy of computing time because the warm-up period only needs to be accommodated once. The end of the observation interval t_f is selected to generate all the data necessary for analysis. However possible autocorrelation of the output data must be dealt with in order to generate the necessary IID data.

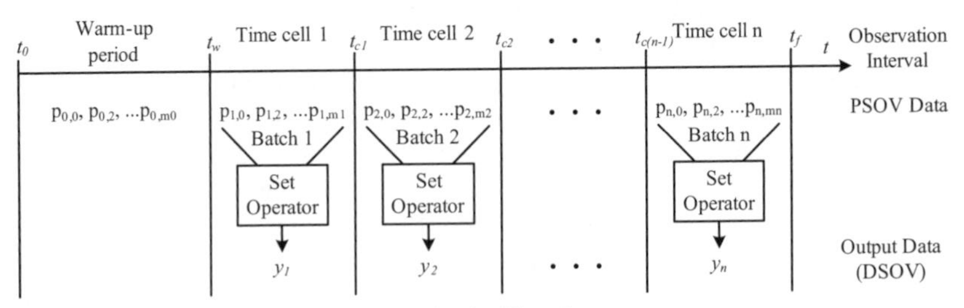

FIGURE 6.6. Output values using method of batch means.

To generate a set of IID values, the observation interval beyond the warm-up period is divided into n time cells as shown in Figure 6.6. The PSOV values that fall into a time cell is called a batch. The end result is a set of n batches. A DSOV output value is then computed for each batch, providing a set of output values y_i for $i = 1, 2, \ldots n$. Equation (6.1) can then be used to obtain a point estimate of the mean value of the distribution of the output variable of interest, together with the corresponding confidence interval.

One of the challenges of the batch means method is the proper selection of the length of the time cells. If the length is too short the y_i values may be correlated. Appropriate checks therefore need to be incorporated. Details about this method (and also other methods which use a single simulation run), can be found in Law and Kelton [6.5].

6.3.3 Experimentation and Data Analysis for the Port Project

The Java program given in Figure 6.7 illustrates how the required experiments for the steady-state study of the Port project can be implemented. The replication–deletion method is being used to generate the necessary output data with a warm-up period of five weeks (previously determined by Welch's method). The major steps include:

1. The *main* method obtains the value of t_f (the right-hand boundary of the observation interval expressed in weeks) from the command line arguments (*arg[0]*). A simulation run termination time expressed in hours is assigned to *endTime*.
2. A set of uncorrelated seeds is generated (to ensure independent replications) and saved into an array. The seeds are reused when carrying out the simulation runs for the alternative case of the port project (this implements the use of common random numbers described in Section 6.4).
3. Two sets of simulation runs are carried out, one set where *numBerths* = 3 (base case), and one set where *numBerths* = 4 (alternate case). This is implemented as a loop that increments *numBerths.*
4. A simulation run consists of instantiating the *PortVer1System* object that is initialised with the start time, number of berths, and random number generator seeds. The termination time of the *PortVer1System* object is first set to *warmUpTime* (using the *setTimef* method) and the simulation program executes for the warm-up period. The output set for the tanker total wait time is cleared. The termination time of the *PortVer1System* object is now set to *endTime* and continues execution until the end of the observation interval.
5. The output data values are then computed by the *ESOuputSet* methods,*computeTrjDSOVs*, and *computePhiDSOVs*. The values are then printed.

For each of the two cases (number of berths equals three and four), output data for each of the output variables (the average group size and tanker total wait time) using t_f equal to 10, 20, and 30 weeks are generated and analysed. In each case results are obtained for a sequence of increasing values of n (number of replications). For each of the output variables, Table 6.5 shows point estimate ($\overline{y}\,n)$), standard deviation ($s(n)$), and confidence interval half length ($\zeta(n)$) values computed from the recorded data using Equation (6.1). The last column shows the ratio of $\zeta(n)/\overline{y}(n)$ which gives a measure of the quality of the point estimate. As expected, increasing the number of runs (n) reduces the confidence interval half length $\zeta(n)$. Increasing the simulation run length also improves the quality of the results. A comparison of the three and four berth options is undertaken in Section 6.4.1 using an appropriate statistical framework.

```
public static void main(String[] args)
{
    double week=(7*24);
    double startTime=0.0;
    double warmUpTime=5*week;
    double endTime
    int NUMRUNS=10000;
    Seeds [] sds = new Seeds[NUMRUNS];
    // Find end time, tf, from command argument
    if(args.length != 1)
    {
       System.out.println("Usage: PortV1Exp2 <endTime>");
       System.exit(1);
    }
    endTime=Double.valueOf(args[0])*week;
    // Lets get a set of uncorrelated seeds
    RandomSeedGenerator rsg = new RandomSeedGenerator();
    for(int i=0 ; i<NUMRUNS ; i++)
       sds[i] = new Seeds(rsg.nextSeed(),rsg.nextSeed(),
                          rsg.nextSeed(),rsg.nextSeed(),
                          rsg.nextSeed());
    // Simulation Runs
    System.out.println("End Time = "+ args[0] +
                        " ("+endTime+")");
    // Run for 3 berths and then 4 berths
    for(int numBerths=3 ; numBerths<=4 ; numBerths++)
    {
       System.out.println("Number of berths = "+numBerths);
       for(int i=0 ; i<NUMRUNS ; i++)
       {
        PortVer1System portSys = new PortVer1System(
                              startTime,numBerths,sds[i]);
        portSys.setTimef(warmUpTime); // end of warmup
        portSys.runSimulation();
        portSys.tankerTW.clearSet();  // clear output set
        portSys.setTimef(endTime); // now run to tf
        portSys.runSimulation();
        // compute DSOV
        portSys.berthGrpN.computeTrjDSOVs(
                             portSys.time0,portSys.timef);
        portSys.tankerTW.computePhiDSOVs();
        System.out.println(portSys.berthGrpN.mean+", "+
                           portSys.tankerTW.mean);
       }
    }
}
```

FIGURE 6.7. Experiments with the Java port simulation program (corresponds to Example 1 of Section 4.3.1).

TABLE 6.5. Results from experiments with the Java port simulation program of Figure 6.7.

(a) Number of berths = 3.

t_f:	10 weeks				20 weeks				30 weeks			
n	$\bar{y}(n)$	s(n)	$\zeta(n)$	$\zeta(n)/\bar{y}(n)$	$\bar{y}(n)$	s(n)	$\zeta(n)$	$\zeta(n)/\bar{y}(n)$	$\bar{y}(n)$	s(n)	$\zeta(n)$	$\zeta(n)/\bar{y}(n)$
					Berth Group Size							
20	1.845	0.138	0.053	0.0290	1.858	0.072	0.028	0.0149	1.866	0.066	0.025	0.0137
30	1.820	0.139	0.043	0.0237	1.833	0.088	0.027	0.0149	1.842	0.078	0.024	0.0132
40	1.824	0.138	0.037	0.0201	1.833	0.095	0.025	0.0138	1.838	0.071	0.019	0.0103
60	1.827	0.147	0.032	0.0173	1.833	0.099	0.021	0.0117	1.828	0.077	0.017	0.0090
80	1.828	0.138	0.026	0.0141	1.834	0.102	0.019	0.0103	1.833	0.076	0.014	0.0077
100	1.823	0.138	0.023	0.0126	1.834	0.101	0.017	0.0091	1.830	0.078	0.013	0.0071
1000	1.834	0.130	0.007	0.0037	1.833	0.094	0.005	0.0027	1.833	0.077	0.004	0.0022
10000	1.826	0.132	0.002	0.0012	1.832	0.095	0.002	0.0009	1.833	0.078	0.001	0.0007
					Tanker Total Waiting Time							
20	7.144	2.387	0.923	0.1292	7.196	1.937	0.749	0.1041	7.256	1.543	0.597	0.0822
30	6.815	2.432	0.754	0.1107	7.042	1.695	0.526	0.0747	7.268	1.394	0.432	0.0595
40	6.703	2.482	0.661	0.0987	7.099	2.020	0.538	0.0758	7.320	1.487	0.396	0.0541
60	7.958	5.733	1.237	0.1554	7.451	2.626	0.567	0.0760	7.324	1.727	0.373	0.0509
80	7.857	5.243	0.976	0.1242	7.454	2.438	0.454	0.0609	7.410	1.762	0.328	0.0442
100	7.844	5.148	0.855	0.1090	7.665	2.498	0.415	0.0541	7.456	1.781	0.296	0.0397
1000	7.613	4.518	0.235	0.0309	7.682	2.817	0.147	0.0191	7.700	2.299	0.120	0.0155
10000	7.449	4.510	0.074	0.0100	7.686	2.926	0.048	0.0063	7.729	2.293	0.038	0.0049

(b) Number of berths = 4.

t_f:	10 weeks				20 weeks				30 weeks			
n	$\bar{y}(n)$	s(n)	$\zeta(n)$	$\zeta(n)/\bar{y}(n)$	$\bar{y}(n)$	s(n)	$\zeta(n)$	$\zeta(n)/\bar{y}(n)$	$\bar{y}(n)$	s(n)	$\zeta(n)$	$\zeta(n)/\bar{y}(n)$
					Berth Group Size							
20	1.859	0.149	0.058	0.0310	1.872	0.075	0.029	0.0154	1.877	0.068	0.026	0.0140
30	1.835	0.146	0.045	0.0247	1.847	0.090	0.028	0.0152	1.854	0.079	0.025	0.0132
40	1.839	0.145	0.039	0.0210	1.847	0.096	0.025	0.0138	1.850	0.071	0.019	0.0103
60	1.844	0.155	0.033	0.0181	1.846	0.101	0.022	0.0118	1.840	0.078	0.017	0.0092
80	1.844	0.146	0.027	0.0147	1.847	0.105	0.019	0.0105	1.845	0.077	0.014	0.0078
100	1.839	0.144	0.024	0.0130	1.847	0.104	0.017	0.0094	1.841	0.080	0.013	0.0072
1000	1.850	0.135	0.007	0.0038	1.846	0.097	0.005	0.0027	1.845	0.079	0.004	0.0022
10000	1.842	0.136	0.002	0.0012	1.845	0.098	0.002	0.0009	1.846	0.080	0.001	0.0007
					Tanker Total Waiting Time							
20	2.533	1.045	0.404	0.1595	2.534	0.641	0.248	0.098	2.533	0.422	0.163	0.0644
30	2.488	0.938	0.291	0.1170	2.556	0.569	0.177	0.069	2.598	0.450	0.140	0.0537
40	2.452	0.885	0.236	0.0962	2.554	0.564	0.150	0.059	2.582	0.422	0.112	0.0435
60	2.769	1.562	0.337	0.1217	2.601	0.705	0.152	0.059	2.547	0.487	0.105	0.0412
80	2.709	1.432	0.267	0.0984	2.582	0.691	0.129	0.050	2.549	0.499	0.093	0.0364
100	2.711	1.379	0.229	0.0845	2.624	0.683	0.113	0.043	2.552	0.496	0.082	0.0323
1000	2.539	1.114	0.058	0.0228	2.557	0.639	0.033	0.013	2.559	0.508	0.026	0.0103
10000	2.501	1.094	0.018	0.0072	2.557	0.662	0.011	0.004	2.564	0.512	0.008	0.0033

6.4 Comparing Alternatives

A frequently occurring requirement among the goals of a modelling and simulation project is the evaluation of several alternate system designs. For example, what reduction in maximum patient waiting time could be expected in the emergency admitting area of a large hospital if an additional orthopaedic specialist were hired or what might be the impact on traffic flow in the downtown core of a large city if a network of one-way streets were implemented? There can be a large number of such design alternatives that need to be evaluated but we first examine the case where there are only two.

In principle the problem solution is straightforward. Develop a simulation program for each of the scenarios (alternatives), obtain a value for some common performance measure applied to each scenario (e.g., a mean value estimate for the distribution of some DSOV), and compare the values obtained. There is, however, a serious complication that emerges, namely, what assurance is there that any observed difference between the performance measure values is a consequence of the design difference being studied and not simply a consequence of the inherent random behaviour within the model?

A number of different approaches have emerged for dealing with this problem and comprehensive discussions can be found in the literature (e.g., Banks et al. [6.1] and Goldsman and Nelson [6.2]). One of the most straightforward is called the *paired-t confidence interval method* The objective here is to first establish a confidence interval for an estimate of the mean of a random variable that is the difference between the output variables associated with each of the scenarios. A decision about relative superiority is then based on the position of the confidence interval relative to zero. Some details are provided below.

6.4.1 Comparing Two Alternatives

Suppose that Y is the output variable to be used for the evaluation and let's assume that we seek as large a value as possible for this variable. The simulation program for each of the design alternatives is replicated n times with appropriate care taken to ensure that in each case the n observations of Y can be assumed to be independent (i.e., by proper management of the random number streams that 'drive' the simulation models). Suppose that y_{1k} is the value of Y obtained for case 1 on the kth replication and suppose that y_{2k} is the value for case 2 on the kth replication. Let:

$$
\begin{aligned}
d_k &= y_{2k} - y_{1k} \qquad k = 1,2,\ldots n \\
\bar{d}(n) &= \frac{1}{n}\sum_{k=1}^{n} d_k \\
s^2(n) &= \frac{\sum_{k=1}^{n}(d_k - \bar{d}(n))^2}{n-1} \qquad (6.3) \\
\zeta &= \frac{t_{n-1,\alpha}\, s(n)}{\sqrt{n}}
\end{aligned}
$$

where $t_{n-1,a}$ is a value from the Student *t*-distribution (see Table A1.4) that corresponds to $(n - 1)$ degrees of freedom and $a = (1 - C)/2$ with *C* the confidence level parameter. Here $\bar{d}(n)$ is a point estimate of the mean of the differences and $s^2(n)$ is the sample variance. (The similarity of these results with those given in Equation (6.1) is worth noting.) The associated confidence interval is $CI(n) = [\bar{d}(n) \pm \zeta]$.

There are three possible outcomes based on $CI(n)$; namely,

a) If $CI(n)$ lies entirely to the right of zero then the result of case 2 exceeds the result of case 1 with a level of confidence given by $100C\%$.
b) If $CI(n)$ lies entirely to the left of zero then the result of case 1 exceeds the result of case 2 with a level of confidence given by $100C\%$.
c) If $CI(n)$ includes zero then at the level of confidence, $100C\%$., there is no meaningful difference between the two cases.

The procedure outlined above is best carried out in conjunction with a technique called *common random numbers* (CRN). When undertaking the comparison of the data that flow from the two simulation programs that embody the two design alternatives, there is reason to be concerned about the extent to which any observed difference is a genuine reflection of the design alternatives or is simply the result of a lack of symmetry in the random phenomena that take place within the respective simulation models.

The common random number technique seeks to establish this symmetry and thus enhance the reliability of the conclusions. The application of the technique corresponds to endeavouring to ensure that, insofar as possible, the random phenomena within the two simulation programs are co-ordinated; for example, comparable entities flowing in the two models are subjected to the same sequence of delays. In principle, this can be achieved by the strict management of the random variate generation procedures within the two programs. This coordination is straightforward for input data models. The coordination task can also be easily achieved with all data models when the simulation model is relatively simple. However, except for input data models, the coordination task can become increasingly more difficult as the simulation model complexity increases. Often the design differences themselves may inhibit a rigorous application of the approach.

The common random number procedure outlined above has the effect of establishing correlation between the output data generated in corresponding simulation runs with the two alternative designs. This, in turn, has a quantitative manifestation; more specifically, the procedure, when operating as intended, should yield the inequality:

$$s^2(n) < s_1^2(n) + s_2^2(n) \ , \qquad (6.4)$$

where $s_1^2(n)$ and $s_2^2(n)$ are the variances for the data obtained for Case 1 and Case 2, respectively and $s^2(n)$ is the value obtained from Equation (6.3).

We return now to our experiments with the port project as outlined in Section 6.3.3. For the two cases where the number of berths is 3 and 4, Table 6.6 shows the output data for each of the output variables (the average group size and tanker total wait time) from a sequence of experiments with t_f = 20 weeks and n = 30. The difference column is obtained as (numBerths=4) – (numBerths=3). The comparison of the two alternatives is carried out using Equation (6.3) and the results are provided at the bottom of Table 6.6 (CI min and CI max are the boundaries of the confidence interval). Some interpretation of the data is as follows.

1. It is clear that increasing the number of berths from three to four does decrease the mean tanker total wait time (by almost 4.5 hours).
2. Although the confidence interval for the berth group size is to the right of zero, the point estimate of the difference is so small relative to the individual point estimates we are obliged to conclude that increasing the berth group size has no meaningful effect on this output variable. This is somewhat counterintuitive but is a consequence of the relative values of tanker arrival rate, the tug's cycle time (time to deberth and berth a tanker), and the tanker loading times. For example, experimentation with the model has shown that when the loading times are increased, the average berth group size does increase. An alternate measure that would be interesting is the percentage of time that all available berths are occupied. The interested reader is encouraged to experiment with the simulation program by exploring the effects of changing these various times in the simulation model (The *PortVer 1* simulation model is available from the textbook Web site).

Table 6.7 shows the data obtained from equivalent experiments which do not use common random numbers (CRN) for the two cases of interest, that is, numBerths=3 and numBerths=4. This was achieved by not using the same seeds for the random number generators that implement the data modules in the experiments. Note that the confidence interval half length $\zeta(n)$ increases for both output variables when compared to the results in Table 6.6. Note also that it can be shown that the Tanker Total Wait Time data in Table 6.7 is not consistent with the inequality of Equation (6.4).

TABLE 6.6. Comparing alternative cases in the port project of Example 1 (with CRN and $n = 30$).

	Berth Group Size			Tanker Total Wait Time		
Run	**numBerths=3**	**numBerths=4**	**Difference**	**numBerths=3**	**numBerths=4**	**Difference**
1	1.721	1.729	0.008	5.981	2.182	-3.799
2	1.847	1.862	0.014	7.433	3.182	-4.251
3	1.911	1.920	0.008	7.594	2.758	-4.836
4	1.876	1.888	0.011	6.299	2.039	-4.260
5	1.897	1.917	0.020	7.177	2.483	-4.694
6	1.833	1.843	0.011	4.589	1.679	-2.910
7	1.866	1.888	0.023	8.234	3.411	-4.823
8	1.864	1.877	0.013	9.310	2.995	-6.315
9	1.790	1.796	0.006	6.398	2.580	-3.819
10	1.949	1.969	0.020	7.131	2.403	-4.728
11	1.952	1.956	0.003	9.387	2.950	-6.438
12	1.830	1.846	0.016	7.776	2.548	-5.229
13	1.720	1.743	0.023	4.764	1.719	-3.046
14	2.004	2.044	0.039	12.640	4.030	-8.610
15	1.811	1.819	0.008	4.678	1.623	-3.055
16	1.834	1.852	0.018	6.126	2.446	-3.680
17	1.932	1.941	0.009	9.013	3.127	-5.887
18	1.836	1.841	0.005	5.782	1.731	-4.050
19	1.828	1.833	0.004	5.795	2.017	-3.778
20	1.856	1.870	0.014	7.816	2.782	-5.035
21	1.921	1.924	0.003	7.344	2.853	-4.490
22	1.744	1.754	0.011	6.092	2.491	-3.601
23	1.636	1.658	0.023	6.183	2.659	-3.524
24	1.823	1.856	0.032	6.797	3.126	-3.672
25	1.747	1.746	0.000	6.619	2.214	-4.405
26	1.843	1.848	0.005	7.074	2.422	-4.652
27	1.668	1.683	0.015	4.597	1.801	-2.795
28	1.699	1.715	0.015	7.429	2.920	-4.509
29	1.928	1.964	0.036	8.780	3.093	-5.686
30	1.828	1.833	0.005	6.410	2.417	-3.993
$\overline{y}(n)$			0.014			-4.486
$s(n)$			0.010			1.221
$\zeta(n)$			0.003			0.379
CI Min			0.011			-4.864
CI Max			0.017			-4.107

TABLE 6.7. Comparing alternative cases in the port project of Example 1 (without CRN and $n = 30$).

	Berth Group Size			Tanker Total Wait Time		
Run	**numBerths=3**	**numBerths=4**	**Difference**	**numBerths=3**	**numBerths=4**	**Difference**
1	1.721	1.834	0.113	5.981	2.943	-3.038
2	1.847	1.881	0.033	7.433	2.591	-4.843
3	1.911	1.887	-0.024	7.594	2.507	-5.087
4	1.876	1.881	0.005	6.299	3.278	-3.021
5	1.897	1.862	-0.035	7.177	2.107	-5.070
6	1.833	1.835	0.002	4.589	3.588	-1.001
7	1.866	1.868	0.002	8.234	2.224	-6.009
8	1.864	1.923	0.059	9.310	2.527	-6.783
9	1.790	1.838	0.048	6.398	1.828	-4.570
10	1.949	1.811	-0.138	7.131	1.895	-5.236
11	1.952	1.905	-0.048	9.387	1.418	-7.969
12	1.830	1.876	0.047	7.776	3.003	-4.774
13	1.720	1.829	0.109	4.764	2.183	-2.581
14	2.004	1.879	-0.126	12.640	2.633	-10.007
15	1.811	1.822	0.011	4.678	2.741	-1.937
16	1.834	1.905	0.071	6.126	4.414	-1.712
17	1.932	1.925	-0.008	9.013	4.340	-4.673
18	1.836	1.854	0.018	5.782	1.658	-4.124
19	1.828	1.865	0.037	5.795	2.158	-3.637
20	1.856	1.878	0.022	7.816	2.346	-5.470
21	1.921	1.844	-0.077	7.344	2.360	-4.984
22	1.744	1.937	0.194	6.092	3.644	-2.448
23	1.636	1.709	0.073	6.183	2.934	-3.249
24	1.823	1.870	0.047	6.797	2.117	-4.680
25	1.747	1.712	-0.034	6.619	2.200	-4.419
26	1.843	1.868	0.026	7.074	2.472	-4.602
27	1.668	1.686	0.017	4.597	2.799	-1.798
28	1.699	1.769	0.070	7.429	1.895	-5.533
29	1.928	1.878	-0.050	8.780	1.786	-6.994
30	1.828	1.909	0.081	6.410	4.787	-1.623
$\overline{y}(n)$			0.018			-4.396
$s(n)$			0.069			2.003
$\zeta(n)$			0.021			0.621
CI Min			-0.003			-5.017
CI Max			0.039			-3.775

6.4.2 Comparing More than Two Alternatives

The paired-t confidence interval method described above can be extended to the case where multiple comparisons need to be carried out. The basis for carrying out this extension is provided by the Bonferroni inequality (sometimes called the Boole inequality). It states that:

$$P[\bigcap_{k=1}^{K} A_k] \geq (1-K) + \sum_{k=1}^{K} P[A_k] \ .$$

In our context, the A_k can be interpreted as the event (in a probability context) that the *k*th confidence interval contains the *k*th mean in a collection of K (pairwise) comparisons, The Bonferroni inequality, in effect, places constraints on the individual comparisons in order to achieve an overall result that has a prescribed level of confidence, $100C\%$. In other words with $100C\%$ confidence, the mean differences all fall into their respective confidence intervals. The (simplified) result that flows from the Bonferonni inequality is that each of the K comparisons should be carried out with a confidence level parameter value of:

$$C_K = 1 - \left(\frac{1-C}{K}\right) \ . \tag{6.5}$$

Note that the result given in Equation (6.5) is overly restrictive because it has imposed the unnecessary (but simplifying) requirement that the confidence level parameter of all constituent comparisons be equal.

The following is a typical scenario. There exists a 'base case' which normally corresponds to the current status of the SUI. The project goals introduce M alternate designs together with the requirement to identify the best of the alternate designs by comparing each alternative to the base case. Thus $K = M$ comparisons need to be made. If an overall confidence level of $100C\%$ is stipulated then the K individual comparisons have to be carried out with a confidence level parameter of C_K as given in Equation (6.5).

It may, on the other hand, be stipulated in the project goals that the M alternative designs not only be compared to the base case but also be pairwise compared to each other. In this case, there is a requirement for $K = M(M + 1)/2$ comparisons. The number of comparisons can easily rise quickly and the reliability of the procedure deteriorate. In addition, of course, the computational overhead can become overwhelming.

Some illustrative results obtained using the multiple alternatives procedure outlined above are given in Table 6.9. The results relate to the Kojo's Kitchen project. We consider a base case (Case 1) which corresponds to the two employees working over the entire business day (10:00 AM – 9:00 PM) and three alternative employee scheduling options (Cases 2, 3, 4). These options allocate different numbers of employees to various segments of the day. The employee scheduling schemes are summarised in Table 6.8. The rightmost column of this Table provides the total number of employee-hours associated with each option. This is relevant in the ultimate selection decision because it represents the 'cost' of the option. The output variable of interest continues to be the percentage of customers who wait more than five minutes before receiving service.

TABLE 6.8. Multiple scheduling alternatives for Kojo's Kitchen.

	Slow (10:00am-11:30am)	Busy (11:30am-1:30pm)	Slow (1:30pm-5:00pm)	Busy (5:00pm-7:00pm)	Slow (7:00pm-9:00pm)	Emp-Hours
Case 1 (Base Case)	2	2	2	2	2	22
Case 2	2	3	2	3	2	26
Case 3	1	3	1	3	1	19
Case 4	1	3	2	3	1	22.5

Table 6.9 provides a summary of the each of the three comparisons. The results shown for Diff21 are obtained by subtracting the results of the base case (Case 1) from Case 2 and applying Equation (6.3), and similarly for Diff31 and Diff41. These results were obtained using the Java simulation program previously discussed in Section 6.2.3. In each case the results are based on data from 100 replications (n = 100) and use of a confidence level parameter value of $C_k = 0.968$ in the determination of the confidence interval for the individual comparisons. This gives a value of $C = 0.904$ using Equation (6.5), that is, a confidence of 90.4% in the conclusions from the comparison. Table 6.9 suggests that the scheduling alternative of Case 2 provides the best improvement over the base case. (Unfortunately it is also the most expensive! Note however that scheduling in Case 4 provides a significant improvement at very little additional cost).

TABLE 6.9. Results for multiple scheduling alternatives (Kojo's Kitchen).

Comparison	Point Estimate $\tilde{y}(n)$	s(n)	ζ	CI Min	CI Max	$\zeta/\tilde{y}(n)$
Diff21	-0.315	0.011	0.024	-0.340	-0.291	-0.076
Diff31	-0.127	0.013	0.028	-0.155	-0.099	-0.220
Diff41	-0.243	0.012	0.025	-0.268	-0.218	-0.105

6.5 Exercises and Projects

6.1 Use the program developed in Problem 5.1 to carry out experiments that provide the values required for the graphs that are stipulated in the goals of the project outlined in Problem 4.1. Write a short report that outlines the problem, the goals of the modeling and simulation project, and the conclusions obtained from the study.

6.2 Use the program developed in Problem 5.2 to carry out experiments that provide the values required for the graphs that are stipulated in the goals of the project outlined in Problem 4.2. Write a short report that outlines the problem, the goals of the modeling and simulation project, and the conclusions obtained from the study.

6.3 Use the program developed in Problem 5.3 to carry out experiments that provide values for the proposed performance measures referred to in part (a) of Problem 4.3. Write a short report that outlines the problem, the goals of the modeling and simulation project, and the conclusions obtained from the study.

6.4 Use the program developed in Problem 5.5 to carry out experiments that provide values for the proposed performance measures referred to in part (a) of Problem 4.3. Write a short report that outlines the problem, the goals of the modeling and simulation project, and the conclusions obtained from the study.

6.5 Use the program developed in Problem 5.7 to carry out experiments that provide values for the proposed performance measures referred to in part (b) of Problem 4.4. Write a short report that outlines the problem, the goals of the modeling and simulation project, and the conclusions obtained from the study.

6.6 Use the program developed in Problem 5.8 to carry out experiments to evaluate the effects of balking introduced in Problem 4.5. Write a short report that outlines the problem, the goals of the modeling and simulation project, and the conclusions obtained from the study.

6.6 References

6.1. Banks J., Carson II, J.S., Nelson, B.L., and Nicol, D.M., (2005), *Discrete-Event System Simulation*, 4th edn., Pearson Prentice Hall, Upper Saddle River, NJ.

6.2. Goldsman, D. and Nelson, B.L., (1998), Comparing Systems Via Simulation, in J. Banks (Ed.), *Handbook of Simulation*, John Wiley, New York, pp. 273–306.

6.3. Goldsman, D. and Nelson, B.L., (2001), Statistical selection of the best system, in B.A. Peters, J.S. Smith, D.J. Medeiros, and M.W. Rohrer, (Eds.), *Proceedings of the 2001 Winter Simulation Conference* IEEE Press, Piscataway, NJ, pp. 139–146.

6.4. Goldsman, D., Schruben, L.W., and Swain J.J., (1994), Test for transient means in simulation time series, *Naval Research Logistics Quarterly*, **41**: 171–187.

6.5. Law, A.M. and Kelton, D.W., (2000), *Simulation Modeling and Analysis*, 3rd edn., McGraw-Hill, New York.
6.6. Robinson S., (2002), A statistical process control approach for estimating the warm-up period, in *Proceedings of the 2002 Winter Simulation Conference*, IEEE Piscataway, NJ, pp. 439–446
6.7. Roth, E., (1994), The relaxation time heuristic for the initial transient problem in M/M/K queuing systems, *European Journal of Operational Research*, **72**: 376–386.
6.8. Welch, P., (1983), The statistical analysis of simulation results, in S. Lavenberg (Ed.), *The Computer Performance Modeling Handbook*, Academic Press, New York, pp. 268–328.

PART 3
CTDS Modelling and Simulation

There are several features that distinguish the modelling and simulation activity within the continuous-time dynamic system (CTDS) domain. Perhaps one of the most important is the dependence of the project's success upon the selection of the behaviour generation tool that is best suited to the nature of the conceptual model. Because the conceptual model in this domain always includes a set of differential equations, the tools in question relate to the numerical procedures for solving these equations (we restrict our discussions to the case where only ordinary differential equations (ODEs) are included in the model).

Many families of approaches for the solution of ODEs can be found in the literature and within each family there generally are numerous specific options. The methods in these families have their characteristic strengths and weaknesses and are often best suited for specific categories of problems. Furthermore, the use of any of these methods usually involves the specification of values for embedded parameters. The range of options is indeed wide and can even become daunting. To embark on a modelling and simulation project in this environment without some appreciation for the issues involved can be foolhardy. Our objective in Part 3 of this textbook is to provide a basic foundation for dealing with these issues.

In Chapter 7 we establish a context for the discussion by formulating a range of simple CTDS conceptual models. For the most part, these have their origins in the portions of the physical world where behaviour can be readily characterised by familiar laws of physics. This central role of the laws of physics is a typical circumstance in the CTDS domain and should not be interpreted as a biased perspective. However, this is not to suggest that CSTD models cannot be formulated in the absence of directly applicable physical laws and we illustrate this point by providing an example of the formulation of some credible CTDS models based entirely on intuitive arguments. The final topic in Chapter 7 is a brief examination of the problem of transforming a conceptual model that has evolved with higher-order differential equations into an equivalent set of first-order differential equations. Such a format is a frequently required by numerical software.

In Chapter 8 we provide an overview of some of the basic numerical tools for solving the ODEs of the CTDS conceptual model. The presentation is relatively informal and is at an introductory level. Features that have practical relevance, especially those that can lead to numerical difficulties, are emphasised.

Because of our assumed absence of random affects in the CTDS models which we treat in this textbook, it is conveniently feasible to include classical optimisation requirements in the project goals. This topic is examined in Chapter 9. The typical objective here is to find values for parameters within the conceptual model which yield a minimum value for a prescribed performance (or criterion) function. Such a function could, for example, correspond to the operating cost of some chemical process. We outline several minimisation procedures that could be applied in a CTDS context.

As a concluding comment in this synopsis, we encourage the reader to examine Annex 3 where we have provided an introduction to Open Desire which is a software tool specifically designed to facilitate simulation experiments with CTDS conceptual models.

Chapter 7 Modelling of Continuous-Time Dynamic Systems

7.1 Introduction

Our concern in this chapter is with exploring the modelling process within the context of continuous-time dynamic systems (CTDS). From our perspective, the essential distinguishing feature of this category of system is the fact that a conceptual model can be formulated as a set of differential equations, possibly augmented with a set of algebraic equations. For the most part, such models emerge from a deductive process that has its basis in physical laws that are known to govern the behaviour we seek to explore, that is, the behaviour of the SUI. This is in contrast to an inductive process whereby a model is developed on the basis of observed (or hypothesised) behaviour, as is the case in the development of almost all models in the realm of discrete-event dynamic systems (DEDS). The deductive model building process is generally associated with systems that have their origins in engineering or in the physical sciences. Because of the availability of 'deep' knowledge provided by relevant physical laws, such models can incorporate subtleties and a level of detail that are not usually possible within the DEDS context. This enhances the scope of project goals that are realistically achievable.

For convenience, we refer to conceptual models that have a differential equation format as *CTDS models*. Although such models arise most commonly from a deductive process, it needs to be stressed that this is not a prerequisite. It is most definitely possible to develop credible and useful CTDS models via an inductive process in certain cases where the SUI falls outside the realm of established physical laws. The fields of biology and economics provide many examples of such an approach.

CTDS models can be formulated in terms of either ordinary or partial differential equations (or both). When the modelling power of partial differential equations is required, the SUI is usually called a *distributed parameter system*. Such systems arise in a wide variety of domains. Included here are: heat transfer, hydrodynamics, electromagnetics, and elasticity. The treatment of models that depend on this formalism is, however, beyond the scope of this textbook. Our considerations are restricted to CTDS models that can be formulated exclusively within the framework of ordinary differential equations. Nevertheless much of the discussion in both this and the following chapter does have relevance to the case of distributed parameter systems.

Frequently random effects are absent in continuous models. Although this is not an essential property, we limit our considerations in this chapter to this restricted (i.e., deterministic) case. One especially significant feature associated with the deterministic context is that a search for operating conditions that yield some prescribed behaviour for the SUI becomes a significantly simpler task because there is no requirement to assess efficacy of a candidate solution over some potentially vast stochastic environment.

The nature of CTDS models as outlined above, suggests a number of differences from the class of DEDS models examined in Parts 1 and 2 in this book. In effect, CTDS models exhibit a "smoothness" property in the sense that the time trajectories of the variables within the model tend to undergo only small changes in response to small changes in parameters or in operating conditions. This feature, combined with the characterisation of system behaviour at a relatively detailed level (resulting from the underlying deep knowledge that is typically available) and the absence of stochastic effects, permits the formulation of project goals that can be more demanding in terms of expected precision and reliability. This, in particular, makes feasible credible optimisation studies whose outcome can, with reasonable confidence, provide the basis for system implementation. We examine this topic in some detail in Chapter 9.

Another important difference between CTDS and DEDS models relates to the nature of the time advance mechanism required in the simulation program. In the case of CTDS models, the fundamental requirement is that of solving the underlying differential equations within the conceptual model and that process intrinsically incorporates a time advance procedure thereby eliminating any need for additional considerations. The mechanisms in question are explored in Chapter 8.

7.2 Some Examples of CTDS Conceptual Models

7.2.1 Simple Electrical Circuit

An electrical circuit consisting of a resistor (R), capacitor (C), inductor (L), and a voltage source ($E(t)$) connected in series (see Figure 7.1) provides an archetypical example of a system whose dynamics can be represented using a CTDS model. An analysis of the circuit based on the application of Kirchoff's voltage law, yields the equation:

$$Lq''(t) + Rq'(t) + \frac{q(t)}{C} = E(t) \ , \tag{7.1}$$

where $q(t)$ is the charge on the capacitor C, and $q'(t)$[1] is the current in the circuit. If we denote by t_0 the left-hand boundary of the observation interval, then it is important to observe here that the solution of Equation (7.1) (i.e., the behaviour generation process) requires the specification of two initial conditions: namely, $q(t_0)$ and $q'(t_0)$ as well as the explicit specification of the function $E(t)$. In other words, $q(t)$ and $q'(t)$ are state variables for the model and $E(t)$ is an input variable.

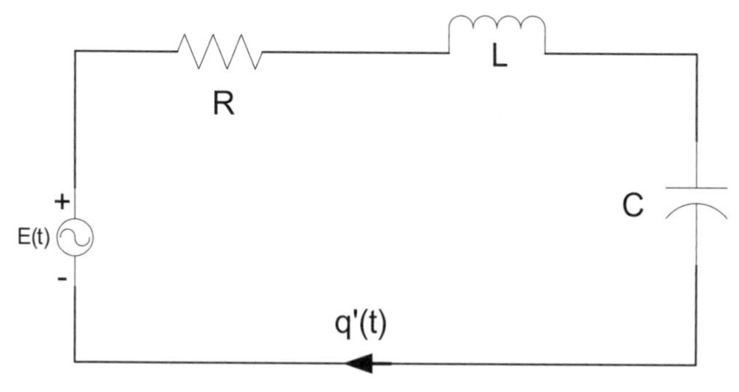

FIGURE 7.1. A simple electrical circuit.

7.2.2 Automobile Suspension System

A vehicle of mass $4M$ is traveling forward at constant velocity over a road which is initially smooth and horizontal. It is in an equilibrium condition and any particular point on the body has a constant vertical displacement from the road surface. The body is connected to each of the four wheels through a spring/shock absorber system and each wheel supports one quarter of the total mass.

At time $t = t_0$ the vehicle begins to travel over a section of the road which has an irregular surface (see Figure 7.2). This causes vertical motion of the vehicle about its equilibrium position. If we use $y(t)$ to represent this vertical displacement, then from the application of Newton's second law, the trajectory of $y(t)$ is defined by:

$$M\, y''(t) + f_b(t) + f_a(t) = 0 \ , \tag{7.2}$$

[1] The prime superscript denotes differentiation with respect to time (d/dt).

where $f_a(t)$ and $f_b(t)$ represent the forces associated with the spring and the shock absorber, respectively. We choose the variable u to represent the vertical irregularities in the road surface, taken with respect to the road's smooth horizontal (equilibrium) condition. Although u is a function of horizontal displacement from some reference point it can, from the perspective of the vehicle moving over it at constant speed, be treated as a function of time; that is, $u = u(t)$. This time function $u(t)$, in fact, represents an input to the CTDS model being formulated. A particular choice for $u(t)$ that matches the presentation in Figure 7.2 is:

$$u(t) = \begin{cases} 0 & \textit{for } t < t_0 \\ \left(\frac{u_{\max}}{2}\right)\left(1 - \cos(\omega(t - t_0))\right) & \textit{for } t \geq t_0 \end{cases}, \tag{7.3}$$

where ω is proportional to the vehicle's horizontal velocity.

For definiteness, we assume that the spring is linear; hence $f_a(t) = k(y(t) - u(t))$ where k is the spring constant. On the other hand let's assume that the shock absorber is nonlinear and that the associated force is:

$$f_b(t) = \Psi \, |v(t)| \, v(t) , \tag{7.4}$$

where $v(t) = (y'(t) - u'(t))$ and Ψ is the shock absorber constant.

If we choose $y(t)$ and $y'(t)$ to be the state variables for the model, then the solution of the second-order differential equation, Equation (7.2), requires the two initial conditions: $y(t_0)$ and $y'(t_0)$. From the definition of $y(t)$ and as a direct consequence of the equilibrium assumption prior to $t = t_0$, both of these values are zero.

A possible goal for a modelling and simulation study associated with the above model could be the determination of values for the spring and shock absorber constants which yield a best value for some prescribed measure of ride quality.

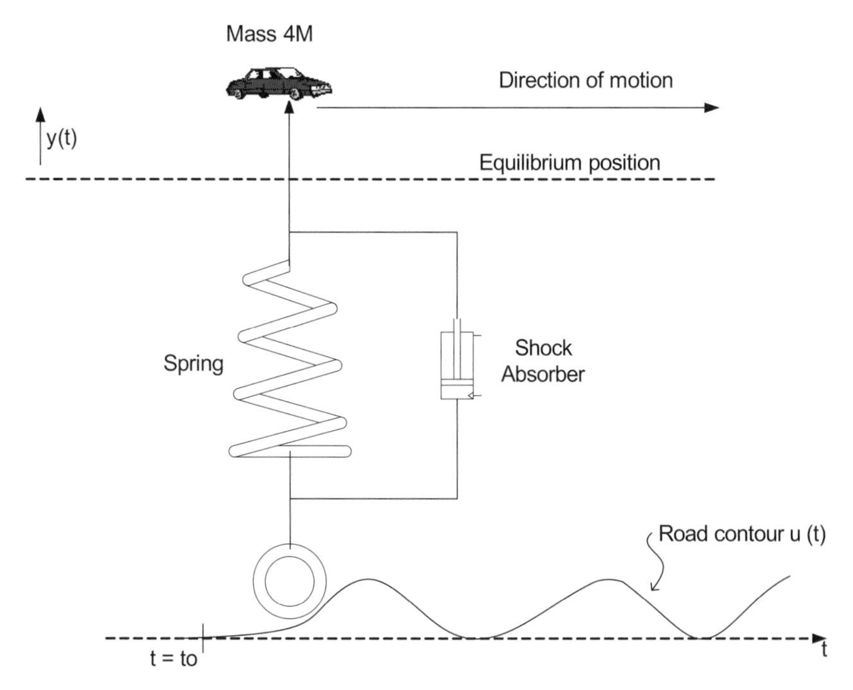

FIGURE 7.2. Automobile suspension system.

7.2.3 Fluid Level Control

The cleaning solution required in an industrial process passes through two holding tanks (see Figure 7.3). Valves control the inflow into each of the tanks and the position of these valves is established by a control strategy based on the height of the liquid in the respective tanks. The rate of change of the volume of liquid in each tank is equal to the difference between the inflow rate and the outflow rate. If we let A_1 and A_2 represent the cross-sectional areas of Tank 1 and Tank 2, respectively; then:

$$\begin{aligned} A_1\, h_1'(t) &= w_0(t) - w_1(t) \\ A_2\, h_2'(t) &= w_1(t) - w_2(t)\,, \end{aligned} \tag{7.5}$$

where $w_0\,(t)$, $w_1(t)$, and $w_2(t)$ are the volume flow rates (e.g., cubic meters per second) into and out of the tanks as shown in Figure 7.3. (Note that the solution of Equation (7.5) requires initial conditions for h_1 and h_2.)

Flow Control Specifications

Tank 1: The valve V_0 opens when the level in Tank 1 is decreasing and falls below a value which corresponds to a half-full tank. More precisely, V_0 moves from a closed to an open position at time t_a where $h_1'(t_a) < 0$ and $h_1(t_a) < H_1/2$. Once open, V_0 stays open until $h_1(t)$ reaches the level H_1 which is the full-tank condition. When V_0 is open, the inflow rate is constant; that is, $w_0(t) = K$.

Tank 2: The control policy for valve V_1 is analogous; that is, V_1 moves from a closed to an open position at time t_b where $h_2'(t_b) < 0$ and $h_2(t_b) < H_2/2$. Once open, V_1 stays open until $h_2(t)$ reaches the level H_2 (the full-tank condition). However, when valve V_1 is open, the inflow rate is given by $w_1(t) = k\, h_1(t)$ where k is a constant. The outflow rate from Tank 2, $w_2(t)$, is given by $w_2(t) = u(t)\, h_2(t)$ where $u(t)$ is a control input to valve V_2 which reflects the external demand for cleaning solution. Because of physical limitations of the piping system, $u(t)$ is constrained; that is, $0 < u(t) < u_{max}$.

The SUI outlined above clearly has a control system context. The conceptual model for the SUI is given by Equation (7.5) together with the (algebraic) equations implicit in the control strategy. A likely project goal here could be the resolution of the design problem of choosing appropriate values for the various parameters within the model (e.g. H_1, H_2 and K) based on assumptions about the external demand, $u(t)$ and some criterion for evaluating performance.

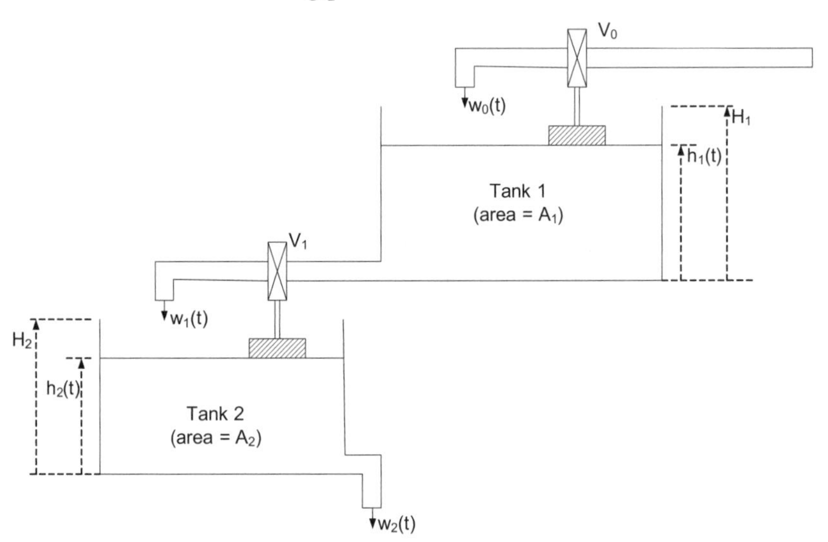

FIGURE 7.3. Fluid level control.

There is a distinctive aspect of this example that is worth noting. The conceptual models for the SUIs outlined in Section 7.2.1 and Section 7.2.2 were formulated entirely on the basis of basic physical laws. In this example, only part of the conceptual model has such 'natural' origins (namely Equation (7.5)). The remainder of the model relates to behaviour that is superimposed by the technological artefact of the control policy (see Flow Control Specifications above). This latter behaviour can be readily altered by the control policy's developer. In fact, its possible modification is likely implicit in the project goals.

7.2.4 Population Dynamics

Often the model associated with a modelling and simulation project in the domain of environmental studies must incorporate a representation of the manner in which the population of various species evolves over time. In many cases the model must reflect the interdependence of several species. Perhaps the best example of the case of interacting populations is given by predator/prey (or host/parasite) situations, for example, wolf/caribou or lynx/hare populations. The characterisation of the behaviour of such populations with a CTDS model implies that the variables representing population values will acquire 'real' (i.e., fractional) values rather than values that are strictly integer. This may appear counterintuitive but with the assumption that the populations are 'large', the fractional parts of real values have little consequence on the general features of the results obtained.

There are no underlying physical laws upon which to base the development of such population models (unlike the circumstances in the examples discussed in Sections 7.2.1, 7.2.2 and 7.2.3). Consequently they are based on essentially intuitive arguments. As demonstrated below, a credible structure for such models can be formulated in a reasonably straightforward manner. However, accommodating the associated data requirements (parameter values) can present a challenge.

We consider first a single population model and let $P(t)$ represent the population at time t. A natural assumption is that the rate of change of population is dependent on two effects: namely, the birth rate $b(t)$ (births per unit time) and the death rate $d(t)$ (deaths per unit time). This yields the basic equation:

$$P'(t) = b(t) - d(t)\,. \tag{7.6}$$

It is reasonable to assume that both $b(t)$ and $d(t)$ are dependent on the current population. If this dependence is linear, that is, $b(t) = k_b\,P(t)$ and $d(t) = k_d P(t)$, then the model becomes:

$$P'(t) = k\,P(t)\,, \tag{7.7}$$

where $k = (k_b - k_d)$. The solution to Equation (7.7) can be easily verified to be

$$P(t) = \exp(kt)\,P_0\,,$$

where P_0 is the population at some (initial) time t_0. Clearly if $k > 0$, the population will grow without bound whereas if $k < 0$, the population will eventually vanish; hence the model is relatively rudimentary.

A possible refinement is to conjecture that k is indeed positive but that there are external effects that prevent the population from exceeding a value of P_{max}. This behaviour can be achieved with a simple modification to the model of Equation (7.7); that is,

$$P'(t) = k\,[1 - (P(t)/P_{max})]\,P(t)\,. \tag{7.8}$$

Now as $P(t)$ approaches P_{max} the growth rate approaches zero.

As an alternative, suppose we choose the dependence in the case of $b(t)$ to be linear but nonlinear in the case of $d(t)$. Specifically, let's choose:

$$b(t) = \alpha\,P(t)$$

$$d(t) = \beta\,P^2(t)\,,$$

where α and β are constants whose values (necessarily positive) remain to be determined as part of the data modelling phase. With the substitution of these relations in Equation (7.8) and with some straightforward manipulation, we obtain:

$$P'(t) = \alpha\,P(t)\,[1 - K\,P(t)]\,. \tag{7.9}$$

Here $1/K = \alpha/\beta$ plays the role of an equilibrium value for the population, $P(t)$. In other words, the solution of Equation (7.8) approaches the value $1/K$ from any initial condition $P_0 = P(t_0)$.

We now extend our considerations to the case of two populations that function in a predator/prey framework. We use P_1 and P_2 to represent the predator and the prey populations, respectively. The behaviour of each of these populations can be assumed to be represented by an equation of the form of Equation (7.9) but suitably augmented by some reflection of the mutual interaction. We assume that the interaction can be characterised by a term that is proportional to the product of the two population sizes. Furthermore, it is reasonable to assume that the interaction is beneficial to the predator population growth rate but is detrimental to the growth rate of the prey population. Under these circumstances we obtain the following CTDS model.

$$P_1'(t) = \alpha_1 P_1(t) [1 - K_1 P_1(t)] + \lambda_1 P_1(t) P_2(t)$$
$$P_2'(t) = \alpha_2 P_2(t) [1 - K_2 P_2(t)] - \lambda_2 P_1(t) P_2(t) \quad , \tag{7.10}$$

where the positive constants λ_1 and λ_2 reflect the 'strength' of the interactions.

A common simplification to the model given in Equation (7.10) is to ignore the effect of 'natural' death rates by setting $\beta_1 = \beta_2 = 0$, which then results in $K_1 = K_2 = 0$. This gives:

$$P_1'(t) = \alpha_1 P_1(t) + \lambda_1 P_1(t) P_2(t)$$
$$P_2'(t) = \alpha_2 P_2(t) - \lambda_2 P_1(t) P_2(t) \ . \tag{7.11}$$

From a validation point of view it is reasonable to require that the predator population $P_1(t)$ approach zero if the prey population vanishes (i.e., if $P_2 = 0$). This requirement can be achieved only if α_1 has a negative value. An equivalent effect can be achieved by replacing α_1 with $-\alpha_1$ (and then taking both α_1 and α_2 to be positive). Our model then becomes:

$$P_1'(t) = -\alpha_1 P_1(t) + \lambda_1 P_1(t) P_2(t) = -\alpha_1 P_1(t) [1 - (\lambda_1/\alpha_1) P_2(t)]$$
$$P_2'(t) = \alpha_2 P_2(t) - \lambda_2 P_1(t) P_2(t) = \alpha_2 P_2(t) [1 - (\lambda_2/\alpha_2) P_1(t)]. \tag{7.12}$$

Equation (7.12) has an equilibrium point given by $P_1^* = \alpha_1/\lambda_1$ and $P_2^* = \alpha_2/\lambda_2$. These values correspond to the case where both $P_1'(t)$ and $P_2'(t)$ are zero. The equilibrium point, however, is unstable and any small perturbation from it leads to an oscillatory trajectory for both $P_1(t)$ and $P_2(t)$ about their respective equilibrium values. Representative trajectories are shown in Figure 7.4.

The CTDS model of Equation (7.12) has been extensively studied and the equations are known as the Lotka–Volterra equations (see, e.g., [7.2]). An interesting study of predator/prey behaviour when harvesting is introduced can be found in [7.1].

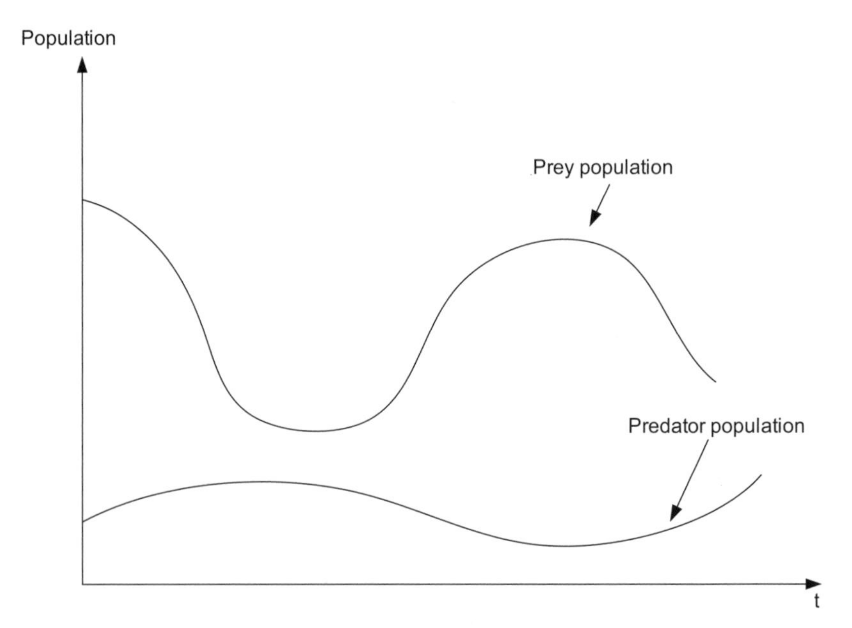

FIGURE 7.4. Predator/prey population.

7.3 Safe Ejection Envelope: A Case Study

Several CTDS models have been presented in the preceding section to illustrate the nature of this family of conceptual models. In this section our focus is on another SUI which gives rise (via a deductive approach) to a CTDS model. However, in this case, we identify a specific goal and, in effect, we formulate a modelling and simulation project.

The problem is one that has been frequently used in the modelling and simulation literature relating to continuous-time dynamic systems. It concerns the safe ejection of a pilot from the cockpit of a disabled fighter aircraft. The specific situation we investigate concerns an aircraft that is flying horizontally at an altitude H, with a constant speed of V_a, when an emergency situation arises and the pilot is obliged to activate the onboard ejection mechanism and abandon the aircraft. Figure 7.5 shows the pilot's general trajectory following ejection.

FIGURE 7.5. Trajectory of the ejected pilot.

The ejection mechanism ensures that the pilot safely leaves the cockpit and once disconnected from the aircraft, the pilot[2] follows a ballistic trajectory that is governed by two forces. One of these is a drag force and the other is the force of gravity which will ultimately return the pilot to the surface of the earth. Notice, however, that once the pilot leaves the aircraft, the aircraft's tail section becomes a projectile that can potentially strike the pilot and cause serious injury. Our concern is with exploring the circumstances that cause such a collision.

A prerequisite for achieving this objective is a model of the dynamic behaviour of the pilot and the aircraft. The modelling perspective which we adopt incorporates two important assumptions, namely,

- The motion is restricted to two dimensions; more specifically the pilot's trajectory stays in the plane defined by the cockpit and tail section (in other words, wind forces that might alter this planar motion are ignored).
- During a free-flight (ballistic) trajectory any object (in this case the pilot) is subjected to a drag force $D = D(t)$ which results from the resistance introduced by air friction. This force acts in a direction opposite to the velocity vector (see Figure 7.10) and we adopt the usual assumption that it can be expressed as

$$D(t) = \mu\, V^2(t)\,, \tag{7.13}$$

[2] For convenience, we usually refer simply to the trajectory of the pilot but it should be recognized that upon leaving the aircraft, the pilot remains connected to the seat and it is the trajectory of the pilot plus seat that is, in reality, being studied. We assume that the seat is jettisoned at some point in time that is beyond the observation interval of interest.

where $\mu = \hat{C}_D \rho$. Here $\hat{C}_D$ is a constant that depends on the physical shape of the moving object and ρ is the local air density which is dependent on altitude. This relationship is known only in terms of a number of data points as provided in Table 7.1.

TABLE 7.1. The altitude/air density relationship.

Altitude (feet)	Air Density (ρ) (lbs/ft^3)
0	2.3777×10^{-3}
1000	2.208×10^{-3}
2000	2.241×10^{-3}
4000	2.117×10^{-3}
6000	1.987×10^{-3}
10,000	1.755×10^{-3}
15,000	1.497×10^{-3}
20,000	1.267×10^{-3}
30,000	0.891×10^{-3}
40,000	0.587×10^{-3}
50,000	0.364×10^{-3}
60,000	0.2238×10^{-3}

There is a variety of factors that influence the form of the pilot's trajectory and hence the possibility of a collision with the tail section; for example, the orientation θ_r of the ejection rail, the ejection velocity V_r, the position of the tail assembly, the velocity V_a of the aircraft, and the altitude H at which the aircraft is flying. Note that the latter is a consequence of the dependence of drag $D(t)$, on air density that, in turn, depends on H.

The specific relationship we undertake to investigate in this project is the one that exists between the constant horizontal velocity of the aircraft (V_a) and a variable we call H_{min}. As is apparent from Equation (7.13), the drag force $D(t)$ acting on the pilot is dependent on the altitude at which the aircraft is flying (indirectly via the air density relationship). Suppose the aircraft is flying at an altitude β_o with horizontal velocity $V_a = \alpha$. A collision will result if the drag force is too high. To avoid a collision at the velocity α, the altitude needs to be increased (air density and hence drag force, both decrease as altitude increases, see Table 7.1) and the least altitude (say β) at which a collision is avoided is the H_{min} value associated with the velocity α. Our project goal is to determine a value of H_{min}, corresponding to each of a selected sequence of values of V_a. A graph of the form shown in Figure 7.6 would be a reasonable means for presenting the data thus acquired.

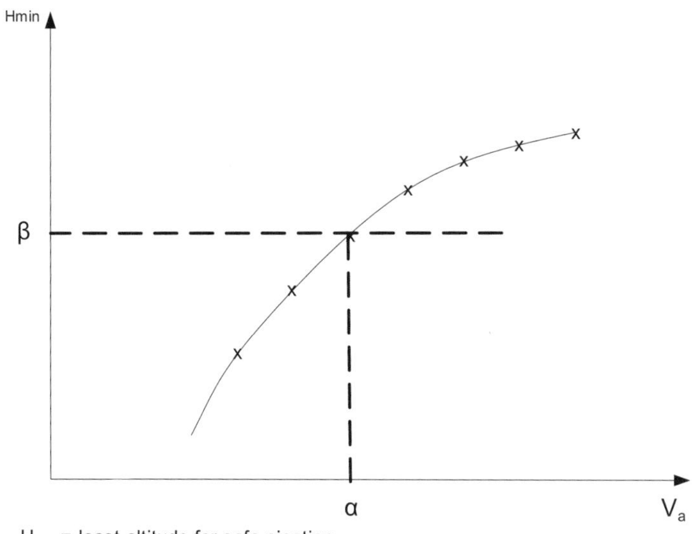

FIGURE 7.6. Generic form of the safe ejection envelope.

The ejection mechanism, once activated (at time $t = 0$), propels the pilot over a short length of rail at a constant velocity V_r. This rail is inclined at an angle θ_r from the vertical (see Figure 7.7). The seat becomes disengaged from the rail after it has risen a vertical distance of Y_r. At that moment (time $t = t_E$) the pilot (and seat) begin a ballistic trajectory that may either pass over or strike the tail section.

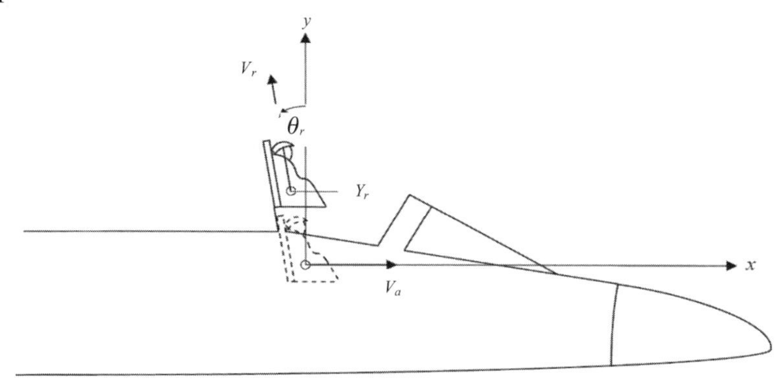

FIGURE 7.7. Initial phase of the ejection trajectory.

There is a variety of ways in which the conceptual model for the dynamic behaviour of interest can be formulated. In our approach, we choose $X_p(t)$ and $Y_p(t)$ to represent the horizontal and vertical displacement, respectively, of the pilot measured relative to a reference point in space A° whose location is fixed in time. A convenient choice for the A° is the point on the aircraft where the seat is initially anchored to the aircraft. If we assume that the ejection process begins at time $t = t_0 = 0$, then $X_p(0) = Y_p(0) = 0$.

We make the simplifying assumption that the leading edge of the tail section is vertical and we let $(X_T(t), Y_T(t))$ be the co-ordinates of the point at the top of the leading edge. This particular point is of interest because it is a reference point for our safe ejection study. We assume that the leading edge of the tail is located a distance BT units behind the point where the seat is anchored. Because the aircraft is moving with a constant horizontal velocity of V_a, it follows that (relative to the fixed point A°), $X_T(t) = (V_a t - BT)$ for $t \geq 0$. Similarly we assume that the top point of the tail section is displaced a distance of HT above the anchor point; thus, $Y_T(t) = HT$ for $t \geq 0$. Both BT and HT are positive constants yet to be specified.

We use t^* to denote the value of time when the pilot is located at the leading edge of the tail section. The value of t^* is implicitly defined by the relation:

$$X_p(t^*) = X_T(t^*) = V_a t^* - BT . \tag{7.14}$$

At $t = t^*$ the pilot is either passing over the leading edge of the tail section ($Y_p(t^*) > Y_T(t^*) = HT$) or is striking it ($Y_p(t^*) \leq HT$). It should also be observed that Equation (7.14), in fact, provides an implicit definition of the right-hand end of the observation interval.

Although, in principle, the collision boundary corresponds to $Y_p(t^*) = HT$ it is realistic to adopt a more conservative criterion (a 'safe miss') which we define to be one where the trajectory passes over the tail section with a vertical displacement of a least $(HT + S_f)$ where S_f is a 'safety factor'. The intent here is to accommodate inherent uncertainties in many of the constants embedded in the dynamic model. Throughout the remaining discussion, references to 'missing the tail' therefore implies $Y_p(t^*) > (HT + S_f)$.

If we denote by $V(t)$ the pilot's velocity vector, then the generic form of the pilot's motion can be represented as shown in Figure 7.8, from which it follows that

$$\begin{aligned} V_x(t) &= V(t)\cos\theta(t) \\ V_y(t) &= V(t)\sin\theta(t) . \end{aligned} \tag{7.15}$$

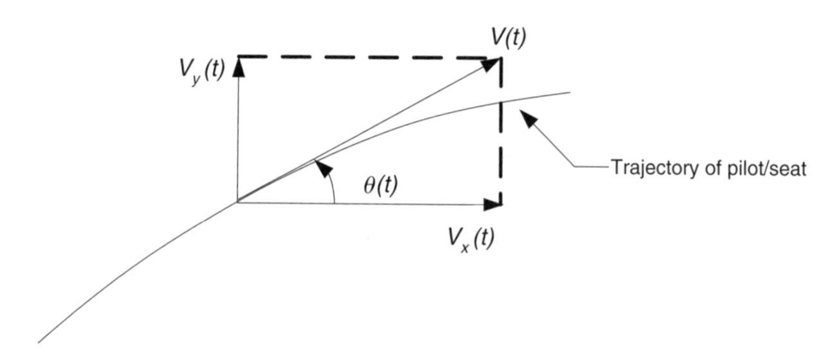

FIGURE 7.8. Generic trajectory of the pilot/seat.

While on the rails, the pilot's velocity vector $V(t)$ is the sum of the constant horizontal velocity of the aircraft V_a, and the constant ejection velocity V_r. The configuration is shown in Figure 7.9 from which it follows directly that

$$\begin{aligned} X_p'(t) &= V_x(t) = V_a - V_r \sin\theta_r \\ Y_p'(t) &= V_y(t) = V_r \cos\theta_r \ . \end{aligned} \tag{7.16}$$

Furthermore because both the magnitude and the orientation of $V(t)$ are constant while the pilot is on the rails, we have that $V_x'(t) = V_y'(t) = 0$.

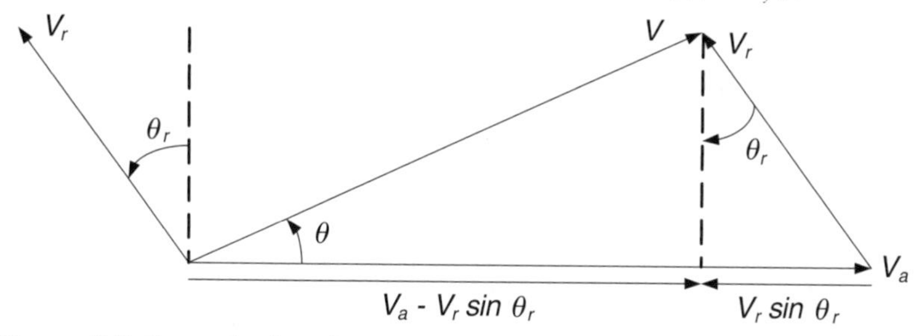

FIGURE 7.9. Constrained motion on rails ($Y_p \leq Y_r$).

Suppose we assume that the pilot/seat leaves the rails at time $t = t_E$. It is straightforward to establish that

$$\begin{aligned} t_E &= Y_r/(V_r \cos\theta_r) \\ X_p(t_E) &= (V_a - V_r \sin\theta_r)\, t_E \\ Y_p(t_E) &= Y_r \\ X_p'(t_E) &= V_x(t_E) = V_a - V_r \sin\theta_r \\ Y_p'(t_E) &= V_y(t_E) = V_r \cos\theta_r \ . \end{aligned} \tag{7.17}$$

Once the pilot/seat is 'disconnected' from the aircraft (i.e., leaves the rails) its motion is governed by two forces, namely, the force of gravity and the drag force $D(t)$ as shown in Figure 7.10. Together these forces create a trajectory that (from the perspective of an observer moving with horizontal velocity of V_a) arcs backwards over the rear of the aircraft (see Figure 7.5).

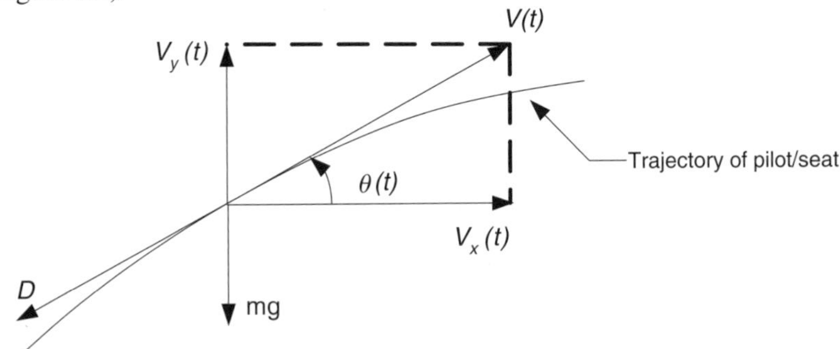

FIGURE 7.10. Free-fall motion (ballistic trajectory).

Because two forces now act upon the pilot, there are acceleration effects introduced as a consequence of Newton's second law. In other words, $V_x(t)$ and $V_y(t)$ are no longer constant. The dynamic model becomes (see Figure 7.10):

$$X_p'(t) = V_x(t) \tag{7.18a}$$

$$Y_p'(t) = V_y(t) \tag{7.18b}$$

$$V_x'(t) = -(D(t)/m)\cos\theta(t) \tag{7.18c}$$

$$V_y'(t) = -(D(t)/m)\sin\theta(t) - g\,. \tag{7.18d}$$

The conceptual model we seek is provided, in its most fundamental form, by Equation (7.18). One shortcoming, however, is the dependence on $V(t)$ (through $D(t)$) and on $\theta(t)$. Two approaches are possible for dealing with this. In the approach we adopt, this explicit dependence is eliminated with some algebraic manipulation that incorporates Equation (7.15) and the specification for $D(t)$ (see Equation (7.13)). Equations (7.18c) and (7.18d) then become:

$$V_x'(t) = -\Psi(t)\, V_x(t)$$

$$V_y'(t) = -\Psi(t)\, V_y(t) - g$$

where: $$\Psi(t) = [\hat{C}_D\, \rho(H + Y_p(t))\, (V_x^2(t) + V_y^2(t))^{0.5}]/m.$$

There now remains the requirement of specifying the observation interval I_o that is pertinent to the project goal. The right-hand end of I_o has, in fact, been established earlier (see Equation (7.14)). The nominal left-hand end of I_0 is the moment when the pilot initiates the ejection process and we have previously associated this with $t = 0$. The values of the four state variables (X_p, Y_p, V_x, V_y) are certainly known at $t = 0$. Notice, however, that values for the state variables are also known at the later time $t = t_E$ (see Equation (7.17)). The fact that there is a severe discontinuity in the derivatives $V_x'(t)$ and $V_y'(t)$ as t passes over the point $t = t_E$ suggests that $t = t_E$ is a more practical choice for the left-hand boundary of I_0 (see Section 8.4.2). In view of this, we choose our conceptual model to be the set of equations given in Equation (7.19).

$$\begin{aligned} X_p'(t) &= V_x(t) \\ Y_p'(t) &= V_y(t) \\ V_x'(t) &= -\Psi(t)\, V_x(t) \\ V_y'(t) &= -\Psi(t)\, V_y(t) - g \end{aligned} \qquad (7.19)$$

where:

$$\Psi(t) = [\hat{C}_D\, \rho(H + Y_p(t))\, V(t)]/m$$
$$V(t) = (V_x^2(t) + V_y^2(t))^{0.5} \;,$$

where the corresponding 'initial' conditions are at $t = t_E$ as prescribed in Equation (7.17). A summary of the various constants associated with the model is given in Table 7.2.

TABLE 7.2. Summary of constants.

Constant	Numerical Value	Units	Role
BT	30	ft	Horizontal displacement of tail section behind origin
$\hat{C}_D$	5		Drag factor
g	32.2	ft/sec^2	Acceleration due to gravity
HT	12	ft	Vertical height of tail section
m	7	slugs	Mass of the pilot/seat combination
S_f	8	ft	Safety factor for avoiding tail section
θ_r	15	degrees	Displacement angle of ejection rails from vertical
V_r	40	ft/sec	Seat velocity while on rails
Y_r	4	ft	Vertical height of rails

It's interesting to also formulate the alternate elaboration of Equations (7.18c) and (7.18d). In this approach we begin with Equation (7.15) from which it follows that

$$V_x'(t) = V'(t)\cos\theta(t) - V(t)\sin\theta(t)\theta'(t) \tag{7.20a}$$

$$V_y'(t) = V'(t)\sin\theta(t) + V(t)\cos\theta(t)\theta'(t) \tag{7.20b}$$

Multiplication of (7.20a) by cos $\theta(t)$ and (7.20b) by $\sin\theta\,(t)$, addition, and substitution of Equations (7.18c) and (7.18d) yields:

$$V'(t) = -\,(D(t)/m) - g\sin\theta(t)\,.$$

Similarly multiplication of (7.20a) by sin $\theta(t)$ and (7.20b) by cos $\theta(t)$, addition, and again substitution of Equations (7.18c) and (7.18d) yields:

$$\theta'(t) = -\,(g\cos\theta(t))/V(t)\,.$$

Thus, an alternate conceptual model for the ballistic trajectory ($t > t_E$) is:

$$\begin{aligned} X_p'(t) &= V(t)\cos\theta(t) \\ Y_p'(t) &= V(t)\sin\theta(t) \\ V'(t) &= -\,(D(t)/m) - g\sin\theta(t) \\ \theta'(t) &= -\,(g\cos\theta(t))/V(t) \end{aligned}$$

$$\text{where: } D(t) = \hat{C}_D\,\rho(H + Y_p(t))\,V^2(t)\,.$$

The safe ejection envelope project is revisited in Section 8.6 where a procedure for its completion is presented together with an Open Desire simulation program which carries out the procedure.

7.4 State Space Representation

7.4.1 The Canonical Form

The differential equations that evolve in the development of a conceptual model for a CTDS can have a variety of formats; for example, they may be linear or nonlinear, they may be a set of first-order equations, they may be equations of higher order, they may be autonomous, or may instead have input functions that reflect pertinent interaction with their environment. Illustrations of these various alternatives can be found in the examples of the previous discussion. The model developed for the electric circuit (Equation (7.1)) is linear, of second order, and is nonautonomous (the voltage source E represents an input). The suspension system model of Equation (7.2) is also a second-order equation but is nonlinear; it also is nonautonomous (the irregular road surface provides the input). The fluid level control model of Equation (7.5) is a pair of first-order equations which are nonlinear (because of the nonlinear dependence of $w_0(t)$ on $h_1(t)$

and $w_l(t)$ on $h_2(t)$) and nonautonomous (the outflow demand represents an input to the model). The population model example presented in Equation (7.12) is a pair of first order non-linear equations that are autonomous.

The above discussion illustrates the wide range of formats in which CTDS conceptual models can evolve. This same variability is certainly present in the realm of DEDS models and it's not surprising to encounter it again. However, CTDS models do have a particularly important feature in this regard; namely, that it is possible to transform all of these formats into a standard (canonical) form. This can be written as

$$\boldsymbol{x}'(t) = \boldsymbol{f}(\boldsymbol{x}(t), \boldsymbol{u}(t), t) \qquad (7.21a)$$

with: $\boldsymbol{x}(t_0) = \boldsymbol{x}_0$

and

$$\boldsymbol{y}(t) = \boldsymbol{g}(\boldsymbol{x}(t)) \,. \qquad (7.21b)$$

Here $\boldsymbol{x}(t)$, $\boldsymbol{u}(t)$, and $\boldsymbol{y}(t)$ are vectors of dimension N, p, and q, respectively, and represent the state, the input, and the output variables, respectively, of the CTDS model. The functions $\boldsymbol{f}$ and $\boldsymbol{g}$ are likewise vectors with dimensions that are consistent with usage. Equation (7.21a) represents a set of N first-order differential equations and, as noted, the initial conditions required for the solution of Equation (7.21a) are assumed to be given. Equation (7.21b) makes provision for the situation where the output variables of the model do not correspond directly to any of the state variables but rather are prescribed functions of the state variables.

The representation given in Equation (7.21) is called a *state space* representation for the particular CTDS model. It has two components, the first are the *state equations*, given by Equation (7.21a) and the second component, given by Equation (7.21b), is called the output equation of the model. Neither, however, is a unique representation for the particular CTDS that is under consideration. Nevertheless, as we shall indicate below, there often are natural choices for the state variables, $x_i(t)$ which form the elements of the state vector, $\boldsymbol{x}(t)$.

The existence of a state space representation for any CTDS model has several important consequences. Among these is the fact that a very substantial body of knowledge about equations of the form of Equation (7.21) has evolved within the domain of applied mathematics. This knowledge is therefore applicable for investigating the properties of CTDS models. Included here are issues that range from the very fundamental, for example, the question of the existence of solutions to the equations that comprise the model, to issues that characterise the properties of the solution, for example, stability. Exploration of these topics is, however, beyond the scope of the considerations in this textbook. The interested reader is encouraged to explore these topics in references such as [7.3] and [7.4].

In addition to important behavioural properties of a CTDS conceptual model that can be explored via its state space representation, there is one very practical benefit also associated with it. Recall that experimentation with any CTDS model requires the means to generate the numerical solution of differential equations. This is a problem that has been extensively studied in the applied mathematics literature and an extensive body of relevant knowledge about the problem exists. But with few exceptions, this body of knowledge addresses the problem of solving a set of differential equations that are of the form of Equation (7.21 a) and likewise the available solution methods apply to this case. Thus the transformation of a CTDS conceptual model into its state space representation is an essential step for purposes of harnessing the numerical tools for solution generation or more specifically, for carrying out simulation activity.

7.4.2 The Transformation Process

If any CTDS conceptual model has a state space representation (i.e., can be transformed into the form of Equation (7.21)), then this must certainly be true for a linear model of the form:

$$\begin{aligned} & y^N(t) + a_{N-1}y^{N-1}(t) + a_{N-1}y^{N-2}(t) + \ldots + a_1 y'(t) + a_0(t) \\ & = b_m u^m(t) + b_{m-1}u^{m-1}(t) + b_{m-2}u^{m-2}(t) + \ldots + b_1 u'(t) + b_0(t) \end{aligned} \tag{7.22}$$

where we assume $m \leq N$ and that $u(t)$ and $y(t)$, $y'(t)$, . . . , $y^{(N-1)}(t)$ are given. In the interest of notational convenience, we assume here that the model has a single input variable $u(t)$ and a single output variable $y(t)$. This general linear case is used to illustrate some features of the transformation process.

First we consider the special case where $m = 0$; that is, the right-hand side of Equation (7.22) contains no derivatives of the input function $u(t)$. (An example of this case is provided by the electrical circuit example, specifically Equation (7.1). The transformation here is particularly straighforward. Let

$$\begin{aligned} x_1(t) &= y(t) \\ x_2(t) &= y'(t) \\ &\;\;\vdots \\ x_N(t) &= y^{(N-1)}(t)\,. \end{aligned} \tag{7.23}$$

The state equations are then:

$$
\begin{aligned}
x_1'(t) &= x_2(t) \\
x_2'(t) &= x_3(t) \\
&\vdots \\
x_N'(t) &= -a_0 x_1(t) - a_1 x_2(t) \cdots - a_{N-1} x_N(t) + b_0 u(t)
\end{aligned}
\qquad (7.24(a))
$$

$$
\textit{with} \quad y(t) = x_1(t) \qquad (7.24(b))
$$

The more conventional compact form for Equation (7.24) is:[3]

$$
\begin{aligned}
\mathbf{x}'(t) &= \mathbf{A}\mathbf{x}(t) + \mathbf{b}u(t) \\
y(t) &= \mathbf{c}^T \mathbf{x}(t)
\end{aligned}
$$

where:

$$
\mathbf{A} = \begin{bmatrix}
0 & 1 & 0 & \ldots & 0 \\
0 & 0 & 1 & \ldots & 0 \\
. & . & . & . & . \\
. & . & . & . & . \\
. & . & . & . & . \\
0 & 0 & 0 & \ldots & 1 \\
-a_0 & -a_1 & -a_2 & \ldots & -a_{N-1}
\end{bmatrix}
$$

$$
\begin{aligned}
\mathbf{b}^T &= [0,0,0,\cdots 0,1] \\
\mathbf{c}^T &= [1,0,0,\cdots 0,0] \\
\mathbf{x}^T &= [x_1(t), x_2(t), \cdots x_N(t)]
\end{aligned}
$$

The initial conditions for the state equations of (7.24) follow directly from the assumptions following Equation (7.22) and the definitions of Equation (7.23).

Let's now consider the case where $m > 0$ in Equation (7.22). A specific example (with $m = 1$) can be obtained from the automobile suspension system model developed earlier if the nonlinear shock absorber is replaced with a linear device. In other words, if we replace the earlier specification for $f_b(t)$ with simply:

$$
f_b(t) = \Psi\, v(t), \quad \text{where } v(t) = y'(t) - u'(t),
$$

then Equation (7.2) can be written as

[3] We use the superscript T to denote the transpose of a vector or matrix.

$$y''(t) + a_1\, y'(t) + a_0\, y(t) = b_1\, u'(t) + b_0\, u(t)\,, \tag{7.25}$$

where $a_1 = b_1 = \Psi/M$, $a_0 = b_0 = k/M$.

Suppose the procedure we outlined earlier is applied; that is, we let $x_1(t) = y(t)$ and $x_2(t) = y'(t)$. The state space representation then becomes:

$$x_1'(t) = x_2(t) \tag{7.26a}$$
$$x_2'(t) = -\,a_1\, x_2(t) - a_0\, x_1(t) + b_1\, u'(t) + b_0\, u(\mathrm{t}) \tag{7.26b}$$

with
$$y(t) = x_1(t)\,. \tag{7.26c}$$

The perplexing outcome here is the explicit reference to the derivative of the input function that appears on the right-hand side of Equation (7.26b). It is not unreasonable to imagine cases of interest where $u(t)$ is not differentiable at all values of t in the observation interval. Recall that for the example that is under consideration, $u(t)$ corresponds to the road surface over which the automobile is travelling. A discontinuity in the road surface could correspond to a hole in the road as shown in Figure 7.11. Because of this discontinuity in $u(t)$, the derivative of $u'(t)$ does not exist at $t = t_b$. Does this mean that Equation (7.25) cannot be solved? Fortunately the answer is 'No'. The dilemma that we have encountered arises because of a poor choice of state variables.

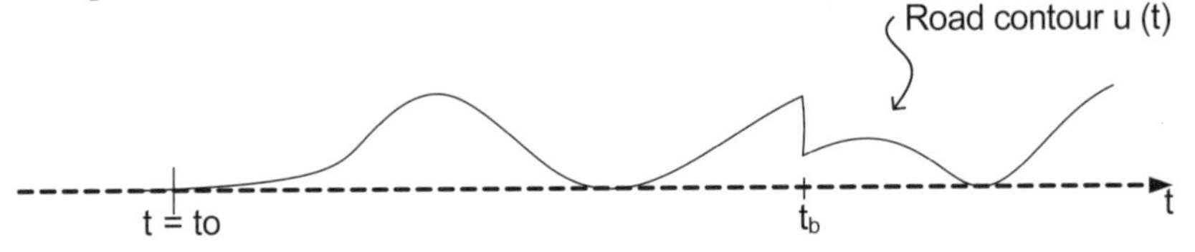

FIGURE 7.11. Discontinuous road surface.

As an alternative candidate for the state space representation, consider:

$$x_1'(t) = x_2(t) \tag{7.27a}$$

$$x_2'(t) = -a_0\, x_1(t) - a_1\, x_2(t) + u(t) \tag{7.27b}$$

with
$$y(t) = b_0\, x_1(t) + b_1 x_2(t)\,. \tag{7.27c}$$

This representation certainly has the desired feature of being independent of any derivatives of the input function, $u(t)$. But is it a valid representation? To confirm that it is, it must be possible to reconstruct the original continuous system model of Equation (7.25) from Equation (7.27) and this can, in fact, be achieved. The process involves straightforward mathematical manipulation that includes successively differentiating Equation (7.27c) and substitutions from Equations (7.27a) and (7.27b) to eliminate derivatives of the state variables $x_1(t)$ and $x_2(t)$.

There is, however, one further issue that needs to be addressed before Equation (7.27) can be accepted as a useful state space representation. This is the matter of initial conditions. Values are provided for $y(t_0)$ and $y'(t_0)$ and these have to be transformed into initial conditions for the state variables x_1 and x_2 so that Equation (7.27a) and Eq. (7.27b) can be solved. The necessary transformation can be developed using Equation (7.27c) together with the result obtained by differentiating Equation (7.27c) and substituting from Equation (7.27b). With t set to t_0 in the resulting equations, we get:

$$y(t_0) = b_0\, x_1(t_0) + b_1\, x_2(t_0) \tag{7.28a}$$

$$y'(t_0) - b_1\, u(t_0) = -a_0\, b_1\, x_1(t_0) + (b_0 - a_1\, b_1)\, x_2(t_0). \tag{7.28b}$$

Equation (7.28) provides two linear algebraic equations for the two unknowns $x_1(t_0)$ and $x_2(t_0)$. A sufficient condition for the existence of a solution to these equations is that the determinant *det* of the coefficient matrix on the right-hand side be nonzero. The value of the determinant is:

$$det = b_0^2 - a_1\, b_0\, b_1 + a_0\, b_1^2\,. \tag{7.29}$$

For the specific case of the (linearised) automobile suspension system, a_0, a_1, b_0, and b_1 have values previously specified (see Equation (7.25)). With these values substituted, $det = (k/M)^2$ and hence is nonzero. Consequently we can conclude that Equation (7.27) is a satisfactory state space representation for Equation (7.25) in that particular context.

In general, however, there is no guarantee that the value of *det* as given in Equation (7.29) is nonzero which means that there is a possibility that the state space representation of Equation (7.25) given by Equation (7.27) may not be acceptable. It can, for example, be easily shown that if $a_1^2 = 4a_0$ and $a_1\, b_1 = 2b_0$ then *det* is identically zero. It is reasonable therefore to wonder about the existence of another state space representation that circumvents this possible flaw. Such an alternative does exist and is given by:

$$x_1'(t) = -\, a_0\, x_2(t) + b_0\, u(t) \tag{7.30a}$$

$$x_2'(t) = x_1(t) - a_1\, x_2(t) + b_1\, u(t) \tag{7.30b}$$

with

$$y(t) = x_2(t)\,. \tag{7.30c}$$

Using the same procedure outlined earlier, it can be demonstrated that Equation (7.25) can be reconstructed from Equation (7.30) and hence Equation (7.30) is a valid representation for Equation (7.25). The equations for the initial conditions follow from Equation (7.30c) and Equation (7.30b) (setting $t = t_0$):

$$y(t_0) = x_2(t_0)$$

$$y'(t_0) - b_1\, u(t_0) = x_1(t_0) - a_1\, x_2(t_0)\,.$$

The determinant of the coefficient matrix for these two algebraic equations has the value –1 and consequently a solution for $x_1(t_0)$ and $x_2(t_0)$ always exists. Specifically:

$$x_1(t_0) = y'(t_0) + a_1\, y(t_0) - b_1\, u(t_0)$$
$$x_2(t_0) = y(t_0)\ .$$

The state space representation given in Equation (7.30) can be extended to general case of Equation (7.22). The form of this representation is given below.

$$\mathbf{x}'(t) = \mathbf{F}\,\mathbf{x}(t) + \mathbf{g}\,u(t)$$
$$y(t) = \mathbf{h}^{\mathrm{T}}\mathbf{x}(t) + b_N\,u(t)\ ,$$

where:

$$\mathbf{F} = \begin{bmatrix} 0 & 0 & 0 & \cdots & 0 & -a_0 \\ 1 & 0 & 0 & \cdots & 0 & -a_1 \\ 0 & 1 & 0 & \cdots & 0 & -a_2 \\ . & . & . & & . & . \\ . & . & . & & . & . \\ 0 & 0 & 0 & \cdots & 0 & -a_{N-2} \\ 0 & 0 & 0 & \cdots & 1 & -a_{N-1} \end{bmatrix}$$

$$\mathbf{g}^T = [b_0 - a_0 b_N, b_1 - a_1 b_N, \cdots, b_{N-1} - a_{N-1} b_N]$$
$$\mathbf{h}^T = [0,0,0,\cdots 0,1]$$
$$\mathbf{x}^T(t) = [x_1(t), x_2(t), \cdots x_N(t)]$$

The vector **g** is shown for the case where $m = N$ in Equation (7.22). The case where $m < N$ is accommodated by setting $b_k = 0$ for $k = (m + 1), (m + 2), \ldots, N$.

Our discussion in this section about the formulation of state space representations for CTDS models has been somewhat limited in scope. Nevertheless many of the key issues have been pointed out and a basis for dealing with them in a broader context has been provided.

7.5 References

7.1. Brauer, F. and Soudack, A.C., (1979), Stability regions and transition phenomena for harvested predator-prey systems, *Journal of Mathematical Biology*, **8**: 55–71.

7.2. Hall, C.A.S. and Day, J.W., (1977). *Ecosystem Modelling in Theory and Practice*, John Wiley, New York.

7.3. Iserles, A., (1996), *A First Course in the Numerical Analysis of Differential Equations*, Cambridge University Press, Cambridge, UK.
7.4. Lambert, J.D., (1991), *Numerical Methods for Ordinary Differential Equations*, Wiley, London.

Chapter 8 Simulation with CTDS Models

8.1 Overview of the Numerical Solution Process

8.1.1 The Initial Value Problem

An implicit requirement associated with modelling and simulation projects within the realm of CTDS models is a means for solving the differential equations embedded in the conceptual model. In very special cases these equations can fall into a category for which closed-form analytic solutions can be developed and this certainly has many advantages. Far more common, however, is the case where the features of the equations preclude such a solution approach. In such situations, numerical approximation procedures provide the only solution alternative. Our concern in this section is with exploring some of these numerical procedures. More specifically, our interest focuses on the means for solving the generic Equation (7.21a) of Chapter 7. (The companion equation (7.21b) is not relevant here because it simply represents a functional relationship defined on the state vector, $\mathbf{x}(t)$).

Our concern, therefore, is with numerical procedures for generating the solution of the equation

$$\mathbf{x}'(t) = \mathbf{f}(\mathbf{x}(t),t) \tag{8.1}$$

over the observation interval $I_0 = [t_0, t_f\ ]$ where t_0 , t_f , and $\mathbf{x}(t_0) = \mathbf{x}_0$ are assumed to be explicitly given. (Note that the explicit dependence of the derivative function f on $u(t)$ that appears in Equation (7.21a) has been suppressed in this representation; the role of $u(t)$ has been merged into the explicit dependence on t.) In general, $\mathbf{x}$ and $\mathbf{f}$ in Equation (8.1) are vectors of dimension N.

The problem stated above is commonly called the initial value problem (IVP). It is distinct from a closely related problem called the boundary value problem (BVP). In both problems at least N pieces of data about the solution are known. In the case of the IVP these are the N components of the N-vector $\mathbf{x}_0$. The situation in the case of the BVP is different because the known values do not occur at the same value of t. The available data could, for example, be:

$$x_1(t_0), x_2(t_0), x_3(t_0), \ldots, x_\eta(t_0), x_{\eta+1}(t_f), \ldots, x_N(t_f),$$

where $1 < \eta \leq N$.

Although a CTDS model almost always incorporates more than one first-order differential equation (i.e., the dimension of the state vector $\mathbf{x}(t)$ is greater than 1), this higher dimensionality introduces unnecessary notational complexity when examining numerical solution methods. Consequently, without loss of generality, we take $N = 1$ throughout most of the discussion that follows.

8.1.2 Existence Theorem for the IVP

The search for a solution of any problem can be undertaken with considerably more confidence when there is assurance that a solution to the problem does indeed exist. With respect to the solution of Equation (8.1) this issue has been extensively studied and substantial knowledge is available. We summarise here some of the most significant results in this regard.

As might be expected, it is the characteristics of the derivative function $f(x,t)$ which play a pivotal role in identification of existence conditions for the solution of Equation (8.1). Our focus therefore is restricted to a function $f(x,t)$ that has two particular features. These are:

a) $f(x,t)$ is defined and continuous in the strip $-\infty < x < \infty$, $t_0 \leq t \leq t_f$, with t_0 and t_f finite
b) There exists a constant L such that for any $t \in [t_0, t_f]$ and any two numbers α and β

$$|f(\alpha, t) - f(\beta, t)| \leq L\,|\alpha - \beta|\ .$$

[a) and b) are called the *Lipschitz conditions* and L is called the *Lipschitz constant*].

Theorem A
Let $f(x\ t)$ satisfy (a) and (b) and let x_0 be any number. Then there exists exactly one function $X(t)$ with the following properties.

i. $X(t)$ is continuous and differentiable for $t \in [t_0, t_f]$.
ii. $X'(t) = f(X(t), t)$ for $t \in [t_0, t_f]$.
iii. $X(t_0) = x_0$.

Remark 1
Theorem A states that under the assumed conditions on the derivative function $f(x,t)$, the IVP of Equation (8.1) not only has a solution, but it is unique.

Remark 2

Suppose $f(x,t)$ has a continuous derivative with respect to x which is bounded in the strip in question (see above); then assumption (a) of Theorem A follows directly whereas (b) follows as a consequence of the mean value theorem; hence the two assumptions of Theorem A are satisfied.

Remark 3

Unless otherwise noted, we assume throughout the remainder of this chapter that the conditions of Theorem A hold. Furthermore we call the function $X(t)$ referred to in Theorem A, the 'true solution' to the IVP under consideration. In some limited circumstances, the true solution may be available as an explicit analytic function. In such circumstances, an 'exact' value can be obtained for the true solution at any value of t within the observation interval I_O, at least to the extent of the precision limitations inherent in the evaluation of the function in question.

8.1.3 What Is the Numerical Solution to an IVP?

A numerical solution to an IVP is a finite set of points; that is,

$$\{(t_n, x_n); n = 0, 2, \ldots, M\},$$

where :

- (t_0, x_0) is the given initial condition.
- x_n is a generated numerical approximation for the true solution value at $t = t_n$, that is, an approximation to $X(t_n)$.
- $t_{n+1} = t_n + h_n$ for $0 \leq n \leq M - 1$ and $t_M = t_f$.

Here h_n is called the step size at t_n. If h_n remains invariant for all values of n, then the solution process is said to be of fixed step size, otherwise it is of variable step size. As becomes apparent in the discussion below, the step size is a critical parameter in the solution process. As might be expected, its value plays a central role in the accuracy of the results obtained. The issues associated with step-size selection are the following: if the step size is to be fixed, then how should the value be selected and if it is to be variable, then what is the procedure for making changes? These are not easy questions to answer but some insight is provided in the discussion that follows. Decisions relating to step-size have to be made by the user of most simulation environments for CTDS models; consequently some familiarity with the underlying issues is essential.

The various notions discussed above are illustrated in Figure 8.1.

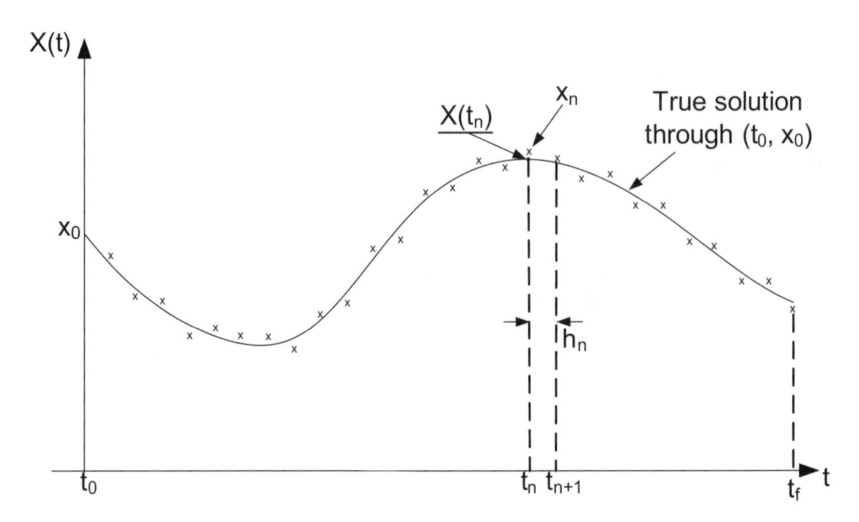

FIGURE 8.1. Numerical solution to an IVP.

An important feature of Figure 8.1 is the implication that the numerical solution rarely coincides with the true solution. In fact, it is only at the starting point $t = t_0$ where there is certainty that the numerical value is identical to the true value. All other numerical values are, in general, different from the true value. This difference, that is, the error, has two basic origins:

a) Truncation (or discretisation) error.

   –This is a property of the solution method.

b) Round-off error.

   –This is a property of the computer program used to implement the solution method. It arises because of the finite precision in number representation.

Although not apparent from Figure 8.1, it is important to appreciate that with all numerical solution methods, each new solution estimate is generated using information from previously generated solution values; in other words it is constructed from data that may already have significant error. This somewhat disturbing fact sets the stage for the propagation of error that, in turn, can lead to instability. In other words, there is the possibility that the size of the error will grow, in an unbounded manner, as new solution values are generated.

Stability is one of several important attributes that can be associated with any solution method. Others are:

- Order (this is closely related to the notion of truncation error introduced earlier).
- Accuracy (this is a reference to the correspondence between the true solution and the numerical solution).
- Local efficiency (this is a measure of the computational effort required to move the generated solution forward from $t = t_n$ to $t = t_{n+1}$; it is typically measured in terms of the number of evaluation of the derivative function f).

In the discussion that follows, we explore these various matters that have vital importance to the simulation phase of a modelling and simulation study within the CTDS realm.

8.1.4 Comparison of Two Preliminary Methods

The Euler Method

The Euler method is the most fundamental of the wide range of approaches that are available for the numerical solution of the IVP. The underlying concept is shown in Figure 8.2.

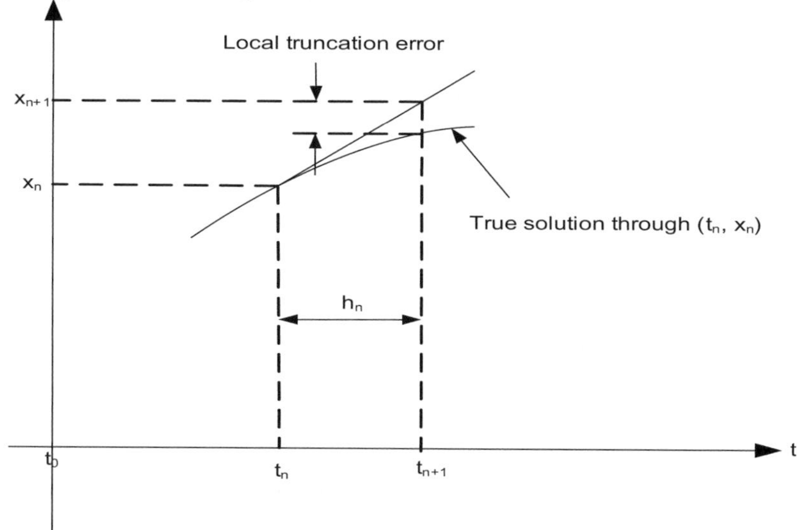

FIGURE 8.2. The Euler method.

The assumption in Figure 8.2 is that the solution process has progressed to $t = t_n$ and the solution value generated at $t = t_n$ is x_n. We denote by f_n the slope of the true solution through the point (t_n, x_n); that is,

$$f_n = f(x_n, t_n) .$$

The solution approximation at $t = t_{n+1} = t_n + h_n$ associated with the Euler method is:

$$x_{n+1} = x_n + h_n f_n . \tag{8.2}$$

Although the approach here is intuitively appealing, it can also be viewed as an approximation arising from the definition of a derivative, namely, from the definition that:

$$x'(t) = \frac{dx}{dt} = \lim_{\Delta \to 0} \frac{x(t+\Delta) - x(t)}{\Delta} .$$

The update formula of Equation (8.2) is then obtained by ignoring the requirement for Δ to approach zero and by making the following associations: $t = t_n$, $\Delta = h_n$, $x(t_n) = x_n$, $x(t_n + h_n) = x_{n+1}$ and $x'(t_n) = f_n$.

The Modified Euler Method (or Trapezoidal Rule)

The Euler method 'moves forward' on the basis of a single derivative function evaluation. This value (namely, $f(x_n, t_n)$) is the slope of the true solution that passes through the solution estimate (t_n, x_n). But because (t_n, x_n) is not generally on the desired solution (the one through (t_0, x_0)) it is reasonable to conjecture that some other slope value might be a better choice. The Modified Euler method creates such an alternate choice by first evaluating the derivative function at the solution estimate produced by the Euler method and then taking an average of the two slope values that are thus available. More specifically,

- Take an Euler step to produce the value $p_{n+1} = (x_n + h_n f_n)$ at $t = t_{n+1}$.
- Let $F_{n+1} = f(p_{n+1}, t_{n+1})$.
- Choose the solution estimate at $t = t_{n+1}$ to be:

$$x_{n+1} = x_n + h_n (f_n + F_{n+1})/2. \tag{8.3}$$

One difference between solution estimates from the Euler and the Modified Euler methods (given by Equations (8.2) and (8.3), respectively) is that the former requires only one derivative function evaluation whereas the latter requires two. It is natural, therefore, to expect some advantage from the added effort. An advantage is certainly present and it is realised in terms of superior error performance, at least at the level of local behaviour. This feature can be explored by examining the Taylor series expansion of the true solution of Equation (8.1) that goes through the point (t_n, x_n). As earlier, we denote this particular solution by $X_n(t)$. A Taylor series expansion gives:

$$X_n(t_n + \delta) = X_n(t_n) + \delta X_n'(t_n) + ½ \delta^2 X_n''(t_n) + O(\delta^3). \tag{8.4}$$

But:

$$X_n(t_n) = x_n$$

and

$$X_n'(t_n) = f(X_n(t_n), t_n) = f(x_n, t_n) = f_n .$$

Then, if we set $\delta = h_n$ in Equation (8.4), we get:

$$X(t_{n+1}) = x_{n+1}^E + O(h_n^2) ,$$

where x_{n+1}^E is the Euler solution estimate of Equation (8.2). This result demonstrates that the Euler method has a local truncation error that is of order h_n^2 which in turn implies that the Euler method is a first-order method.

A similar analysis with the Modified Euler method gives the result that:

$$X(t_{n+1}) = x_{n+1}^{ME} + O(h_n^3) ,$$

where x_{n+1}^{ME} is the solution estimate provided by Equation (8.3). Thus the local truncation error of the Modified Euler method is of order h_n^3 and the method is a second-order method. This, in particular, demonstrates that the additional derivative function evaluation required by the Modified Euler method provides the benefit of realising a higher-order method.

The order of a solution method is one of its most important characterising features. As illustrated above, this feature relates to the nature of the error between the solution value produced by the method over a single step relative to the true solution, when both begin at a common starting point. This error estimate evolves from the Taylor series expansion of the true solution around the starting point and reflects the degree of correspondence between the series expansion and the generated solution value.

A practical interpretation of the meaning of a solution method of order r is that such a method generates solution values that have zero error for the case where the true solution is a polynomial of order r or less (provided zero round-off error is assumed). Thus a second-order method will produce exact solution values for the IVP:

$$x'(t) = a_1 + 2\, a_2 (t - t_0); \quad x(t_0) = a_0$$

because the solution to this equation is the quadratic function:

$$x(t) = a_0 + a_1 (t - t_0) + a_2 (t - t_0)^2 .$$

8.2 Some Families of Solution Methods

The most common numerical solution methods for the IVP fall into two broad classes. We examine each of these in turn beginning with the Runge–Kutta family.

8.2.1 The Runge–Kutta Family

There are two representations for the Runge–Kutta family and these are referred to as explicit and implicit representations. We restrict our considerations to the explicit representation. The explicit s-stage Runge–Kutta formula is given below:

$$x_{n+1} = x_n + h \sum_{i=1}^{s} b_i g_i \,, \tag{8.5}$$

where:

$$\begin{aligned}
g_1 &= f(x_n, t_n) \\
g_2 &= f(x_n + h\, a_{21}\, g_1, t_n + c_2\, h) \\
g_3 &= f(x_n + h\, (a_{31}\, g_1 + a_{32}\, g_2), t_n + c_3\, h) \\
&\;\;\vdots \\
g_s &= f(x_n + h\, (a_{s1}\, g_1 + a_{s2}\, g_2 + \ldots\; a_{s\,s-1}\, g_{s-1}), t_n + c_s\, h)\,.
\end{aligned}$$

Remarks

- The s-stage formula requires s evaluations of the derivative function f to advance one step (of length h) along the t-axis.
- The s-stage formula has $S = (s^2 + 3s - 2)/2$ free parameters, namely, the collection of b_i's, a_{ij}'s, and c_k's. Numerical values for these parameters are determined by a procedure that undertakes to establish an equivalence between the computed value x_{n+1} and the first r terms in a Taylor series expansion for the true solution $X(t)$ passing through (t_n, x_n). This creates a formula of order r. It is always true that $r \le s$. In essentially all cases, there are many ways to select values for the S parameters in order to achieve a formula of order $r \le s$.
- The general formula given in Equation (8.5) is explicit because the solution value x_{n+1} evolves directly without the need for the resolution for further numerical issues. Observe also that no past solution information is needed to generate x_{n+1}. As we show in the discussion that follows, these features are not always provided by other methods.

It is interesting to observe that both methods introduced earlier in Section 8.1.4 are members of the Runge–Kutta family. The first-order Euler method corresponds to the case where $s = 1$, and $b_1 = 1$ and the second-order Modified Euler method corresponds to the case where $s = 2$, $b_1 = b_2 = 1/2$, $a_{21} = c_2 = 1$. An alternate second-order method, frequently called the Heun form, is given by $s = 2$, $b_1 = 1/4$, $b_2 = 3/4$, $a_{21} = c_2 = 2/3$.

Third- and fourth-order Runge–Kutta formulas are often used and a representative of each of these classes is provided below. The third-order formula given below is often called the Heun form.

$$x_{n+1} = x_n + \frac{1}{4} h\,[g_1 + 3\,g_3]$$

$$g_1 = f(x_n, t_n)$$

$$g_2 = f(x_n + \frac{1}{3} h\,g_1, t_n + \frac{1}{3} h)$$

$$g_3 = f(x_n + \frac{2}{3} h\,g_2, t_n + \frac{2}{3} h)\,.$$

The fourth-order formula given below is often called the Kutta (or the 'classic') form.

$$x_{n+1} = x_n + \frac{1}{6} h\,[g_1 + 2\,g_2 + 2\,g_3 + g_4]$$

$$g_1 = f(x_n\,, t_n)$$

$$g_2 = f(x_n + \frac{1}{2} h\,g_1, t_n + \frac{1}{2} h)$$

$$g_3 = f(x_n + \frac{1}{2} h\,g_2, t_n + \frac{1}{2} h)$$

$$g_4 = f(x_n + h\,g_3, t_n + h)\,.$$

8.2.2 The Linear Multistep Family

Specific methods in this family are constructed from the following generic formula,

$$x_{n+1} = \sum_{i=1}^{k} \alpha_i x_{n+i-1} + h \sum_{i=0}^{k} \beta_i f_{n+i-1}, \tag{8.6}$$

where $f_j = f(x_j\,, t_j)$. Notice that an essential difference from the Runge–Kutta family is the reliance on past values of the numerical solution and on the slope of the solution at those values, that is, on the derivative function f evaluated at those past values. Several special cases can be identified:

a) If $k = 1$, then we have a single-step method (reliance on past values is restricted to values at the current time point t_n.
b) If $\beta_0 = 0$, then we have an explicit/open/predictor method; if $\beta_0 \neq 0$, then we have an implicit/closed/corrector method.

The implicit case ($\beta_0 \neq 0$) gives rise to a 'circular' situation where the generation of the solution value, that is, x_{n+1}, requires data that directly depend on x_{n+1}, namely, f_{n+1}. This introduces an accessory problem that needs to be addressed before a practical solution procedure is realised.

It is important also to observe here that the dependence of linear multistep methods on past solution values implies a fundamental shorcoming, namely, a 'start-up' problem. Past solution values are needed to initiate the solution procedure and these can only be obtained by reliance on some ancillary method that is not similarly constrained. Typically Runge–Kutta methods are used in practice to provide these preliminary values.

8.2.2.1 Predictor–Corrector Methods

The predictor–corrector methods represent the standard implementation approach for the linear multistep family. The underlying idea is to first use an explicit formula to project forward (i.e., to predict) a solution value estimate and then, as a second refinement (or corrector) step, an implicit formula is used to create a tentative solution value. This tentative value may or may not be accepted; in the latter case one or more iterations may follow. This procedure, in effect, deals with the underlying issue introduced by the implicit formula.

Values for the coefficients in a linear multistep formula of order r are established via the same approach used to develop specific members of the Runge–Kutta family, namely, by establishing an equivalence between the computed value x_{n+1} and the first r terms in a Taylor series expansion for the true solution $X(t)$ passing through (t_n, x_n). As an example, we give the formulas for the Adams fourth-order predictor–corrector process:

Predictor (Adams–Bashforth)

$$x_{n+1} = x_n + h(55 f_n - 59 f_{n-1} + 37 f_{n-2} - 9 f_{n-3})/24 . \tag{8.7a}$$

Corrector (Adams–Moulton)

$$x_{n+1} = x_n + h\,(9 f_{n+1} + 19 f_n - 5 f_{n-1} + f_{n-2})/24 . \tag{8.7b}$$

Notice that the predictor is an explicit formula whereas the corrector is implicit. The associated procedure is summarised below.

a) Use Equation (8.7a) to generate: $x_{n+1}^{(p)}$.

b) Use Equation (8.7b) to generate: $x_{n+1}^{(c)}$ (with $x_{n+1}^{(p)}$ used to compute f_{n+1}).

If $|x_{n+1}^{(c)} - x_{n+1}^{(p)}| < \epsilon$ then $x_{n+1} = x_{n+1}^{(c)}$; otherwise do step (c).

c) Replace $x_{n+1}^{(p)}$ with $x_{n+1}^{(c)}$ and repeat from step (b).

Here ϵ is a predefined operational parameter that provides accuracy control; its value is usually set by the user. Note also that if the error check at step (b) is successful on the first iteration then the new solution value x_{n+1} is generated after only two derivative function evaluations (assuming past derivative values have been stored). This is a significant improvement over the four evaluations required by a fourth-order Runge–Kutta method. In other words this predictor–corrector method is potentially significantly more efficient (in terms of derivative function evaluations) than a Runge–Kutta method of like order. This can have important consequences in a simulation project where the conceptual model has many equations and/or the derivative functions are particularly complex.

8.3 The Variable Step-Size Process

Thus far our discussion has implicitly assumed that the step-size h used in the numerical solution procedure for the IVP remains invariant. Such solution procedures are certainly widely used in the simulation (i.e. experimentation) phase of modelling and simulation projects that have a CTDS context. Such procedures can, however, be inefficient because the nature of the solution may be such that a small value of h is required only over a minor portion of the observation interval whereas larger values can be used elsewhere without the danger of compromised solution quality. This gives rise to the need for automatic step-size adjustment.

The realisation of such a variable step-size procedure needs to address two basic issues: how to determine when a step-size change is needed (either increase or decrease) and how to carry out a meaningful change in the value of the step-size. It is reasonable to assume that the criterion for step-size change ought to be based on the size of the local truncation error (or an estimate of this error, say, E_{est}) relative to some (user-specified) error tolerance E_{tol}.

The specification of a variable step-size process within such a context can be summarised as shown in Figure 8.3. Each repetition of the process moves the solution forward by one time step and each begins with the current solution value (t_n, x_n) and a nominal step-size h_n.

1. Compute a solution estimate x_{n+1} at $t_{n+1} = t_n + h_n$
2. Compute E_{est}, an estimate of the magnitude of the local truncation error at x_{n+1}
3. Compute E_{tol}, the upper bound for the admissible value for E_{est}
4. If $E_{est} > E_{tol}$, then
   - 4.1 reduce the value of h_n

   else
   - 4.2 accept the solution estimate x_{n+1} and set $t_{n+1} = t_n + h_n$
   - 4.3 compute a "best estimate" for the next step-size, h_{n+1}
   - 4.4 replace n+1 with n
5. Repeat from step 1

FIGURE 8.3. The variable step-size process.

Several of the steps in Figure 8.3 require some elaboration and this is provided below.

Step 2: Obtaining an estimate for the local truncation error is a key aspect of the variable step-size process. A variety of approaches has emerged but for the most part their comprehensive development depends on the exploration of issues in numerical mathematics that are beyond the scope of this textbook (the interested reader can find relevant discussion in [8.5, 8.14]). The general nature of a few of these approaches is provided in the brief summaries given in the discussion that follows.

a) The half-step approach
The idea here is to obtain two estimates for the solution at time t_{n+1}: the first obtained on a single step with step-size h_n and the other obtained using two half-steps, each of size $h_n/2$. If we denote these two solution estimates x_{n+1} and x^*_{n+1}, respectively, then (with certain assumptions) it can be shown that a reasonable estimate of the local truncation error at t_{n+1} is:

$$E_{est} = \lambda_r (x^*_{n+1} - x_{n+1}) \ ,$$

where $\lambda_r = 2^r/(2^r - 1)$ and r is the order of the solution method. A notable feature of this approach is that it has general applicability inasmuch as it is not linked to any particular solution method. Clearly a significant disadvantage is a substantial efficiency penalty because there is a threefold increase in the number of derivative function evaluations that would otherwise be required to advance the solution by one step.

b) The embedded approach
A good illustration of this approach is provided by the Runge–Kutta–Fehlberg method that is given in Equation (8.8). The underlying idea here is the development of two Runge–Kutta formulas that differ in order by

one and can be constructed from a shared collection of derivative function evaluations. In the Runge–Kutta–Fehlberg method two solution estimates x_{n+1} and x^*_{n+1} of order 4 and 5, respectively, are produced at each step. Their difference $(x^*_{n+1} - x_{n+1})$ provides a good estimate of the local truncation error in the lower-order result. Notice that six derivative function evaluations are required and if the fourth-order result is used, then there is a 50% overhead incurred here in obtaining the error estimate, relative to the 'classic' fourth-order Runge–Kutta formula given earlier.

$$
\begin{aligned}
x_{n+1} &= x_n + h\left[\frac{25}{216}g_1 + \frac{1408}{2565}g_3 + \frac{2197}{4104}g_4 - \frac{1}{5}g_5\right] \\
x^*_{n+1} &= x_n + h\left[\frac{16}{135}g_1 + \frac{6656}{12825}g_3 + \frac{28561}{56430}g_4 - \frac{9}{50}g_5 + \frac{2}{55}g_6\right] \\
g_1 &= f(x_n, t_n) \\
g_2 &= f(x_n + \frac{h}{4}g_1, t_n + \frac{h}{4}) \\
g_3 &= f(x_n + \frac{h}{32}(3g_1 + 9g_2), t_n + \frac{3h}{8}) \\
g_4 &= f(x_n + \frac{h}{2197}(1932g_1 - 7200g_2 + 7296g_3), t_n + \frac{12}{13}h) \\
g_5 &= f(x_n + h(\frac{439}{216}g_1 - 8g_2 + \frac{3680}{513}g_3 - \frac{845}{4104}g_4), t_n + h) \\
g_6 &= f(x_n + h(-\frac{8}{27}g_1 + 2g_2 - \frac{3544}{2565}g_3 + \frac{1859}{4104}g_4 - \frac{11}{40}g_5), t_n + \frac{1}{2}h) \\
E_{est} &= x^*_{n+1} - x_{n+1} = h\left[\frac{1}{360}g_1 - \frac{128}{4275}g_3 - \frac{2197}{75240}g_4 + \frac{1}{50}g_5 + \frac{2}{55}g_6\right]
\end{aligned}
\tag{8.8}
$$

c) A predictor–corrector approach

As the name suggests, this approach is specific to predictor–corrector methods. With suitable assumptions, the underlying analysis shows that a reasonable estimate of the local truncation error at t_{n+t} has the form:

$$E_{est} = \lambda_r (x_{n+1} - x^*_{n+1}),$$

where x_{n+1} and x^*_{n+1} are the corrector and predictor values, respectively (necessarily of the same order), and the constant λ_r is dependent on the order of the method.

Step 3: A standard format for the bound on the local truncation error is: $E_{tol} = (K_1 + K_2 |x_{n+1}|)$ where K_1 and K_2 are user-specified parameters. The first term (K_1) provides an 'absolute' contribution to the tolerance bound and the second term $(K_2 |x_{n+1}|)$ provides a relative contribution; that is, if the solution value itself is large, then the error tolerance increases.

Step 4.1: The result of the analysis leading to a meaningful formula for reducing the value of h_n is surprising simple (the analysis itself, however, is outside the scope of our present interest; relevant discussion can be found in [8.14]). The general form of the update formula is:

$$h_n \Leftarrow c\,\left(\frac{E_{tol}}{E_{est}}\right)^{\frac{1}{r+1}} h_n \;, \tag{8.9}$$

where r is the order of the solution value x_{n+1} and c is a 'safety factor' that is typically incorporated (a common value is 0.9). A reduction in size results because $E_{est} > E_{tol}$ at Step 4.1.

Step 4.3: The situation represented at this step corresponds to the case where $E_{est} \leq E_{tol}$. This can be interpreted as reflecting a step-size that is overly conservative and therefore could possibly be increased on the subsequent phase of the solution process. The underlying analysis shows that the appropriate update formula for h_n is again given by Equation (8.9).

8.4 Circumstances Requiring Special Care

Thus far in this chapter we have explored features of the most important numerical tools commonly used to solve the IVP and hence to carry out simulation studies with CTDS models. As with all tools these likewise have inherent limitations and restrictions on their applicability and it is prudent for tool users to be aware of these. Our goal in this section is to provide some insight into this important topic.

8.4.1 Stability

The notion of stability is concerned with the existence of upper bounds on the magnitude of the step-size h used in the solution-generating process. In-depth investigation of this important feature is, of necessity, carried out in the context of linear systems because extensive analysis is possible only in this restricted context. Nevertheless these results can often be extended to the general case of nonlinear models by observing that linear approximations can be constructed for nonlinear models around any particular point on the solution trajectory. Although relevance of such approximations is restricted to a small region about the chosen point, useful insights into behaviour can nevertheless be obtained.

The essential point can be illustrated by considering the following simple linear IVP.

$$\begin{aligned} u'(t) &= -c_1 u(t) + v(t); \quad u(0) = \alpha_1 \\ v'(t) &= -c_2 v(t); \quad v(0) = \alpha_2 \ , \end{aligned} \tag{8.10}$$

where c_1 and c_2 are positive constants. It can be easily verified (e.g., by direct substitution) that the true solution of Equation (8.10) is:

$$\begin{aligned} u(t) &= (\alpha_1 + \gamma)\exp(-c_1 t) - \gamma \exp(-c_2 t) \\ v(t) &= -\alpha_2 \exp(-c_2 t) \ , \end{aligned}$$

where $\gamma = \alpha_2/(c_1 + c_2)$. Observe that both $u(t)$and $v(t)$ approach 0 as $t \to \infty$ independent of the specific values chosen for c_1, c_2, α_1, and α_2.

Suppose now that a fixed step-size Euler method is applied to generate a numerical solution to Equation (8.10). The iterative process that results can be expressed as

$$\begin{aligned} u_{n+1} &= u_n + h\,(-c_1 u_n + v_n) = (1 - c_1 h)\, u_n + h\, v_n \\ v_{n+1} &= v_n + h\,(-c_2 v_n) = (1 - c_2 h)\, v_n \ . \end{aligned}$$

Clearly if the numerical solution is to have any credibility whatsoever, a fundamental requirement is that both $u_n \to 0$ and $v_n \to 0$ as $n \to \infty$. The necessary and sufficient conditions for this to occur are:

$$\begin{aligned} &|1 - c_1 h| < 1; \quad \text{that is, } -1 < (1 - c_1 h) < 1 \\ \text{and} \quad &|1 - c_2 h| < 1; \quad \text{that is, } -1 < (1 - c_2 h) < 1 \ , \end{aligned}$$

which, in turn, implies: $h < \min[2/c_1, 2/c_2]$. In other words, there is a very practical constraint on how large a value can be assigned to the step-size h. If this upper bound is exceeded, then the numerical solution is simply unstable and has no relationship to the true solution.

This result clearly raises several important questions; for example, are all solution methods subject to such step-size constraints and is there anything special (generalisable) about the nature of the specific constraint obtained above? With respect to the first of these questions, it is certainly true that such a constraint does exist for all members of the Runge–Kutta family. However, the constraint does not apply to all solution methods. This can be illustrated by considering a method called the backward Euler method which is a special case of the linear multistep family given in Eq. (8.6). This method is a single-step implicit method ($k = 1$ and $\beta_0 \neq 0$). The updating formula for the backward Euler method is:

$$x_{n+1} = x_n + f(x_{n+1}, t_{n+1}) \ .$$

When this formula is applied to our test case of Equation (8.10), the iterative process that results is:

$$u_{n+1} = \frac{u_n}{(1+c_1 h)} + \frac{h\,v_n}{(1+c_1 h)(1+c_2 h)}$$

$$v_{n+1} = \frac{v_n}{(1+c_2 h)}$$

It is easy to conclude here that the necessary and sufficient conditions to ensure that both $u_n \to 0$ and $v_n \to 0$ as $n \to \infty$ are:

$$|1 + c_1 h| > 1 \quad \text{and} \quad |1 + c_2 h| > 1.$$

Both these conditions are satisfied for any (positive) value of h (recall our original assumption that both c_1 and c_2 are positive). Hence we have an example of a method that does not place a bound on the size of the step-size h.

Let's return now to our earlier observation of the instability that results when an unacceptably large value of step-size is used to solve Equation (8.10) with the Euler method. Are there more general conclusions that can be identified? The answer most certainly is 'Yes'. To proceed, we generalise our test case to an IVP that is the set of N linear first equations; that is,

$$\mathbf{x}'(t) = \mathbf{A}\,\mathbf{x}(t) \tag{8.11}$$

with $\mathbf{x}(t_0) = \mathbf{x}_0$. We assume here the simplest case where the $N \times N$ coefficient matrix $\mathbf{A}$ has real, distinct, and negative eigenvalues.[1] In this circumstance, it can be shown that the true solution of Equation (8.11) approaches zero independent of the initial value $\mathbf{x}_0$. If the Euler method is used to generate the solution of Equation (8.11) then it can be demonstrated that the stability requirement (namely, the requirement that the computed solution likewise approaches zero), is $h < 2/|\lambda_{max}|$ where λ_{max} is the largest (in absolute value) of the eigenvalues of $\mathbf{A}$. We leave as an exercise for the reader to confirm that our earlier stability conclusion for the special case of Equation (8.10) is entirely consistent with this general result. (Hint: show that the eigenvalues of the coefficient matrix in Equation (8.10) are $-c_1$ and $-c_2$.)

The general result above is restricted to the most fundamental of the methods in the Runge–Kutta family. One might reasonably wonder about the nature of the stability requirement for other members of this family.

[1] The eigenvalues of the $N \times N$ matrix $\mathbf{A}$ are the N solutions, $\lambda_1, \lambda_2, \ldots, \lambda_N$ to the equation $\det(\lambda\mathbf{I} - \mathbf{A}) = 0$, where det( ) represents the determinant.

This is a topic that has been extensively investigated in the numerical mathematics literature and information can be found in textbooks such as [8.8], [8.9], and [8.12]. In this regard, we note that the stability bound for the fourth-order Kutta form given earlier is $h < 2.78/|\lambda_{max}|$ under the assumed conditions on the coefficient matrix **A** in Equation (8.11).

8.4.2 Stiffness

Stiffness is a property of some CTDS models. It is of particular importance because it interacts with the step-size constraint that is intrinsic to many numerical solution methods in a manner that seriously deteriorates the efficiency of the solution process. The background prerequisites for a comprehensive presentation of the topic are substantial and hence its treatment is beyond the scope of this textbook. Nevertheless, the essential nature of the problem can be readily illustrated by examining a straightforward example. (The interested reader is encouraged to explore the issue in the numerical mathematics literature, e.g., [8.5].)

Consider the following two simple linear CTDS models.

Model A:

$$\begin{aligned} u'(t) &= -u(t) + 2; u(0) = 0 \\ v'(t) &= -v(t) + 2;\ v(0) = 0 \quad . \end{aligned} \tag{8.12}$$

Model B:

$$\begin{aligned} p'(t) &= -500.5\,p(t) + 499.5\,q(t) + 2;\ \ p(0) = -0.1 \\ q'(t) &= -499.5\,p(t) - 500.5\,q(t) + 2;\ \ q(0) = 1 \quad . \end{aligned} \tag{8.13}$$

It is easy to confirm (e.g., by direct substitution) that the solution to Equation (8.12) is:

$$u(t) = v(t) = 2\,(1 - \exp(-\,t)) \tag{8.14}$$

and that the solution to Equation (8.13) is:

$$\begin{aligned} p(t) &= u(t) - \delta(t) \\ q(t) &= v(t) + \delta(t) \ , \end{aligned}$$

where $\delta(t) = 0.1\ \exp(-1000\ t)$. Observe that the solutions to Equations (8.12) and (8.13) are essentially identical for $t > 0.02$ because $\delta(t)$ has almost vanished.

It's now important to consider what might constitute a reasonable value for the right boundary of the observation interval I_0 for these two simple models (the left boundary has already been set to 0). This can easily be inferred from Equation (8.14) from which it is apparent that the solution in both cases is dominated by the term exp($-t$) which tends towards zero as t increases. Inasmuch as this term has effectively vanished after $t = 10$, a reasonable choice for the right boundary of I_0 is 10. In other words, it is unlikely that an interest in the behaviour of either of these models would extend beyond $t = 10$.

Let's now examine what impact the stability constraint of a numerical solution method would have. On the basis of our earlier considerations, let's assume a constraint of the form $h < K/|\lambda_{max}|$ where K could be in the range between 2 and 3. To proceed we require the eigenvalues of the two linear models given in Equations (8.12) and (8.13). For model A it is easily seen that the two eigenvalues of the coefficient matrix are both equal to –1. For model B it can be demonstrated that the eigenvalues are –1 and –1000. The surprising result that now flows from the stability constraint is that even though the true solutions for both models are 'almost' identical (at least for $t > 0.02$) a maximum step-size of K would be permitted in studying model A whereas the step-size would have to be restricted to less than $K/1000$ when studying model B! Apart from the computational burden that is thus imposed upon the investigation of model B, the unavoidable roundoff errors that could accumulate during the relatively large number of steps needed to traverse the observation interval could seriously deteriorate solution quality. The study of model A would not encounter either of these difficulties.

This rather unexpected result has its origins in the wide separation between the largest and smallest eigenvalues of model B. This property is called *stiffness*. As might be expected, it has been extensively studied in the numerical mathematics literature and a considerable body of knowledge about it has emerged, for example, [8.7] and [8.5]. These studies are often in the context of linear systems because of the convenience of analysis that linearity provides. The phenomenon nevertheless does arise in nonlinear systems which can always be linearly approximated in suitably small regions. The underlying difficulty arises simply because the smallest (in magnitude) eigenvalue generally determines the right boundary of the observation interval whereas the largest (in magnitude) eigenvalue can introduce a size constraint on the step-size h. As we have illustrated above, these two effects have conflicting and undesirable impacts on the numerical solution process.

It needs to be stressed, however, that solution methods specifically designed to accommodate stiffness have been developed and should be used in any simulation experiment where there is a possibility that the CTDS model may exhibit stiffness (see, e.g., [8.5]). These methods do involve additional computational overhead and are not recommended for general usage.

One might be tempted to conjecture that stiff systems are no more than curiosities intended mainly to provide a platform for mathematical analysis. It is easy to demonstrate that this is not the case. Consider, for example, the automobile suspension system that was introduced in Section 7.2.2 and subsequently linearised in Equation (7.25). Suppose we assign the specific values $k = 0.5$ (newtons/m) and $M = 0.5$ (kg) to the spring constant and the mass, respectively; then Equation (7.25) becomes:

$$y''(t) + 2\,\psi\, y'(t) + y(\mathrm{t}) = 2\,\psi\, u'(t) + u(t) \;,$$

where ψ (kg/sec) is the stiffness parameter of the shock absorber. The state variable representation of Equation (7.17) becomes:

$$x_1{}'(t) = -x_2(t) + u(t)$$
$$x_2{}'(t) = x_1(\mathrm{t}) - 2\,\psi\, x_2(t) + 2\,\psi\, u(t)$$

with $\quad y(t) = x_2(t)$.

It can be easily established that the two eigenvalues λ_1 and λ_2 of the coefficient matrix are the solutions to the algebraic equation:

$$\lambda^2 + 2\,\psi\,\lambda + 1 = 0 \;;$$

that is, $\lambda_1 = -\psi + \mathrm{sqrt}(\psi^2 - 1)$ and $\lambda_2 = -\psi - \mathrm{sqrt}(\psi^2 - 1)$. Now assume that ψ is large; in particular, that it is much greater than 1. With this assumption the value -2ψ is a reasonable approximation for λ_2. To obtain a helpful approximation for λ_1 we note that for small δ, a first-order Taylor series approximation for the function $R(z) = \mathrm{sqrt}(z)$ is:

$$R(z + \delta) = R(z) + 0.5\,\delta\,/R(z) \;.$$

Consequently (bearing in mind the assumption that ψ is much larger than 1):

$$\mathrm{sqrt}(\psi^2 - 1) = \psi\,\mathrm{sqrt}(1 - 1/\psi^2) \approx \psi[\mathrm{sqrt}\,(1) - 0.5/(\psi^2\,\mathrm{sqrt}(1))] = \psi - (0.5/\psi)$$

and so an approximate value for λ_1 is $-0.5/\psi$. Thus when the shock absorber constant ψ is large (relative to the spring constant k), there is a significant spread between the two eigenvalues; in particular, $|\,\lambda_2/\lambda_1| = 4\psi^2$ (which, for exampLe equals 900 when $\psi = 15$).

In practical terms, a large value for ψ (relative to k) means that a ride over an uneven road surface would be very bumpy for the passengers in the automobile because the suspension system would appear to be very stiff. The need to investigate such a circumstance could very well arise if the project goals included assessment of an automobile's dynamic behaviour in extreme conditions, such as evaluation of the impact of a shock absorber failure which could correspond to ψ becoming very large, hence the need to deal with a CTDS model that has the stiffness property.

A meaningful and generally accepted formal definition of stiffness has proved to be elusive. Instead it is simply regarded as a property of CTDS models that imposes upon some numerical solution procedures the requirement for an unusually small step-size over a substantial portion of the observation interval. As we have demonstrated above, in the special case of a linear system whose coefficient matrix has distinct real eigenvalues, this property is present when there is a significant spread between the smallest and the largest eigenvalues.

8.4.3 Discontinuity

CTDS models frequently contain discontinuities. Unless special precautions are taken in handling these, it is almost certain that the solution trajectories that are obtained will be flawed. In some cases these flawed solutions may still be adequate within the context of the goals of the modelling and simulation project whereas in other cases these flaws cannot be tolerated and specialised numerical procedures need to be used.

Two of the examples previously considered have embedded discontinuities: namely the bouncing ball project (Section 2.2.4) and the pilot ejection project (Section 7.3). In the case of the bouncing ball, the discontinuity occurs each time the ball strikes the ice surface and bounces. The bounce really corresponds to an instantaneous change in both the horizontal and the vertical velocities of the ball. The latter case is especially severe inasmuch as both the direction and magnitude of the ball's vertical velocity changes. In the case of the model for the pilot ejection project, the discontinuity occurs at the moment when the pilot/seat leaves the rails. At that moment there is an instantaneous change in the rate of change of both the horizontal and vertical velocities of the pilot/seat (while on the rails both $V_x'(t)$ and $V_y'(t)$ are zero but this changes instantaneously at the moment when the pilot/seat leaves the rails).

A discontinuity occurs when one or more state variables or the derivatives of state variables undergo an instantaneous change. Such an occurrence is usually called an 'event'. Events fall into two categories, namely, time events and state events. The distinguishing feature of a time event is that the time at which it occurs is explicitly known. The time of occurrence of a state event is known only implicitly through some functional specification that involves the state variables. For example, in the case of the bouncing ball there is a sequence of state events and the time of occurrence of each corresponds to the condition $y = 0$ (vertical displacement is zero; i.e., the ball is striking the ice surface).

The fact that the time of occurrence of a time event is known is very significant because it enables a simple circumvention of the numerical difficulty that, as we outline below, is otherwise present. More specifically, if it is known that a time event occurs at $t = t^*$ then the obvious practical approach is simply to execute the solution procedure up to $t = t^*$, carry out the change(s) associated with the event, and then continue the solution either to the next time event or to the right boundary of the observation interval, whichever occurs first. This approach preserves the integrity of the solution and requires only a minor disruption in the normal flow of the solution procedure. Handling time events, therefore, is relatively straightforward.

It is interesting to observe that in the case of the pilot/seat model, the simple analysis that yields Equation (7.19) effectively transforms the apparent state event into a time event. Because of the constant velocities that prevail while the seat is on the rails, the time when the seat leaves the rails is easily determined to be $t_E = Y_r/(V_r \cos \theta_r)$. Furthermore there is nothing in the goals of the project that necessitates trajectory information prior to t_E and consequently the situation becomes even more straightforward; that is, simply initiate the numerical solution at the event time t_E (or more precisely, incrementally beyond the event time).

The circumstances in the case of the bouncing ball are quite different; the state events that occur at the bounces cannot be circumvented. What then is the numerical issue that emerges? To address this question we need to reflect on the program code requirements that are necessitated by the discontinuity. As the following discussion points out, to deal with the state event the simulation model itself must now acquire a facet that is beyond the simple programming of the algebraic expressions that constitute the derivative functions of the model.

The basic requirement here is clearly a means for locating the occurrence of the state event so that the changes associated with it can be carried out. This is usually achieved by the introduction of *switch functions*. One such function is created for each state event that needs to be accommodated in the CTDS model. The key requirement in defining these switch functions is to capture, in a simple way, the implicit specification of the time of occurrence of the state event. A standard approach is to define the switch function so that its algebraic sign changes when the state event occurs. In other words the zero of the switch function signals the occurrence of the state event. For example, in the case of the bouncing ball model, an appropriate switch function is $\varphi_1(t) = y_1(t)$ (recall that $y_1(t)$ represents the vertical position of the ball above the ice surface). In the general case, we assume the existence of m such switch functions associated with the CTDS model being studied; for example, $\varphi_1(t)$, $\varphi_2(t)$, . . . , $\varphi_m(t)$.

There are in fact two distinct problems that need to be solved. These are called the detection problem and the location problem. The task of the detection procedure (which resolves the detection problem) is to identify an interval in which it is certain that at least one zero crossing of a switch function occurs. With this interval as its input, the location procedure then has the task of locating the time of the leftmost of these crossings; this constitutes the solution of the location problem.

To correctly deal with known discontinuities, a CTDS simulation model should incorporate, in some form or another, the equivalent of the following pseudocode. Step 1 in this code corresponds to the detection procedure and steps 2 and 3 correspond to the location procedure. This code needs to be executed at the completion of each successive time-step over the course of the underlying solution procedure. For definiteness, let's assume that the current solution step has moved the solution from $t = t_a$ to $t = t_b$.

1. For each i in the range 1 through m, determine if φ_i signals the occurrence of event i and if so place i in $\check{\mathbf{I}}$.
2. For each $i \in \check{\mathbf{I}}$ determine t_i^* such that $\varphi_i(t_i^*) = 0$ and place t_i^* in $\check{\mathbf{T}}$.
3. Determine t^{**}, the least value in $\check{\mathbf{T}}$.
4. Restart the solution process at $t = t_a$ and solve to t^{**}.
5. Carry out the changes required at the event(s) occurring at t^{**}.
6. Continue the solution process from t^{**}.

Correct and robust implementation of the pseudocode outlined above is not a trivial undertaking because the resolution of both the detection problem and the location problem requires considerable care. Various approximations are typically accepted but these can introduce substantial error and/or numerical misbehaviour.

Consider, for example, the bouncing ball model; in this case $m = 1$ because there is only one state event that needs to be monitored and $\varphi_1(t) = y_1(t)$. It is reasonable to conjecture that in the neighbourhood of an event time t^*, $\varphi_1(t)$ would have the form shown in Figure 8.3 where we assume that t_a and t_b are adjacent solution points resulting from a fixed step-size solution process. The signal for the occurrence of the state event (the bounce) could be taken simply to be the observation that $\varphi_1(t_a)$ and $\varphi_1(t_b)$ have opposite algebraic signs. Having thus established that a state event has occurred, we now need to identify t^*, the time of its occurrence. A gross, but very convenient, assumption is simply to take $t^* = t_b$. Because the solution process is currently at t_b, it is very straightforward to modify the values of horizontal velocity (x_2) and vertical velocity (y_2) to reflect the changes required by the state event. As a final and entirely artificial change to reflect the intended reality, y_1 can also be set to zero.

The procedure outlined above significantly compromises the accuracy of the solutions for the ball's trajectory and hence the accuracy of the results obtained for the underlying modelling and simulation project. But this is not to say that the results are unacceptable. There was no 'accuracy specification' included with the project description and possibly some latitude is permitted. Note, in fact, that the experiments with the bouncing ball carried out in Annex A3 are undertaken with these same rough approximations.

The approach taken above in handling the location problem is certainly primitive (namely, taking $t^* = t_b$). In the case where the switch function $\varphi_1(t)$ can safely be assumed to have the form shown in Figure 8.4, (i.e., a single crossing between t_a and t_b) a relatively simple bisection procedure can be used to solve the location problem in a more credible manner. The idea is simply to half the length of the interval that is known to contain the point of zero crossing on each of a sequence of iterations. This sequence ends either when the interval length is reduced to a sufficiently small size or until the value of φ_1 at the midpoint of the current interval is sufficiently close to zero. A specification of this bisection procedure based on the latter termination criterion is given below:

$t_c = (t_a + t_b)/2$
while ($|\varphi_1(t_c)| > \epsilon$)
    if ($\varphi_1(t_a) * \varphi_1(t_c) < 0$) $t_b = t_c$
    else $t_a = t_c$
    $t_c = (t_a + t_b)/2$
endwhile
$t^* = t_c$.

Here ϵ is a parameter that controls the accuracy of the final result that is generated. It should also be appreciated that each evaluation of φ_1, (at time t_c) requires that the underlying solution procedure re-solve the model equations from time t_a to time t_c. Computational overhead has clearly increased!

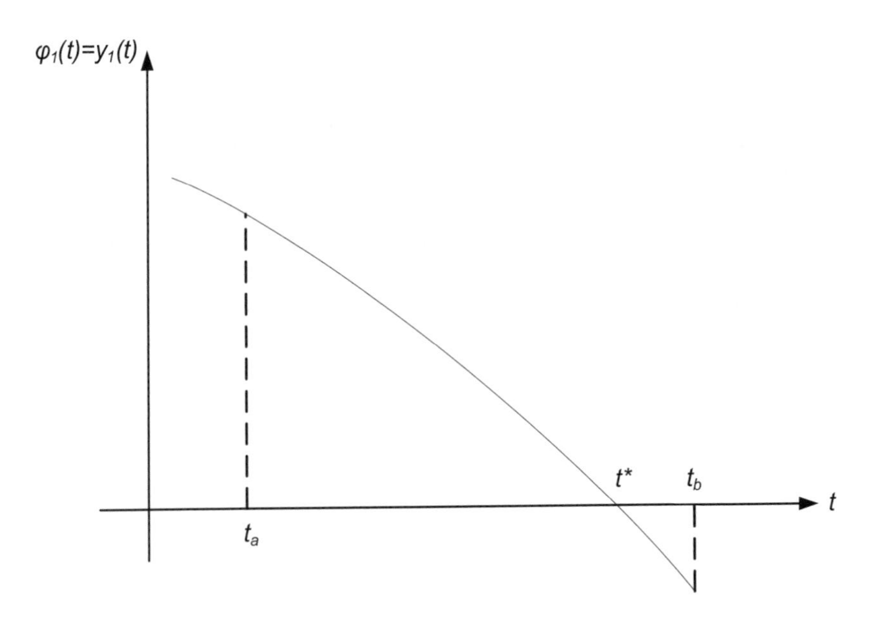

FIGURE 8.4. Locating the state event for the bouncing ball.

Note also that in general, there is no assurance that there is only a single zero crossing in the interval identified by the detection procedure. For example, the behaviour of a switch function (but not the one we have been discussing for the bouncing ball) might have the form shown in Figure 8.5. Because there are multiple zeros in the interval the bisection method outlined above would be an inappropriate choice for the location procedure. A more robust approach would need to be formulated.

The situation in handling discontinuities acquires a different (but nonetheless challenging) perspective when a variable step-size procedure is used as the equation-solving tool.

Some interesting investigations of this challenging numerical problem in handling CTDS models with discontinuities can be found in [8.1], [8.3], [8.6], and [8.13] (a variety of interesting example problems is likewise provided). A current and comprehensive discussion can be found in Cellier and Kofman [8.2].

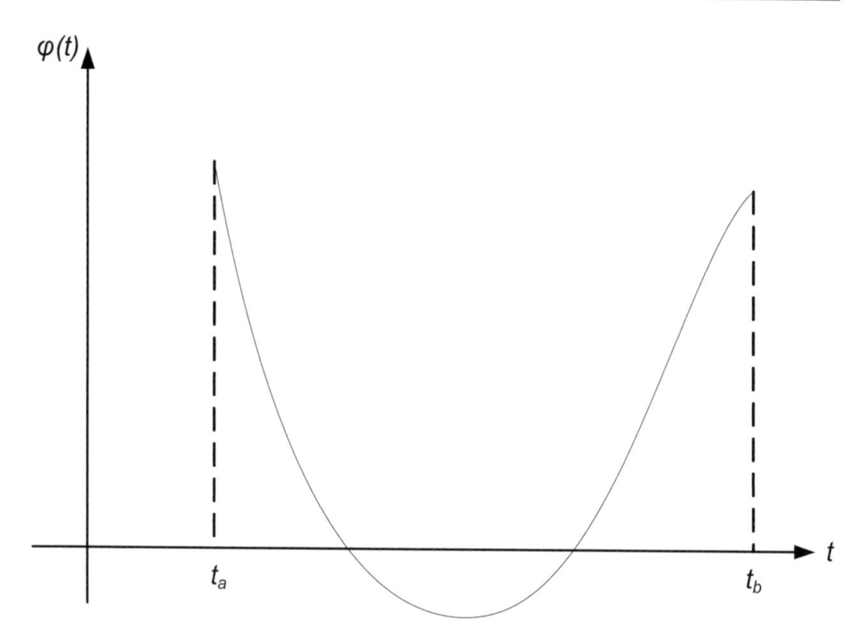

FIGURE 8.5. A switch function with multiple crossings.

8.4.4 Concluding Remarks

The main purpose of the discussion in Section 8.4 has been to demonstrate that the numerical tools required to carry out the simulation phase of a modelling and simulation project in the CTDS domain need to be used with some degree of caution. There are potential pitfalls and these are not always made clear to the users of the many simulation software products that are available in the marketplace. What may appear to be interesting dynamic behaviour in a CTDS model may simply be the reflection of numerical anomalies.

Mechanisms to detect such anomalies and bring them to the attention of the user are rarely provided. Thus it is important for the user to be alert and to have reasonable background knowledge and insight in order to be able to assess curious behaviour that may arise. Unfortunately there are very few guaranteed checks that can be applied to reveal the existence of problems. Nevertheless, one simple option that is always worth considering is the use of an alternate solution method whenever there is some reason to suspect that the numerical solution process is being compromised. Large inconsistencies in the results obtained provide a reasonable signal of underlying difficulty.

It is appropriate finally to stress also that robust solution methods for efficiently handling the differential equation of a CTDS conceptual model continue to evolve, especially in a modelling and simulation context. Readers interested in exploring such developments will find relevant topics in the recent work of Cellier and Kofman [8.2].

In this regard it is particularly interesting to note the work described by Kofman and Junco [8.10] and further elaborated in [8.2]. Traditional numerical methods for ODEs discretise the time axis as the underlying mechanism for driving the solution forward. The work referenced above takes the alternate approach of discretising the state space. This introduces an entirely new landscape which is, in particular, well suited to a unified treatment (at the computational level) of models that have DEDS and CTDS components.

8.5 Options and Choices in CTDS Simulation Software

A wide variety of software products/environments is available for carrying out the simulation (i.e. experimentation) phase of any modelling and simulation project in the CTDS realm. Some of these are commercial products (e.g., Dymola [8.4]) whereas others are in the public domain (e.g., Open Desire [8.11])). By and large, each has a relatively distinctive manner for specifying the conceptual model that is to be studied and often has, as well, many distinctive capabilities. Such distinctive capabilities (e.g., matrix inversion, eigenvalue calculation, discontinuity handling, animation, etc.) can be especially relevant to a particular project and thus provide a basis for making a selection from among available alternatives.

From the discussion in Sections 8.2 through 8.4 it is reasonable to suggest that a practical requirement for any CTDS simulation product is a solution engine that provides a variety of numerical solution methods. This is especially important when the conceptual model is large (many differential equations) and/or complex (i.e., derivative function evaluation is time consuming) because in such cases solution efficiency can become a matter of concern. The availability of solution method alternatives gives the user the option of making tradeoffs between computational overhead and accuracy.

Quite apart from a choice from among solution methods, there are still decisions to be made with respect to embedded parameters. The most fundamental, of course, is the step-size *h*. In the absence of other guidelines or insights, a rule-of-thumb often used when the solution method is of fourth-order, is to assign *h* the value $10^{-3}|I_0|$ (where $|I_0|$ is the length of the observation interval). In the case where a predictor–corrector method has been selected, the parameter ϵ that provides some accuracy control (see Section 8.2.2.1) may be available for assignment by the user. When a variable step-size method is selected, several associated parameters typically emerge (e.g., the error tolerance parameters K_1 and K_2 introduced in Section 8.3) and these must be assigned meaningful values by the user.

Making prudent value choices for these various embedded parameters is not an easy matter for a novice because very little guidance is available. Fortunately, with 'well-behaved' conceptual models it is usually a noncritical task. However with ill-behaved situations these value assignments can have a significant impact and improper assignments may even jeopardise the success of the modelling and simulation project.

8.6 The Safe Ejection Envelope Project Revisited

In Chapter 7 a CTDS conceptual model was formulated to provide the data required to establish the safe ejection envelope for a pilot forced to abandon a disabled fighter aircraft. Briefly, the objective is to determine for each of a range of horizontal aircraft velocities, the least altitude at which the ejection mechanism will yield an ejection trajectory that avoids the aircraft's tail assembly by a suitable margin of safety. The conceptual model is given by Equation (7.19) with initial conditions given by Equation (7.17).

The envelope we seek is, in fact, a graph of (V_a, H^*) pairs where H^* is the least 'safe altitude' associated with the horizontal velocity V_a. The procedure makes use of the fact that if ejection at a particular altitude is unsafe (i.e., results in a trajectory that does not clear the tail assembly by a sufficient distance) then increasing the altitude will eventually locate a safe value. This is a consequence of the fact that the drag force due to air density diminishes as altitude increases.

A procedure for generating the data required to create a graph of the form shown in Figure 7.6 is given in Figure 8.6. This procedure assumes the existence of a verified simulation program based on the conceptual model of Equation (7.19). Several parameters have been introduced to define the boundaries of the study; these are summarised in Table 8.1.

TABLE 8.1. Parameters used in the safe ejection envelope study.

Parameter	**Interpretation**	**Value**
V_{start}	Initial horizontal velocity	100 ft/sec
H_{start}	Initial altitude	0 ft
V_{limit}	Largest horizontal velocity	950 ft/sec
Δ_h	Increment in altitude	500 ft
Δ_v	Increment in velocity	50 ft/sec

An Open Desire simulation program that carries out this task is given in Figure 8.7.[2] The resulting safe ejection envelope is given in Figure 8.8 (however, some enhancement of the original Open Desire presentation has been carried out).

$V_a \leftarrow V_{start}$
$H \leftarrow H_{start}$
repeat
  while (miss $< S_f$)
    * $H \leftarrow H + \Delta_h$
    * solve ode's of the model from $t = t_E$ to $t = t^*$ where t^* is first occurrence of $X_p(t^*) \le V_a t^* - BT$
    * miss $= Y_p(t^*) - HT$
  endwhile
record (V_a, H^* with $H^* \leftarrow H$)
$V_a \leftarrow V_a + \Delta_v$
until ($V_a > V_{limit}$)
Plot the collected (V_a, H^*) pairs

FIGURE 8.6. Generating the envelope data.

[2] An overview of this particular simulation environment can be found in Annex 3. Readers unfamiliar with Open Desire are urged to review Annex 3 in order to better appreciate the simulation program in Figure 8.7.

```
---Safe Ejection Envelope Project
-----------------------------------------------------------
---CONSTANTS
g=32.2 |  ---acceleration due to gravity (ft/sec^2)
m=7 |  ---mass of pilot and seat (slugs)
BT=30 |  ---horizontal displacement of tail section (ft)
HT=12 |  ---vertical height of tail section (ft)
Cdhat=5 |  ---drag coefficient (ft-sec^2)
Sf=8 |  ---safety factor for avoiding tail (ft)
thetaD=15 |  ---angle of ejection rails (degrees)
thetaR=thetaD*(PI/180) |  ---angle of ejection rails
                                                (radians)
Vr=40 |  ---seat velocity while on rails (ft/sec)
Yr=4 |  ---vertical height of rails (ft)
Va=100 |  ---initial aircraft (horizontal)velocity
                                                (ft/sec)
H=0 |  ---initial aircraft altitude (ft)
---TABLE: Relative Air Density, RHO, versus altitude
dimension RHO[24]
data 0,1E+3,2E+3,4E+3,6E+3,1E+4,1.5E+4,2E+4,3E+4,4E+4
data 5E+4,6E+4, 2.377E-3,2.308E-3,2.241E-3
data 2.117E-3,1.987E-3,1.755E-3,1.497E-3, 1.267E-3
data 0.891E-3,0.587E-3,0.364E-3,0.2238E-3
---Storage Arrays for crossplot data
dimension VaV[20],HV[20]
---EXPERIMENT
read RHO
t=Yr/(Vr*cos(thetaR))
                   ---left hand end of observation interval
TMAX=3 |  ---right-hand end of observation interval
irule=3 |  ---fixed stepsize RK4
DT=0.004 |  ---integration step size
Xp=(Va-(Vr*sin(thetaR)))*t
                  ---horizontal position when leaving rails
Yp=Yr |  ---vertical position when leaving rails
Vx=Va-Vr*sin(thetaR)
                  ---horizontal velocity when leaving rails
Vy=Vr*cos(thetaR) |---vertical velocity when leaving rails
---setup display for pilot trajectories
display W300,80
display 2 |  display A |  display R
display C17 |  display N11
NN=1000 |  scale=1
knt=0
```

FIGURE 8.7. Open Desire simulation program for safe ejection envelope.

```
---main loop follows
write "Va      ";"H      ";"Miss       ";"Time"
repeat
  drun  |  miss=(Yp-HT)  |  stop=t  |  reset
  while miss<Sf
    H=H+500
    drun  |  miss=(Yp-HT)  |  stop=t  |  reset
   end while
  write Va;"  ";H;"  ";miss;"  ";stop
  knt=knt+1  |  VaV[knt]=Va  |  HV[knt]=H
  Va=Va+50  |  Vx=Vx+50  |  Xp=Xp+50*t
  until Va>950
write '>>>type "go" to continue'  |  STOP
-----------------------------------------------------
---OUTPUT(the safe ejection envelope (H vs Va))
-----------------------------------------------------
display F  |  NN=knt
drun ENVELOPE
-----------------------------------------------------------
DYNAMIC
-----------------------------------------------------------
HplusYp=H+Yp  |  func rho=RHO(HplusYp  |  ---compute air
                                                density
PSI=(Cdhat*rho*sqrt(Vx*Vx+Vy*Vy))/m
d/dt Xp=Vx  |  d/dt Yp=Vy
d/dt Vx=-PSI*Vx  |  d/dt Vy=-PSI*Vy-g
-----------------------------------------------------------
OUT
XTail=Va*t-BT
term XTail-Xp
term -Yp
-----------------------------------------------------------
---OUTPUT(pilot trajectories)
SQ=((XTail-Xp)+15)/15  |  SYp=0.075*Yp-1
                                 ---scaling for trajectories
dispxy SQ,SYp
-----------------------------------------------------------
label ENVELOPE
get Va=VaV
get H=HV
SVa=0.002*Va-1  |  SH=0.000025*H-0.999  |  ---scaling for
                                                envelope
dispxy SVa,SH
```

FIGURE 8.7. Open Desire program for safe ejection envelope (continued).

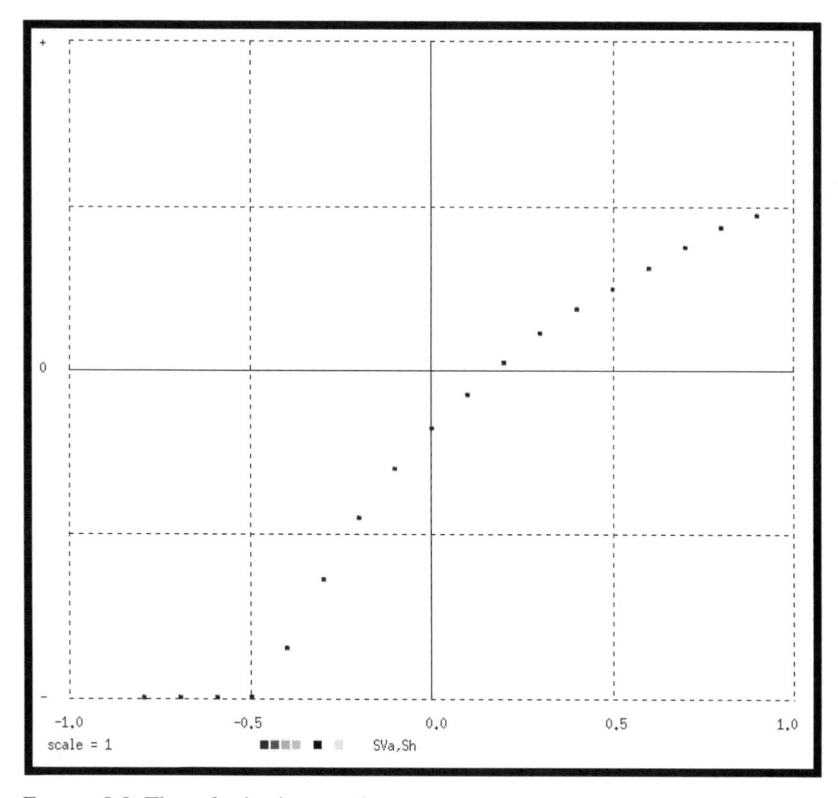

FIGURE 8.8. The safe ejection envelope.

8.7 Exercises and Projects

8.1 The Foucault pendulum was proposed in 1851 by Léon Foucault as a means of demonstrating the earth's rotation. Implementations of the Foucault pendulum can be found in science museums throughout the world. The special feature of this pendulum is that the pivot point can turn and consequently the plane in which the swinging bob moves can change in both the x- and y-directions. In fact, because of the earth's rotation, the plane of the swing will continuously change. Because of a complex interaction of forces, the rate at which the plane of the swing changes is dependent on the latitude λ where the observer is located. For example, at either of the poles ($\lambda = \pm 90$ degrees), it requires 24 hours for a complete rotation of 360 degrees (i.e., angular rate of $2\pi/24$ radians per hour) and this period decreases as the observer moves toward the equator ($\lambda = 0$) where the angular rate is zero. The equations that govern this behaviour (with the assumption that air friction effects can be ignored) are:

$$x''(t) - 2\omega \sin(\lambda)\ y'(t) + K^2 x(t) = 0$$

$$y''(t) + 2\omega \sin(\lambda)\ x'(t) + K^2 y(t) = 0\ ,$$

where ω represents the earth's rotational velocity (7.3 radian/sec) and $K = g/L$ where g is the acceleration due to gravity and L is the pendulum length (necessarily large, e.g., 50 meters) and λ is the latitude of the observer.

a) Formulate a modelling and simulation project based on the conceptual model given above, to determine the angular velocity (radians per hour) of the pendulum's plane of swing for each of the following values of latitude, λ: 5, 10, 15, . . . , 80 and 85 degrees. (Hint: Observe the graph that results when $x(t)$ is plotted against $y(t)$).
b) Determine from a search in the available literature (e.g., the Web), what the relation should be and confirm the validity of the results obtained in part (a).

8.2 A bumblebee colony represents an example of a 'stratified population', that is, one in which the total population is made up of different forms of the same species. Only impregnated females survive the winter to found a new colony in the spring. She prepares a simple nest and begins laying eggs at the rate of 12 eggs per day. The lifecycle is as follows.

a) An egg takes 3 days to hatch and what emerges is a larva.
b) The larva grows for 5 days and then turns into a pupa.
c) The pupa exists for 14 days and then turns into an adult/worker.
d) The adult lives for 5 weeks.

Formulate a modelling and simulation project whose goal is to obtain insight into how the population of the colony reacts to the death of the queen bee. Suppose, in particular, that the queen dies after T_0 days. As a result the population of the colony will eventually diminish to zero. Suppose this happens T_1 days after the death of the queen. The value of T_1 depends on the size of the population at the time T_0 which in turn depends on T_0 itself. Using an appropriate CTDS model, obtain sufficient data to produce a graph of T_1 versus T_0 with T_0 in some suitable range that adequately illustrates the pertinent aspects of the behaviour of interest.

Note that there are four state variables associated with the colony; namely,

a) $N_e(t)$, the egg population at time t
b) $N_r(t)$, the larva population at time t
c) $N_p(t)$, the pupa population at time t
d) $N_a(t)$, the adult population at time t.

In formulating the model, assume that t has the units of days. The fact that an egg exists for 3 days means that 1/3 of the egg population moves from the egg population to the larva population each day. Similarly 1/5 of the larva population moves out of the larva population each day. As a consequence, two of the four equations of the conceptual model are:

$$N_e'(t) = 12 - \frac{1}{3} N_e(t)$$

$$N_r'(t) = \frac{1}{3} N_e(t) - \frac{1}{5} N_r(t)$$

8.3 In this project we consider the motion of two masses moving horizontally on frictionless surfaces as shown in Figure 8.9. Mass m_1 is a block that rolls (without friction) on a horizontal surface and mass m_2 is a wheel that rolls on top of mass m_1 (again without friction). Each of these masses is individually connected with a spring to a vertical wall. The spring that connects m_1 has a spring constant of k_1 and the spring that connects m_2 has a spring constant of k_2. We assume that up until $t = 0$ this system has been resting in an equilibrium state. The lengths of the two springs are such that at equilibrium the wheel rests at the midpoint of the block whose width is $2w$. We take $m_1 = m_2 = 5$ kg, $k_1 = k_2 = 15$ Newton/meter and $w = 1.6$ meter.

If we let $x_1(t)$ and $x_2(t)$ represent the horizontal positions of the two masses relative to their respective equilibrium positions, then the CTDS conceptual model for the system is:

$$0.5\, m_2\, x_2''(i) - (m_1 + 0.5\, m_2)\, x_1''(i) = k_1\, x_1(t)$$
$$0.5\, m_2\, x_2''(i) - 1.5\, m_2\, x_1''(i) = k_2\, x_2(t)$$

At $t = 0$ the block is moved to the right by a distance $\alpha = 1.5$ meters and then released (the wheel on the other hand remains at its equilibrium position); thus, $x_1(0) = \alpha$, $x_2(0) = 0$, $x_1'(t) = 0, x_2'(t) = 0$. The goal of this modeling and simulation project is to gain insight into the circumstances that cause the wheel to fall off the surface of the block.

a) Determine if the wheel will fall off the block for the parameter values and the initial conditions that are given.
b) It is reasonable to assume that there are regions in the (positive) k_1–k_2 plane for which the ball will fall off the block and conversely regions where the wheel will not fall off the block (with the given values for the various parameters). Carry out experiments to determine these regions.
c) Determine how the regions found in part (b) are affected by changes in the value of α.

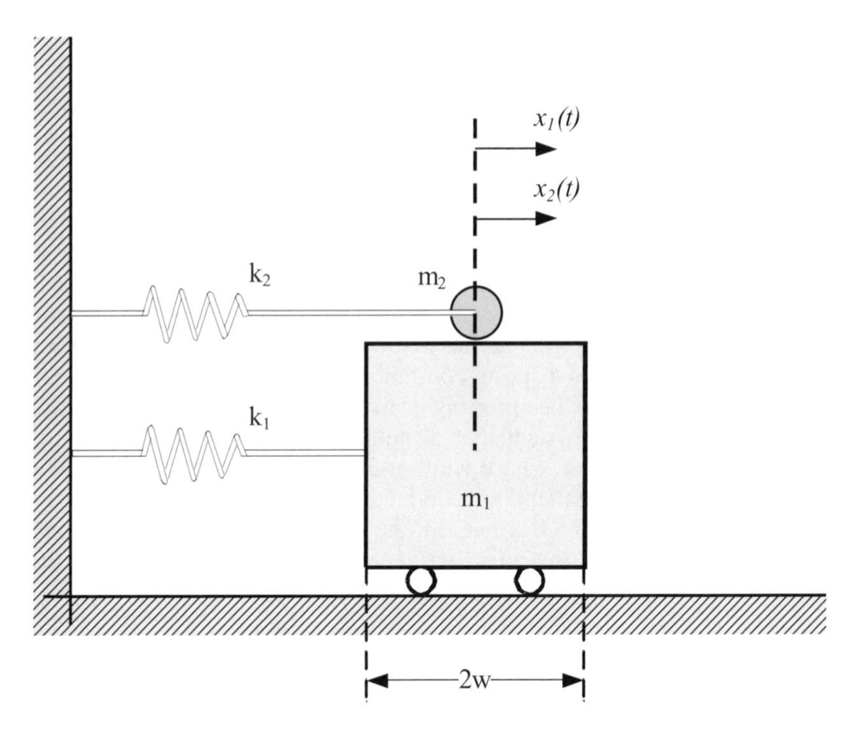

FIGURE 8.9. Rolling masses.

8.4 In this study a proposed system for halting an aircraft that might otherwise overshoot the runway during its landing manoeuvre is to be investigated. The system has particular utility in the context of an aircraft carrier. The configuration of the upper half of the system is shown in Figure 8.10. The complete system is symmetric about the center line; that is, an identical configuration to that shown in Figure 8.10 exists below the center line.

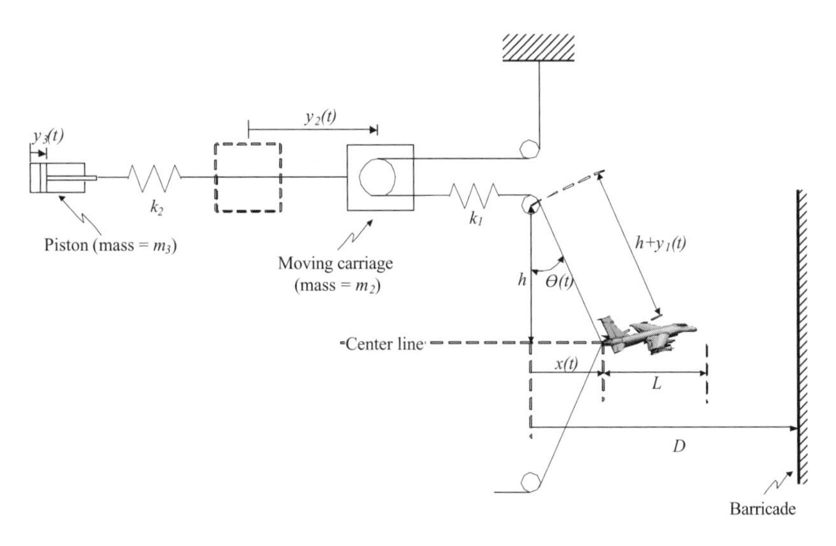

FIGURE 8.10. Schematic representation of aircraft arresting mechanism.

The springs shown as k_1 and k_2 are fictitious. They are intended to represent the elastic properties of the steel cables which are the connecting members. In particular, this means that these springs cannot be compressed. If, for example, y_2 becomes less than y_3 the cable connecting the piston and the moving carriage simply goes limp.

An appropriate analysis of the elements of the system yields the following conceptual model.

$$m_1 \ddot{x}(t) = -2 f_1(t) \sin\theta(t)$$

$$\sin\theta(t) = \frac{x(t)}{\sqrt{h^2 + x^2(t)}}$$

$$y_1(t) = \sqrt{h^2 + x^2(t)} - h$$

$$f_1(t) = \begin{Bmatrix} k_1(y_1(t) - 2y_2(t)) & for\, y_1 > 2y_2 \\ 0 \quad otherwise & \end{Bmatrix}$$

$$m_2 \ddot{y}_2(t) = 2 f_1(t) - f_2(t)$$

$$f_2(t) = \begin{Bmatrix} k_2(y_2(t) - y_3(t)) & for\, y_2 > y_3 \\ 0 \quad otherwise & \end{Bmatrix}$$

$$m_3 \ddot{y}_3(t) = f_2(t) - f_d(t)$$

The force $f_d(t)$ is a consequence of the shock absorber effect of the piston which is moving through a cylinder filled with water. Its value is dependent on the square of the velocity $\dot{y}_3(t)$; that is, $f_d(t) = k_3 \dot{y}_3^2(t)$. The drag coefficient k_3 furthermore is dependent on y_3 and its value, as established from experimental data, is given in Table 8.2.

TABLE 8.2. Drag coefficient of the piston.

y_3 (Meters)	k_3 (Newtons/(m/sec)^2)
0	1720
9	1340
18	1100
37	1480
46	1480
55	1480
64	1720
73	1960
82	2500
86	3000
90	3650
93	4650
95	5400
100	7800

The values of the various constants in the model are summarized in Table 8.3.

TABLE 8.3. Summary of constants.

Constant	Value
m_1	25,000 kg
m_2	1300 kg
m_3	350 kg
k_1	115,000 Newtons/m
k_2	430,000 Newtons/m
h	30 m
L	15 m
D	300 m

The specific system to be investigated has a relatively solid barricade located $D = 300$ meters from the contact point ($x = 0$) which will bring the aircraft to a full stop provided it is not traveling faster than 5 m/sec when it strikes the barricade. There are two specific issues that need to be investigated. The first is to determine V^*, where V^* is the largest initial velocity of the aircraft such that its velocity, when the front of the aircraft strikes the barrier will not exceed 5 m/sec. In addition, it is of interest to obtain some insight into the relationship between this maximum initial velocity and the mass of the aircraft. For this purpose, it is required to obtain a graph of V^* versus aircraft mass (m_1) for m_1 in the range 20,000 kg to 30,000 kg.

8.8 References

8.1. Birta, L.G., Ören, T.I., and Kettenis, D.L., (1985), A robust procedure for discontinuity handling in continuous system simulation, *Transactions of the Society for Computer Simulation*, **2**: 189–205.

8.2. Cellier, F.E. and Kofman, E., (2006), *Continuous System Simulation*, Springer-Verlag, New York.

8.3. Ellison, D., (1981), Efficient automatic integration of ordinary differential equations with discontinuities, *Mathematics and Computation in Simulation*, **23**: 12–20.

8.4. Elmquist, H., (2004), Dymola – Dynamic modeling language, user's manual, Version 5.3, DynaSim AB, Research Park Ideon, Lund, Sweden.

8.5. Gear, C.W., (1971), *Numerical Initial Value Problems in Ordinary Differential Equations*, Prentice-Hall, Englewood Cliffs, NJ.

8.6. Gear, C.W. and Osterby, O., (1984), Solving ordinary differential equations with discontinuities, *ACM Transactions on Mathematical Software*, **10**: 23–44.

8.7. Hairer, E. and Wanner, G., (1996), *Solving Ordinary Differential Equations II: Stiff and Differential-Algebraic Problems*, 2nd edn., Springer-Verlag, Berlin.

8.8. Iserles, A., (1996), *A First Course in the Numerical Analysis of Differential Equations*, Cambridge University Press, Cambridge, UK.

8.9. Kincaid, D. and Cheng, W., (2002), *Numerical Analysis: Mathematics of Scientific Computing*, 3rd edn., Brooks/Cole, Pacific Grove, CA.

8.10. Kofman, E. and Junco, S., (2001), Quantized state systems: A DEVS approach for continuous system simulation, *Transactions of the SCS*, **18**(3): 123–132.

8.11. Korn, G.A., (1998), *Interactive Dynamic-system Simulation Under Windows 95 and NT*, Gordon Breach, London.

8.12. Lambert, J.D., (1991), *Numerical Methods for Ordinary Differential Equations*, Wiley, London.

8.13. Shampine, L.F., Gladwell, I., and Brankin, R.W., (1991), Reliable solutions of special event location problems for ODEs, *ACM Transactions on Mathematical Software*, **17**: 11–25.
8.14. Watts, H.A., (1984), Step-size control in ordinary differential equation solvers, *Transactions of the Society for Computer Simulation*, **1**: 15–25.

Chapter 9 Optimisation

9.1 Introduction

Optimisation studies are frequently embedded within the goals of a modelling and simulation project. In some cases this optimisation aspect may simply be a preliminary requirement in the development of the model that is to be subsequently used in the simulation study. In other cases it may constitute the main aspect of the project goals. We refer to these two alternatives as the model refinement problem and the strategy formulation problem, respectively.

As an example of the model refinement problem, consider a situation where there exists a general model of how a particular drug that is required in the treatment of some illness, dissipates through the human body. However, before the model can be used it must be adapted ('calibrated') to the particular patient undergoing treatment. In other words, the values for various parameters within the model have to be established so that it 'best fits' the patient. This could be achieved by minimising the difference between one or more of the model's output variables and clinical data obtained from the patient. Once optimised in this sense, the model is available for use by the physician to assist in establishing a proper continuing dosage level for the drug.

As an example of the strategy formulation problem, consider a model of a chemical process which has been developed using known properties of the chemical kinetics that are involved in the process. Suppose one of the model's outputs represents the cost of production over the period defined by the observation interval. A goal of a modelling and simulation project might be to determine a minimum value for this output by the optimum selection of parameters embedded within an operating policy that is represented by one of the inputs to the model.

It can be reasonably assumed that in both these examples the system under investigation is a continuous-time dynamic system. In fact, our considerations throughout this chapter are restricted to this domain. This is not to suggest that optimisation studies cannot be undertaken with models from the DEDS domain. However, the tools required for handling the optimisation task in that domain need to deal with the inherent stochastic nature of DEDS models. This superimposes another level of complexity that is beyond the scope of our considerations in this textbook. Nevertheless, a variety of approaches for handling the problem has been developed and descriptions can be found in the literature (e.g., [9.4], [9.16], [9.23], and [9.28]). A comprehensive overview of a range of commercially available optimisation packages, their vendors, and the heuristic search procedures that they use, can be found in [9.12].

It is also fair to suggest that in the DEDS environment, the results obtained from the optimisation process generally need to be treated with some caution. For example, they are rarely precise enough to permit decisions to be confidently made at a detailed design level. Nevertheless the results obtained can provide valuable insight which, after all, is an essential aspects of the modelling and simulation activity.

9.2 Problem Statement

There exist two groups of relevant variables when an optimisation problem is superimposed on a model of a continuous-time dynamic system. As in our previous considerations with the CTDS domain, there is the N-vector $\mathbf{x} = \mathbf{x}(t)$ which we use to represent the state vector for the model. The model, in fact, is the set of first-order differential equations:

$$\mathbf{x}'(t) = \mathbf{f}(\mathbf{x}(t), t)$$

with $\mathbf{x}(t_0) = \mathbf{x}_0$ and, as well, with a specified observation interval $I_0 = [t_0, t_f]$. We now introduce the m-vector of parameters $\mathbf{p}$, whose values are to be optimally selected. The presence of the parameter vector $\mathbf{p}$, needs to be reflected in the specification of the model and this can be simply achieved by rewriting our model as:

$$\mathbf{x}'(t) = \mathbf{f}(\mathbf{x}(t), t; \mathbf{p}) \ .$$

It is frequently convenient to explicitly indicate the dependence of the state vector x on the parameter vector $\mathbf{p}$ by writing it as $\mathbf{x}(t;\mathbf{p})$.

To guide the selection of the best value for the parameter vector $\mathbf{p}$, there is necessarily associated with the problem a scalar, real-valued criterion function which we denote $J = J(\mathbf{p})$. The objective, then, is to find a value $\mathbf{p}^*$ for $\mathbf{p}$ which yields an extreme value for J. This may be either a maximum or a minimum but for definiteness in our discussions, we assume the latter (note that the maximisation of J is equivalent to the minimisation of $-J$). Thus we seek to find $\mathbf{p}^*$ (the minimising argument) such that:

$$J(\mathbf{p}^*) \leq J(\mathbf{p}) \quad \text{for all } \mathbf{p} \,\varepsilon\, \Phi \ .$$

In general, not all possible m-vectors are permitted candidates for $\mathbf{p}^*$ and consequently the minimisation of J could be restricted to a particular subset of admissible values which is denoted Φ. Such restrictions may be explicit; for example, the first component p_1, of **p** must be positive. Alternately, the restrictions may be implicitly defined via a collection of functional constraints; for example, $\Phi_j(\mathbf{x}(t;\, \mathbf{p})) \leq 0$ for $j = 1, 2, \ldots, c_1$ and $\Phi_j(\mathbf{x}(t;\, \mathbf{p})) = 0$ for $j = c_1 + 1, c_1 + 2, \ldots, c_2$. Such a functional constraint would arise, for example, in the case of a manufacturing process where the tensile strength of a plastic material that is being produced is compromised if the rate of cooling at a particular phase of the process is excessive. In such a circumstance only those values of **p** that do not create the unacceptable cooling conditions would be allowed.

As might be expected, the existence of restrictions on the permitted values for **p** (the constrained problem) introduces additional complexity upon the solution task. One approach that can be effectively used is called the *penalty function method.* Here the constraints are manipulated into a special form and appended to the criterion function to produce an 'augmented' criterion function whose basic feature is that it penalises violation of the constraints. The minimisation of this augmented criterion function is therefore undertaken without the burden of having to explicitly restrict the search space. The constrained problem is thus transformed into an unconstrained problem (more correctly, there is a requirement for the solution of a sequence of unconstrained problems). In other words, this approach allows the constrained problem to be treated with the same numerical tools as the unconstrained problem. Elaboration of this approach as well as other methods for handling the constrained optimisation problem can be found in [9.3], [9.8], and [9.10].

The specific form of the criterion function J evolves from the nature of the problem to be solved. The only requirement is that $J(\mathbf{p})$ have a real scalar value for each value of the m-vector **p**. Note, however, that because the parameter vector **p** is embedded in a CTDS conceptual model the evaluation of J, for any given **p**, requires the solution of a set of differential equations. This is, in principle, of no particular consequence for any optimisation process, however, it can have significant practical consequences in terms of computational overhead.

Some typical forms for the criterion function J are:

$$\text{(a)}\quad J = g(\mathbf{x}(t_f;\, \mathbf{p}))$$

$$\text{(b)}\quad J = \sum_{j=1}^{s} g(\mathbf{x}(t_j;\mathbf{p}))$$

$$\text{(c)}\quad J = \int_{t_0}^{t_f} g(\mathbf{x}(t;\mathbf{p}))dt\ .$$

In each of these cases g is some scalar function of its argument $\mathbf{x}$. An example where (a) would be an appropriate choice is provided by the bouncing ball problem that was considered earlier (see Section 2.2.4). Recall that the task is to find an initial release angle which results in the ball falling through a hole in the ice surface. The release angle represents the parameter (there is only one) and g could be selected to be the square of the distance between the point where the ball strikes the surface and the location of the hole. The implicit assumption that the problem has a solution means that g has a minimum value of zero; that is, the ball falls through the hole. A successful search for the release angle that minimises g will therefore provide the solution to the problem.

A criterion function of the form shown in (b) could have relevance to the model refinement problem outlined earlier. The calibration process in question could, for example, be based on the manner in which blood sugar is absorbed following an injection of insulin. In this case, the s time points, t_j, $j = 1, 2, \ldots, s$ that are referenced could be the points in time where blood sugar measurements are taken from the patient and the function g could be the absolute value of the difference between the measured data from the patient and the value acquired by some particular output variable of the model. Finding values for the set of model parameters that yield a minimum value for J would then correspond to the calibration process.

The criterion function form shown in (c) maps directly onto a classic control system design problem. The feedback controller for a continuous time dynamic system (e.g., an aircraft autopilot) typically has several parameters whose values need to be chosen in a way that, in some sense, optimises system performance. A frequently used performance measure is 'integral-square-error', that is, the integral of the square of the deviation between a desired system output and the output that actually occurs when the system has a prescribed input. Assuming that a CTDS conceptual model is available for the system and its controller, the goal of finding best values for the controller parameters would be based on using the model in the minimisation of a criterion function of the form shown in (c). In this case $g = (y - \hat{y})^2$ where y is the output of interest (some function of the model's state vector **x**) and $\hat{y}$ is the desired value for y.

As might be expected, the difficulty of the minimisation task is very much dependent of the geometric nature of the criterion function $J(\mathbf{p})$. In particular, there is the very serious issue of multiple local minima. Most minimisation procedures are unable to distinguish such 'false' minima and consequently may converge upon such a point, thereby yielding an erroneous result. Another geometric feature that is poorly accommodated by most procedures is the existence of a 'long' gently sloping valley. Such a situation can cause premature termination of a minimisation procedure and the presentation of an inferior result. Unfortunately these difficult circumstances are not uncommon.

By way of illustration we show in Figure 9.1 a representative criterion function that is dependent on two parameters.[1] The multiplicity of local extreme values and the existence of sloping valleys are apparent.

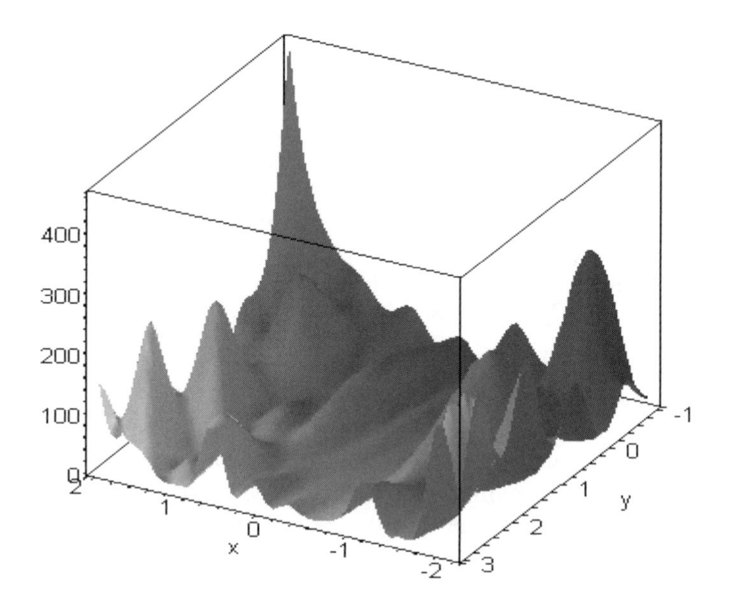

FIGURE 9.1. A response surface for a two-dimensional criterion function.

[1] Figure 9.1 has been taken from Pinter et al. [9.19] with the permission of the authors.

9.3 Methods for Unconstrained Minimisation

The discussion in Section 9.2 has stressed that optimisation problems embedded in the goals of a modelling and simulation project in the CTDS domain are essentially the same as the 'classical' problem that is treated in the numerical optimisation literature. There is, however, one important distinctive feature, namely, that the evaluation of the criterion function at any particular argument value **p** requires the solution of a set of differential equations (i.e., the conceptual model). This can, at least in principle, simply be regarded as part of the computational overhead.

A wide range of methods for dealing with the unconstrained function minimisation problem have been developed. A comprehensive review of these is well beyond the scope of our interest in this textbook. Our intent here is simply to provide an introduction to some of the basic ideas upon which these methods are based. It is strongly recommended that the reader who needs to carry out an optimisation study probe deeper into the topics that are introduced in the discussion that follows. Relevant information can be found in the numerous textbooks that deal specifically with numerical optimisation (e.g., [9.5], [9.8], [9.15], [9.18]).

There is a variety of ways for categorising the relatively large number of available function minimisation methods. Perhaps the most fundamental is whether gradient information is required by the procedure. Methods not requiring gradient information are often referred to as heuristic methods because their basis of operation is primarily based on intuitive notions. In the two sections that follow, we outline a representative member of both the heuristic and the gradient-dependent categories.

9.3.1 The Nelder–Mead Simplex Method

The Nelder–Mead method first appeared in the optimisation literature in 1965 (see [9.14]) and continues to be of practical value and of theoretical interest [9.11], [9.24]. One of its features is the absence of any need for gradient information hence it can be classed as a heuristic method. In a modeling and simulation context, this is especially significant as becomes apparent in our discussion in Section 9.3.2.

The process begins with the specification of a regular simplex which is defined in terms of $(m + 1)$ points in m-space (recall that our parameter vector **p** is a vector of m dimensions). When $m = 2$, the simplex is a triangle. The defining points for the initial simplex are part of the initialisation procedure. Generally a (priming) point $\mathbf{p}^0$ which represents a 'best' solution estimate is prescribed; the remaining m points of the initial simplex are generated by a simple procedure that uses the priming point.

The minimisation procedure consists of a sequence of operations referred to as reflection, expansion, and contraction. Each step begins with a reflection operation which is then followed by either an expansion operation or a contraction operation. These operations produce a sequence of simplexes that change shape and move through the m-dimensional parameter space until (one hopes) they encompass, and then contact upon, the minimising argument $\mathbf{p}^*$.

Let $\{\mathbf{p}^0, \mathbf{p}^1, \mathbf{p}^2, \ldots, \mathbf{p}^m\}$ be the vertices of the current simplex. Let $\mathbf{p}^L$ be the vertex that yields the largest value for J, $\mathbf{p}^G$ be the vertex that yields the next largest value for J, and $\mathbf{p}^S$ be the vertex that yields the smallest value for J. Correspondingly, let $J_L = J(\mathbf{p}^L)$, $J_G = J(\mathbf{p}^G)$, and $J_S = J(\mathbf{p}^S)$. The centroid of the simplex with $\mathbf{p}^L$ excluded is given by:

$$\mathbf{p}^C = \frac{1}{m}\,[(\sum_{k=0}^{m}\mathbf{p}^k) - \mathbf{p}^L\,]\ .$$

A reflection step (Figure 9.2a) is carried out by reflecting the worst point $\mathbf{p}^L$ about the centroid, to produce a new point $\mathbf{p}^R$, where

$$\mathbf{p}^R = \mathbf{p}^C + \alpha\,(\mathbf{p}^C - \mathbf{p}^L)\ .$$

Here α is one of three user-assigned parameters associated with the procedure; the requirement is that $\alpha > 1$ and it is typically chosen to be 1. One of three possible actions now take place depending on the value of $J_R = J(\mathbf{p}^R)$. These are:

i. If $J_G > J_R > J_S$, then $\mathbf{p}^R$ replaces $\mathbf{p}^L$ and the step is completed.

ii. If $J_R < J_S$ then a new 'least point' has been uncovered and it is possible that further movement in the same direction could be advantageous. Consequently an expansion step (Figure 9.2b) is carried out to produce $\mathbf{p}^E$ where

   $$\mathbf{p}^E = \mathbf{p}^C + \gamma\,(\mathbf{p}^R - \mathbf{p}^C) \quad (\gamma > 1 \text{ and is typically } 2)\ .$$

   If $J_E = J(\mathbf{p}^E) < J_S$ then $\mathbf{p}^L$ is replaced with $\mathbf{p}^E$; otherwise $\mathbf{p}^L$ is replaced with $\mathbf{p}^R$. In either case, the step is completed.

iii. If $J_R > J_G$ then a contraction step is made to produce the point $\mathbf{p}^D$ where

   $$\mathbf{p}^D = \mathbf{p}^C + \beta\,(\tilde{\mathbf{p}} - \mathbf{p}^C) \quad (0 < \beta < 1 \text{ and is typically } 0.5)\ .$$

   Here $\tilde{\mathbf{p}}$ is either $\mathbf{p}^R$ or $\mathbf{p}^L$ depending on whether J_R is smaller or larger than J_L (see Figures 9.2c and d). If $J_D = J(\mathbf{p}^D) < J_G$ then the step ends. Otherwise the simplex is shrunk about $\mathbf{p}^S$ by halving the distances of all vertices from this point and then the step ends.

Either of two conditions can termination the search procedure. One is based on the relative position of the vertices of the current simplex; that is, if they are sufficiently closely clustered then $\mathbf{p}^S$ can be taken as a reasonable approximation of the minimising argument $\mathbf{p}^*$. Alternately, the termination can be based on the variation among the values of the criterion function J at the vertices of the simplex. If these values are all within a prescribed tolerance, then again $\mathbf{p}^S$ can be taken as a reasonable approximation of the minimising argument $\mathbf{p}^*$.

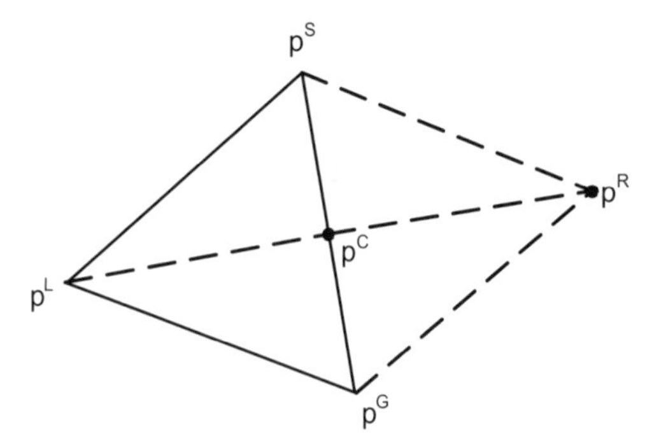

FIGURE 9.2a. Reflection step.

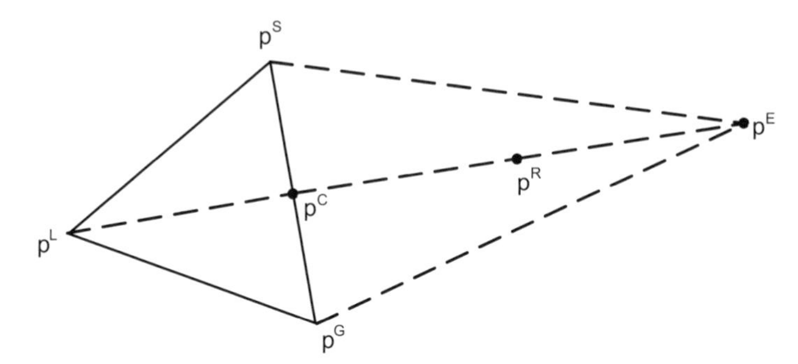

FIGURE 9.2b. Expansion step.

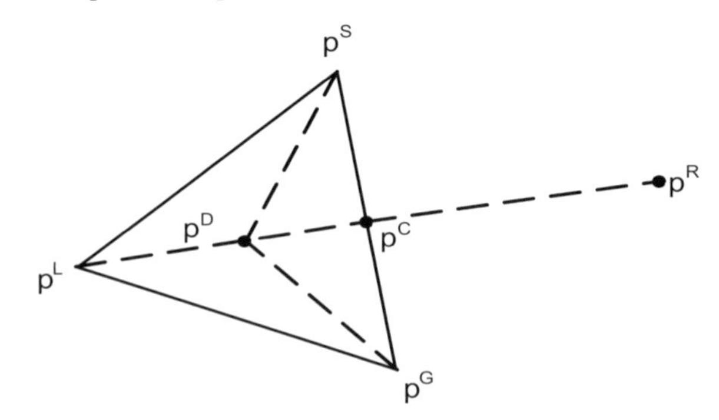

FIGURE 9.2c. Contraction step ($J_L < J_R$).

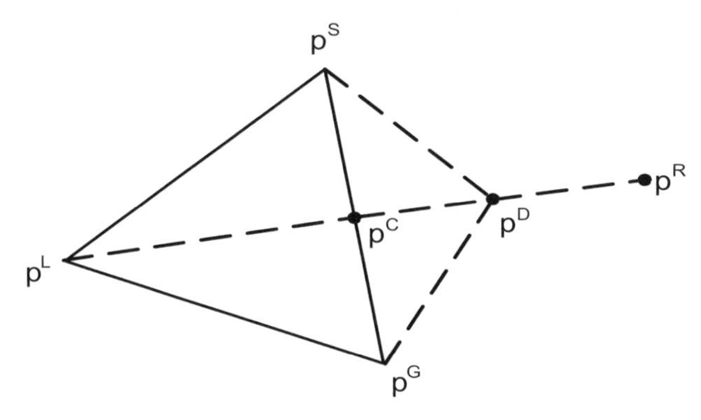

FIGURE 9.2d. Contraction step ($J_R < J_L$).

9.3.2 The Conjugate Gradient Method

We begin this section with a brief review of the notion of the gradient, specifically, the gradient of the criterion function $J = J(\mathbf{p})$. Inasmuch as $\mathbf{p}$ is a vector of dimension m, the gradient of J is likewise a vector of dimension m. The kth component of this vector is the partial derivative of J with respect to p_k, that is, with respect to the kth component of $\mathbf{p}$. The gradient of $J(\mathbf{p})$ is denoted $\mathbf{J}_p(\mathbf{p})$. Suppose, for example, that $J(\mathbf{p}) = 10(p_2 - p_1^2)^2 + (1 - p_1)^2$. Then,

$$\mathbf{J}_p(\mathbf{p}) = \begin{bmatrix} \dfrac{\partial J}{\partial p_1} \\ \dfrac{\partial J}{dp_2} \end{bmatrix} = \begin{bmatrix} -(40\, p_1(p_2 - p_1^2) + 2\,(1 - p_1)) \\ 20\,(p_2 - p_1^2) \end{bmatrix} .$$

The gradient vector is especially relevant in function minimisation for two reasons:

a) If $\overline{\mathbf{p}}$ is a point in m-space, then the negative gradient vector evaluated at $\overline{\mathbf{p}}$ has the property that it points in the direction of greatest decrease in the function J. In other words, for suitably small but fixed ε, $J(\overline{\mathbf{p}} + \varepsilon\, \mathbf{v})$ is smallest when $\mathbf{v} = -\mathbf{J}_p(\overline{\mathbf{p}})$.
b) When J is continuously differentiable, a necessary (but not sufficient) condition for $\mathbf{p}^*$ to be a local minimum for the function $J(\mathbf{p})$ is that $\mathbf{J}_p(\mathbf{p}^*) = \mathbf{0}$.

A concept that has played an important role in the development of numerical minimisation procedures is that of *conjugate directions*. The concept relates to a specified symmetric positive definite matrix **A** of dimension m. Specifically, a set of η ($\eta \leq m$) nonzero m-vectors (or equivalently, 'directions') $\mathbf{r}^0, \mathbf{r}^1, \ldots, \mathbf{r}^{\eta-1}$ is A-conjugate if $(\mathbf{r}^j)^T \mathbf{A}\, \mathbf{r}^k = 0$ for $j \neq k$ and $j, k = 0, 1, \ldots, (\eta - 1)$.

A-conjugate directions have a variety of interesting properties which include the feature that any such collection of directions is linearly independent. There is one property that is especially relevant from the point of view of function minimisation but before outlining it, the notion of a linear or line search needs to be introduced.

Suppose $\bar{\mathbf{p}}$ is a given point in m-space and $\bar{\mathbf{r}}$ is a given m-vector (direction). For any positive value of the scalar α, the m-vector $(\bar{\mathbf{p}} + \alpha\, \bar{\mathbf{r}})$ can be regarded as a point in m-space reached by moving a distance of α away from $\bar{\mathbf{p}}$ in the direction $\bar{\mathbf{r}}$. Suppose now that J is a given scalar valued criterion function whose value depends on the m-vector $\mathbf{p}$, that is, $J = J(\mathbf{p})$ and suppose we substitute $(\bar{\mathbf{p}} + \alpha\, \bar{\mathbf{r}})$ for $\mathbf{p}$. Because both $\bar{\mathbf{p}}$ and $\bar{\mathbf{r}}$ are fixed, J becomes simply a function of the scalar α and consequently we can write $J = J(\alpha)$. Furthermore, it is reasonable to assume that there is a value of α (which we denote α^*) that yields a minimum value for $J(\alpha)$. Finding the value of α^* is called the line (or linear) search problem. In effect, the line search problem corresponds to locating a minimum of J in a specific plane (or 'slice') of the parameter space. This is illustrated in Figure 9.3. Note the possible existence of multiple local minima.

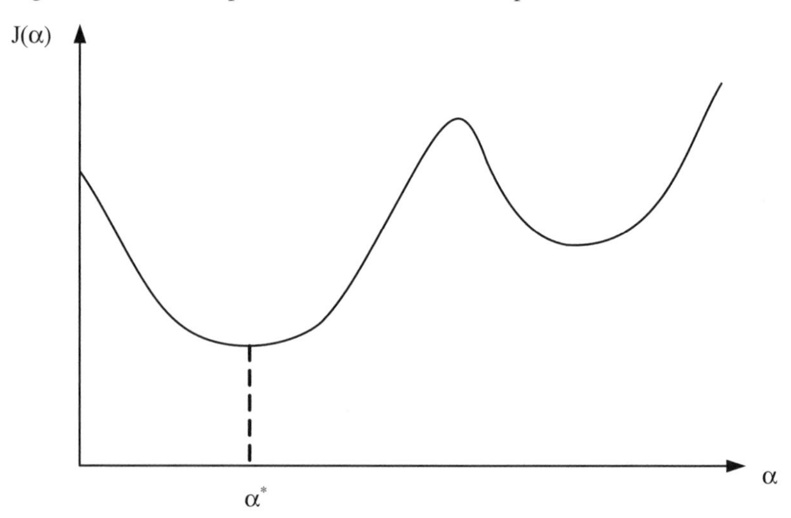

FIGURE 9.3. Illustration of the line (linear) search problem.

The following result is the essential property of conjugate directions from the point of view of function minimisation.

The CD Lemma

Let:

- $J(\mathbf{p}) = \frac{1}{2}\ \mathbf{p}^T \mathbf{A}\ \mathbf{p} + \mathbf{b}^T \mathbf{p} + c$ with **A** symmetric and positive definite and **p** an m-vector
- $\mathbf{p}^0$ be a given initial point
- the m-vectors $\mathbf{r}^0, \mathbf{r}^1, \ldots \mathbf{r}^{\eta-1}$, $(\eta \leq m)$ be a set of A-conjugate directions
- the kth point, $\mathbf{p}^k$ in the sequence $\mathbf{p}^1, \mathbf{p}^2, \ldots \mathbf{p}^\eta$ be generated by carrying out a line search from $\mathbf{p}^{k-1}$ along $\mathbf{r}^{k-1}$; that is, $\mathbf{p}^k = \mathbf{p}^{k-1} + \alpha^* \mathbf{r}^{k-1}$ where $J(\mathbf{p}^k) = \min_{\alpha} J(\mathbf{p}^{k-1} + \alpha \mathbf{r}^{k-1})$.

Then:

i) $\mathbf{J}_p(\mathbf{p}^k)$ has the property that $(\mathbf{J}_p(\mathbf{p}^k))^T \mathbf{r}^j = 0$ for $j = 0, 1, \cdots (k-1)$

ii) the same point $\mathbf{p}^k$ is reached independent of the order in which the directions $\mathbf{r}^j$ are used in the sequence of line searches.

The general quadratic form for the criterion function, $J(\mathbf{p})$, considered in the CD Lemma is clearly very specialised. Nevertheless the Lemma is important because any criterion function has a quadratic approximation in a sufficiently small neighbourhood around its minimum. Consequently any implications flowing from this Lemma are relevant in such a neighbourhood.

There is, in fact, one especially important consequence of the Lemma, namely, when $\eta = m$, there must exist an index $K \leq m$ such that $\mathbf{p}^K = \mathbf{p}^*$, the minimising argument of J. This follows from the linear independence of the m-vectors $\mathbf{r}^0, \mathbf{r}^1, \ldots, \mathbf{r}^{m-1}$ and outcome (i) of the CD Lemma. More specifically, (i) states that the gradient of J at $\mathbf{p}^K$ (i.e., $\mathbf{J}_p(\mathbf{p}^K)$) is orthogonal to each of a set of m linearly independent m-vectors which, in turn, implies that $\mathbf{J}_p(\mathbf{p}^K)$ must be zero (recall that the zero m-vector is the only one that can be simultaneously orthogonal to each of a set of m linearly independent m-vectors). Because of the assumed special structure of J, the condition $\mathbf{J}_p(\mathbf{p}^K) = 0$ is both necessary and sufficient for $\mathbf{p}^K = \mathbf{p}^*$, the minimising argument of J. Note that the case where $K = m$ is a 'worst' case; for certain choices of the initial point $\mathbf{p}^0$, it can occur that $K < m$. In other words, the minimising argument of J will be located in at most m steps of the procedure.

The fundamental prerequisite for implementing any function minimisation method that is based on conjugate directions is, of course, the availability of the necessary set of directions. Furthermore it must be borne in mind that any such approach is, at least in principal, relevant only to the minimisation of a quadratic function because the directions are, after all, 'A-conjugate' where **A** is the matrix that defines (at least in part) the specific quadratic function of interest. Thus the whole undertaking may appear somewhat pointless inasmuch as the minimum of a quadratic function can easily be obtained without the need for a numerical search process. (For the quadratic criterion function assumed in the CD Lemma above, the minimising argument is given by $\mathbf{p}^* = -\mathbf{A}^{-1}\,\mathbf{b}$.)

The escape from this apparent dilemma is via the observation made earlier that any criterion function has a quadratic approximation in some suitably small region around its minimising argument $\mathbf{p}^*$. Thus if a minimisation process can move into this region then the properties of the conjugate directions will result in rapid convergence upon $\mathbf{p}^*$. But it needs to be appreciated that in the general case, the specific quadratic function is never known hence any practical conjugate directions method needs to internally generate directions that will ultimately be A-conjugate even though there is no knowledge of the characterising matrix, **A**. Although this may appear to be a formidable task, numerous such procedures have been developed. The family of conjugate gradient methods is included among these procedures.

The original function minimisation procedure in the conjugate gradient family was proposed by Fletcher and Reeves [9.7]. The *k*th step in the procedure ($k \geq 1$) begins with the current estimate of the minimising argument $\mathbf{p}^{k-1}$ and a search direction $\mathbf{r}^{k-1}$. There are two tasks carried out during the step. The first generates a new estimate for the minimising argument denoted $\mathbf{p}^k$, where

$$\mathbf{p}^k = \mathbf{p}^{k-1} + \alpha^* \mathbf{r}^{k-1} \quad \text{and} \quad J(\mathbf{p}^k) = \min_{\alpha} J(\mathbf{p}^{k-1} + \alpha\,\mathbf{r}^{k-1}).$$

In other words, $\mathbf{p}^k$ is the result of a line search from $\mathbf{p}^{k-1}$ in the direction $\mathbf{r}^{k-1}$.

The second task carried out on the kth step is the generation of a new search direction, denoted $\mathbf{r}^k$, where

$$\mathbf{r}^k = -\mathbf{J}_p(\mathbf{p}^k) + \beta_{k-1}\,\mathbf{r}^{k-1} \quad \text{with} \quad \beta_{k-1} = \frac{\|\mathbf{J}_p(\mathbf{p}^k)\|}{\|\mathbf{J}_p(\mathbf{p}^{k-1})\|}\,.$$

(In the above, for an *m*-vector $\mathbf{v}$ we use $\|\,\mathbf{v}\,\|$ to represent the square of the Euclidean length of $\mathbf{v}$ which is given by $\mathbf{v}^T\mathbf{v}$).

For the first step in this procedure, (i.e., when $k = 1$), $\mathbf{p}^0$ is an initial 'best' estimate of the minimising argument and $r^0 = -\,\mathbf{J}_p(\mathbf{p}^0)$. The sequence of steps ends when some predefined termination criterion is satisfied (e.g., a point $\mathbf{p}^k$ is located at which the length of the gradient vector; i.e., sqrt($\|\mathbf{J}_p(\mathbf{p}^k)\|$), is sufficiently small.

The significant feature of this procedure is that when the criterion function $J(\mathbf{p})$ is quadratic then the search directions $\mathbf{r}^0, \mathbf{r}^1, \ldots, \mathbf{r}^k$ that are generated are A-conjugate. Consequently it follows from the CD Lemma that the minimising argument of J will be located in at most m steps (or m line searches).

A number of variations on this original procedure have been proposed. Several of these have suggested alternate values for β_{k-1} and others have tried to better accommodate the reality of nonquadratic criterion functions. For example, Polack and Ribière [9.20] have proposed

$$\beta_{k-1} = \frac{(\mathbf{J}_p(\mathbf{p}^k))^T(\mathbf{J}_p(\mathbf{p}^k) - \mathbf{J}_p(\mathbf{p}^{k-1}))}{\|\mathbf{J}_p(\mathbf{p}^{k-1})\|} \quad ,$$

and Sorenson [9.25] recommends

$$\beta_{k-1} = \frac{(\mathbf{J}_p(\mathbf{p}^k))^T(\mathbf{J}_p(\mathbf{p}^k) - \mathbf{J}_p(\mathbf{p}^{k-1}))}{(\mathbf{p}^{k-1})^T(\mathbf{J}_p(\mathbf{p}^k) - \mathbf{J}_p(\mathbf{p}^{k-1}))} \quad .$$

It's perhaps worth observing that if β_{k-1} is set to zero, then the procedure outlined above becomes the classic steepest descent process. The practical performance of that approach, however, is poor and its selection is not recommended, especially in view of the far superior alternatives that are conveniently available.

Suggestions have also been made for 'restarting' the conjugate gradient procedure in some cyclic fashion, in other words, abandoning the collection of search directions that have been generated and reinitiating the procedure (which usually means choosing the negative gradient as the search direction). The procedure's m-step property when applied to a quadratic function suggests that after a cycle of m-steps (or line searches), the procedure could be reinitialised. Although the natural choice for the restart direction is the negative gradient, Beale [9.1] has shown that the finite termination property on the quadratic function can be maintained even when the first search direction is not the negative gradient. Based on this observation, a restart strategy that incorporates a novel specification for the search directions was proposed. The approach suggested by Beale was further developed by Powell [9.21].

The line search problem is one which, on first glance, appears deceptively simple to resolve (see Figure 9.3). After all, there is only a single parameter α that needs to be considered and it is usually known that the minimising value of α is positive. There is even an easily established orthogonality condition that the minimizing argument α^* must satisfy; namely,

$$(\mathbf{J}_p(\mathbf{p}^{k-1} + \alpha^* \mathbf{r}^{k-1}))^T \mathbf{r}^{k-1} = 0 \ .$$

Nevertheless, obtaining an accurate solution to the problem can be a challenging numerical task. Note also that there is an implicit requirement for efficiency because a line search problem needs to be solved on each step of the procedure and indeed, the solution of these subproblems consumes a substantial part of the computational effort in solving the underlying criterion function minimisation problem.

A variety of approaches can be considered for solving the line search problem. The first that usually comes to mind is a polynomial fitting process. For example, by evaluating $J(\alpha)$ at three 'test' values of α, it is possible to obtain a quadratic approximation for J whose minimum can be readily determined. That value can be taken as an approximation (albeit rather crude) for α^*. Various refinements of this approach are clearly possible, for example, obtaining a new quadratic approximation using 'test points' that are in the region of the previous minimum or incorporating a higher order polynomial (possibly cubic).

If it can be assumed that there is available a known interval $\hat{I}$ which contains α^* and that $J(\alpha)$ is unimodal[2] in $\hat{I}$ then an interval reduction technique can be used. This involves the judicious placement of points in a sequence of intervals of decreasing length where decisions to discard portions of each interval in the sequence are made on the basis of the relative size of $J(\alpha)$ at the selected points. The decisions that are made ensure that the retained interval segment contains α^*. The process ends when the current interval length is sufficiently small and then its midpoint is typically chosen to be α^*. Arguments based on maximising the usefulness of each evaluation of J give rise to the placement of points in a manner that is related either to the golden section ratio or to the Fibonacci sequence. A discussion of the underlying ideas can be found in [9.6].

The significance and practical value of carrying out exact line searches is a topic that has received considerable attention in the optimisation literature (e.g. [9.2] and [9.3]). It can, for example, be easily shown that when the line search is not exact the Fletcher–Reeves formula could generate a search direction $\mathbf{r}^k$ that is not a descent direction. A variety of conditions has been proposed for terminating the line search when a sufficient decrease has occurred in the value of the criterion function (e.g., the Wolfe conditions [9.27]). Many of these are outlined in [9.17].

We end the discussion in this Section by addressing a distinctive and important feature of the optimisation problem that has specific relevance to the CTDS realm that is of interest in this textbook.

---

[2] Within the present context, this implies that while α is in $\hat{I}$, $J(\alpha)$ always increases as α moves to the right from α^* and likewise $J(\alpha)$ always increases as α moves to the left from α^*.

The conjugate gradient method (and indeed a large number of other powerful numerical optimisation methods) requires the gradient of the criterion function. In our case the criterion function J is not an analytic function for which required derivative information can be obtained simply by differentiation. Consequently some alternate approach is required.

Recall that the kth component of the gradient vector $\mathbf{J}_p$, evaluated at the specific point $\mathbf{p} = \boldsymbol{\omega}$, is (by definition) given by:

$$\left.\frac{\partial J}{\partial p_k}\right|_{\mathbf{p}=\boldsymbol{\omega}} = \lim_{\varepsilon \to 0} \frac{J(\boldsymbol{\omega} + \varepsilon \mathbf{e}^k) - J(\boldsymbol{\omega})}{\varepsilon} ,$$

where $\mathbf{e}^k$ is the kth column of the $m \times m$ identity matrix. The obvious numerical approximation to this formal definition is:

$$\left.\frac{\partial J}{\partial p_k}\right|_{\mathbf{p}=\boldsymbol{\omega}} \approx \frac{J(\boldsymbol{\omega} + \Delta \mathbf{e}^k) - J(\boldsymbol{\omega})}{\Delta} ,$$

where Δ is a suitably small positive scalar. With this approach, each of the m components of the gradient vector can be individually approximated. Determination of $\mathbf{J}_p(\boldsymbol{\omega})$ requires m evaluations of J where each such evaluation corresponds to a small perturbation in one of the components of the reference point $\boldsymbol{\omega}$. (We assume here that the value of J at the reference point $\boldsymbol{\omega}$, i.e., $J(\boldsymbol{\omega})$, is already known.) Selecting the most appropriate value for the perturbation Δ requires careful consideration because 'small' is a highly ambiguous notion. If, for example, Δ is 'too small' then the result obtained can become hopelessly corrupted by numerical noise. Nevertheless, with proper care the approach can usually be sufficiently accurate to enable an effective implementation of a conjugate gradient minimisation procedure or, indeed, the implementation of any procedure requiring gradient information.

We note nevertheless that one particular case that would merit special caution in this respect is the circumstance where a discontinuity is known to exist in the conceptual model. As pointed out in the discussion in Section 8.4.3, dealing with such models has inherent numerical difficulties and the errors introduced could undermine the success of the gradient approximation outlined above.

9.4 An Application in Optimal Control

Typically an optimal control problem involves the determination of the time trajectory for one or more control inputs to a continuous-time dynamic system in a manner that minimises a prescribed criterion function. This problem appears, on first glance, to be beyond the scope of our interest in this chapter because the determination of entire time trajectories was never part of the intended considerations. We note, however, that a substantial body of literature relating to the solution of this generic problem is available (see, e.g., [9.13], [9.26]) and among the important results that have emerged is the Pontriagin minimum principle. This, in particular, provides a basis for transforming the optimal control problem into a boundary value problem which can then be reformulated as a function minimisation problem. In this section we illustrate this process with a straightforward example.

Our concern in this example is with the control of a first-order irreversible exothermic chemical reaction carried out in a stirred tank reactor. Control of the process is achieved by injecting coolant through a valve into a cooling coil inserted into the reactor. The conceptual model is based on characterising the perturbations around a steady-state condition. It is relatively simple but highly nonlinear. The model is given in Equation (9.1):

$$
\begin{aligned}
x_1'(t) &= -(1+2x_1(t)) + R(t) - S(t) \\
x_2'(t) &= 1 - x_2(t) - R(t)
\end{aligned}
\tag{9.1}
$$

where

$$R(t) = 0.5 + (x_2(t) + 0.5)\exp(y(t))$$

$$y(t) = \frac{25x_1(t)}{x_1(t)+2}$$

$$S(t) = u(t)\,(x_1(t) + 0.25)\ .$$

Here $x_1(t)$ and $x_2(t)$ represent deviations from steady-state temperature and concentration, respectively, and $u(t)$ is the control input. We take $x_1(t_0) = 0.05$ and $x_2(t_0) = 0$ and for convenience we assume that $t_0 = 0$. The objective is to rapidly return the reactor to steady-state conditions ($x_1 = x_2 = 0$) while at the same time avoiding excessive usage of coolant. Choosing $u(t)$ to minimise the following function reflects these objectives:

$$\mathrm{P} = \int_0^1 (x_1^2(t) + x_2^2(t) + 0.1\,u^2(t))dt\ .$$

The application of the minimum principle gives rise to an auxiliary set of differential equations; namely,

$$
\begin{aligned}
v_1'(t) &= v_1(t)\,(2 + u(t)) - Q(t)\,(v_1(\mathrm{t}) - v_2(\mathrm{t})) - 2\,x_1(\mathrm{t}) \\
v_2'(t) &= v_2(\mathrm{t}) - (v_1(\mathrm{t}) - v_2(\mathrm{t}))\exp(y(\mathrm{t})) - 2\,x_2(\mathrm{t})\ ,
\end{aligned}
\tag{9.2}
$$

where

$$Q(t) = \frac{50(x_2(t) + 0.5)exp(y(t))}{(x_1(t) + 2)^2} \qquad (9.3a)$$

and

$$u(t) = 5\, v_1(t)\, (x_1(t) + 0.25)\ . \qquad (9.3b)$$

The solution to Equations (9.1) and (9.2) (for the case where $v_1(1) = v_2(1) = 0$) provides the necessary conditions for the optimality of $u(t)$ as prescribed in Equation (9.3b).

The difficulty that arises here is that initial conditions are given for $x_1(t)$ and $x_2(t)$ (i.e., conditions at $t = 0$) whereas the boundary conditions on $v_1(t)$ and $v_2(t)$ are specified at $t = 1$. In other words there is a need to solve a two-point boundary value problem. Such problems have been extensively studied in the numerical mathematics literature and a variety of methods is available. One approach is to recast the problem as a criterion function minimisation problem within the class considered in this chapter.

In this reformulation, the CTDS model of interest is the group of four differential equations given by Equations (9.1) and (9.2) together with Equation (9.3). We assume the set of initial conditions:

$$x_1(0) = 0.05,\ x_2(0) = 0,\ v_1(0) = p_1,\ v_2(0) = p_2\ ,$$

where p_1 and p_2 are parameters. The values of we seek for p_1 and p_2 are those which yield a minimum value for the criterion function:

$$J(p_1, p_2) = v_1^2(1) + v_2^2(1)\ .$$

Then, provided that the minimisation process yields a minimum value of zero for J (implying $v_1(1) = 0 = v_2(1)$), the value of $u(t)$ which results will be the solution to the original optimal control problem.

The approach which is illustrated in this example has general applicability to a wide range of optimal control problems and is, at least in principle, equally applicable to boundary value problems in general.

9.5 Exercises and Projects

9.1 The general quadratic function of dimension m can be written as:

$$J(\mathbf{p}) = ½\, \mathbf{p}^T \mathbf{A}\, \mathbf{p} + \boldsymbol{b}^T \mathbf{p} + c\ ,$$

where $\mathbf{p}$ is an m-vector, $\mathbf{A}$ is an $m \times m$ positive definite symmetric matrix, and $\mathbf{b}$ is a m-vector. Consider the point $\mathbf{p}^o$ and a search direction $\mathbf{r}$. Show that if α^* solves the line search problem in the direction **r** from the point $\mathbf{p}^o$, that is, α^* has the property that

$$J(\mathbf{p}^o + \alpha^*\mathbf{r}) = \min_{\alpha} J(\mathbf{p}^o + \alpha\,\mathbf{r}) \ ,$$

then

$$\alpha^* = -\frac{\mathbf{r}^T J_p(\mathbf{p}^0)}{\mathbf{r}^T \mathbf{A}\,\mathbf{r}} \ .$$

9.2 Develop a computer program that implements an efficient line search procedure which is based on the golden section search. Details about this approach can be found in [9.10] or [9.22] or at the Wikipedia site: //en.wikipedia.org/wiki/Golden_section_search). Test your program on a variety of quadratic functions by comparing your results with the analytic result given in Problem 9.1.

9.3 The bouncing ball project was introduced in Chapter 2 (Section 2.2.5). The goal is to find a release angle θ_0 that results in the ball's trajectory entering the hole in the ice surface. This task can be formulated as a line search problem in the following way. Consider the criterion function $J(\theta_0) = (H - \hat{x}_k)^2$ where H is the location of the hole and $\hat{x}_k$ is the ball's horizontal position when the kth collision with the ice surface occurs. J has a minimum value of zero when $\hat{x}_k = H$, that is, when the ball falls through the hole on the kth 'bounce'. Because the criterion function J depends only on a scalar parameter (namely, θ_0), the minimisation problem is one-dimensional (hence a line search problem).

The solution requirements also stipulate that there must be at least one bounce before the ball passes through the hole; that is, $k > 1$. This can be handled (somewhat inelegantly) by first finding a value for θ_0 for which the second and third bounces straddle the hole. This value can then be used as the starting point for the line search process.

Embed a syntactically compatible version of the program developed in Problem 9.2 into the Open Desire simulation model for the bouncing ball given in Figure A3.6 and make appropriate changes in the Experiment segment of the program so that it finds a suitable value for θ_0.

9.4 Probably the most intuitively appealing approach for locating the minimising argument of a criterion function $J(\mathbf{p})$ is a succession of line searches along the co-ordinate axes. This implies that the searches are along the directions $\mathbf{e}_1, \mathbf{e}_2, \ldots, \mathbf{e}_m$ where $\mathbf{e}_k$ is the kth column of the $m \times m$ identity matrix ($\mathbf{e}_k$ is an m-vector whose entries are all 0 except for the entry in the kth position which is 1). One notable feature of this procedure (usually called the univariate search) is that it does not require gradient information. It can be viewed as a series of iterations where each iteration begins at the point $\mathbf{p}^{o}$ and ends at the point $\mathbf{p}^{m}$ which is the point that is reached after a sequence of m line searches along the co-ordinate axes. The procedure is illustrated in Figure 9.4 for the two-dimensional case.

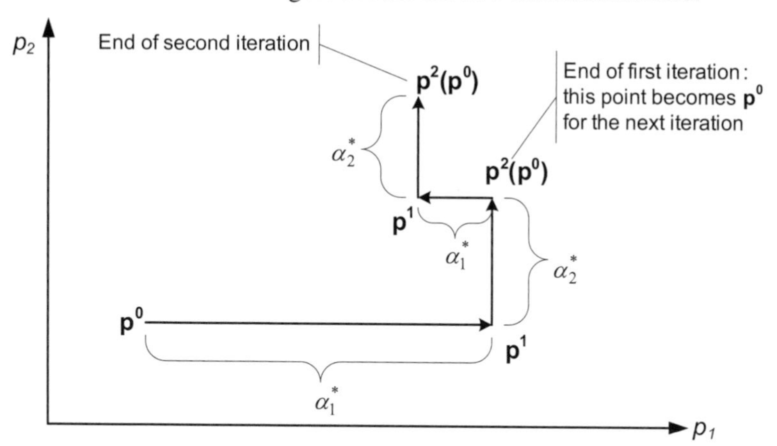

FIGURE 9.4. The univariate search in two dimensions.

The procedure for the univariate search can be written in the following way. Choose a value for the termination parameter ε and an initial estimate $\hat{\mathbf{p}}$ for the minimising argument of J and set $\mathbf{p}^{m} = \hat{\mathbf{p}}$.

repeat

$k = 0$

$\mathbf{p}^{o} \leftarrow \mathbf{p}^{m}$

while $(k < m)$

$k \leftarrow k + 1$

Find α^{*}_{k} such that $J(\mathbf{p}^{k-1} + \alpha^{*}_{k}\, \mathbf{e}^{k}) = \min\limits_{\alpha} J(\mathbf{p}^{k-1} + \alpha \mathbf{e}^{k})$

$\mathbf{p}^{k} \leftarrow \mathbf{p}^{k-1} + \alpha^{*}_{k}\, \mathbf{e}^{k}$

endwhile

$J_{max} \leftarrow \max(|J(\mathbf{p}^{o})|, |J(\mathbf{p}^{m})|)$

until $\left| \dfrac{J(\mathbf{p}^{0}) - J(\mathbf{p}^{m})}{J_{max}} \right| < \varepsilon$

The value of $\mathbf{p}^{m}$ upon termination of the repeat/until loop is the accepted estimate for the minimising argument.

Show that the univariate procedure will converge to the minimising argument of the general quadratic function given in Problem 9.1 if exact line searches are carried out.

HINT: Consider what must be true if the procedure makes no progress on some particular iteration and then use the fact that the only m-vector that can be simultaneously orthogonal to m orthogonal m-vectors is the zero vector.

In many situations the performance of the univariate procedure outlined in Problem 9.4 can be significantly improved by incorporating a slight modification. This simply involves an additional line search in the direction $\mathbf{s} = (\mathbf{p}^m - \mathbf{p}^o)$. This modified procedure (which we call the extended univariate search) is illustrated in Figure 9.5.

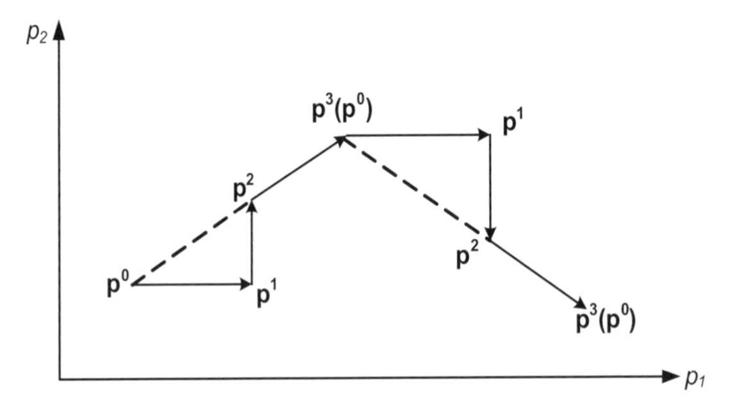

FIGURE 9.5. The extended univariate serach in two dimensions.

a) Modify the procedure given in Problem 9.4 so that it represents the extended univariate search as described above.
b) Formulate an argument that demonstrates that the extended univariate search will also locate the minimising argument of the general quadratic function.

9.5 a) Write a program that implements the univariate search procedure as presented in Problem 9.4. Incorporate the line search program that was developed in Problem 9.2.

b) Test the effectiveness of the program using the following two test problems.

i. $J(\mathbf{p}) = 100(p_2 - p_1^2)^2 + (1 - p_1)^2$.

ii. $J(\mathbf{p}) = (p_1 + 10p_2)^2 + 5(p_3 - p_4)^2 + (p_2 - 2p_3)^4 + 10(p_1 - p_4)^4$.

Use initial estimates (0,1) and (1,0) for test problem (i) and initial estimates (1, 0, 1, 0) and (–1, 0, 0, 1) for test problem (ii). The minimum value of the criterion function for both test problems is zero. The termination parameter ε should be set to a value no larger than 10^{-5}.

9.6 Repeat Problem 9.5 for the case of the extended univariate search.

9.7 Develop an Open Desire simulation program to solve the optimal control problem that is outlined in Section 9.4. Use a syntactically compatible version of the extended univariate search program developed for Problem 9.6 to solve the criterion function minimisation problem. Use (2, 2) as an initial estimate of the minimising argument of the criterion function.

9.6 References

9.1. Beale, E.M.L., (1972), A derivation of conjugate gradients, in F.A. Lottsma (Ed.), *Numerical Methods for Non-Linear Optimization*, Academic Press, London, pp. 39–43.

9.2. Al-Baali, (1985), Descent property and global convergence of the Fletcher-Reeves method with inexact line search, *IMA Journal on Numerical Analysis*, **5**: 121–124.

9.3. Bertsekas, D.P., (1996), *Constrained Optimization and Lagrange Multiplier Methods*, Athena Scientific, Nashua, NH.

9.4. Bhatnager, S. and Kowshik, H.J., (2005), A discrete parameter stochastic approximation algorithm for simulation optimization, *Simulation*, **81**(11).

9.5. Bonnans, J.F., Gilbert, J.C., Lemaréchal, C., and Sagastizabal, C.A., (2003), *Numerical Optimization: Theoretical and Practical Aspects*, Springer-Verlag, Berlin.

9.6. Cormen, T.H., Leisserson, C.E., and Rivest, R.L., (1990), *Introduction to Algorithms*, MIT Press, Cambridge, MA.

9.7. Fletcher, R. and Reeves, C.M., (1964), Function minimization by conjugate gradients, *Computer Journal*, **7**: 149–154.

9.8. Fletcher, R., (1987), *Practical Methods of Optimization*, 2nd edn., John Wiley, New York.

9.9. Gilbert, J. and Nocedal, J., (1992), Global convergence properties of conjugate gradient methods for optimization, *SIAM Journal on Optimization*, **2**: 21–42.

9.10. Heath, M.T., (2000), *Scientific Computing, An Introductory Survey*, 2nd edn., McGraw-Hill, New York.

9.11. Lagarias, J.C., Reeds, J.A., Wright, M.H., and Wright, P.E., (1998), Convergence properties of the Nelder-Mead simplex method in low dimensions, *SIAM Journal of Optimization*, **9**: 112–147.

9.12. Law, A.M. and Kelton, D.W., (2000), *Simulation Modeling and Analysis*, (3rd edn.), McGraw-Hill, New York.

9.13. Lewis, F.L. and Syrmos, V.L. (1995), *Optimal Control*, 2nd edn., John Wiley and Sons, New York.
9.14. Nelder, J. and Mead, R., (1965), A simplex method for function minimization, *Computer Journal*,, **7**: 308–313.
9.15. Nocedal, J. and Wright, S.J., (1999), *Numerical Optimization*, Springer-Verlag, New York.
9.16. Olafason, S. and Kim, J. (2002), Simulation optimization, in *Proceedings of the 2002 Winter Simulation Conference*, pp. 79–84.
9.17. Oretega, J.M. and Rheinboldt, W.C., (1970), *Iterative Solution of Nonlinear Equations in Several Variables*, Academic Press, New York.
9.18. Pedregal, P., (2004), *Introduction to Optimization*, Springer-Verlag, New York.
9.19. Pinter, Janos D., Linder, David and Chin, Paulina (2006), Global Optimization toolbox for maple: an introduction with illustrative applications, *Optimization Methods and Software,* vol 21(4), pp. 565-582.
9.20. Polack, E. and Ribière, G., (1969), Note sur la Convergence de Méthodes de Directions Conjuguées, *Revue Française d'Informatique et de Recherche Opérationnelle*, **16**: 35–43.
9.21. Powell, M.J.D., (1978), Restart procedures for the conjugate gradient method, *Mathematical Progress, 12*: 241–254.
9.22. Press, W.H., Teukolsky, S.A., Vetterling, W.T., and Flannery, B.P., (1999), *Numerical Recipes in C, The Art of Scientific Computing*, 2nd edn., Cambridge University Press, Cambridge, UK.
9.23. Rubinstein, R. and Shapiro, A., (1993), *Discrete Event Systems: Sensitivity Analysis and Stochastic Optimization by the Score Function Method*, John Wiley, New York.
9.24. Rykov, A., (1983), Simplex algorithms for unconstrained optimization, *Problems of Control and Information Theory*, **12**: 195–208.
9.25. Sorenson, H.W., (1969), *Comparison of Some Conjugate Directions Procedures for Function Minimization*, J Franklin Institute, vol. 288, pp. 421–441.
9.26. Seierstad, A. and Sydstaeter, K., (1987), *Optimal Control Theory with Economic Applications*, North Holland, Amsterdam.
9.27. Wolfe, P., (1969), Convergence conditions for ascent methods, *SIAM Review*, **11**: 226–235.
9.28. Zabinsky, Z.B., (2003), *Stochastic Adaptive Search for Global Optimization*, Kluwer Academic, Dordrecht.

Annex 1 Probability Primer

A1.1 Motivation

The exploration of a number of important issues relating to the development of DEDS-based models and to the correct interpretation of the results acquired from experiments with simulation programs developed for such models (i.e., 'simulation experiments') requires some familiarity with the fundamental notions of probability theory. Our purpose in this annex is to briefly review these so that these discussions (e.g., in Chapters 3 and 6) will have a reasonable foundation. Our presentation is relatively informal; more extensive and mathematically comprehensive treatments can be found in references such as [A1.1], [A1.2], or [A1.3].

A1.2 Random Experiments and Sample Spaces

We begin with the basic notion of a *random experiment*, which is an experiment whose outcome is uncertain or unpredictable. Although unpredictable, it is nevertheless assumed that the outcome of such an experiment will fall within some known set of possible outcomes called the *sample space*. A sample space that consists of a finite number, or an infinite sequence, of members is called a discrete sample space. A sample space may also be an interval or a collection of disjoint intervals from the real line, in which case the sample space is said to be continuous.

Consider, for example, the random experiment that corresponds to rolling two dice, one red and the other green. If we regard the outcome of this experiment to be the pair of integers (n_1,n_2) which corresponds to the number of dots on the red and green dice, respectively, then the sample space is the set

$$\hat{S} = \{(1,1), (1,2), \ldots, (6,6)\} ,$$

which is discrete and has 36 members. An alternate representation of $\hat{S}$ which provides visual features that are helpful in later examples is shown below.

$$\hat{S} = \begin{Bmatrix} (1,1) & (1,2) & (1,3) & (1,4) & (1,5) & (1,6) \\ (2,1) & (2,2) & (2,3) & (2,4) & (2,5) & (2,6) \\ (3,1) & (3,2) & (3,3) & (3,4) & (3,5) & (3,6) \\ (4,1) & (4,2) & (4,3) & (4,4) & (4,5) & (4,6) \\ (5,1) & (5,2) & (5,3) & (5,4) & (5,5) & (5,6) \\ (6,1) & (6,2) & (6,3) & (6,4) & (6,5) & (6,6) \end{Bmatrix} . \qquad \text{(A1.1)}$$

Or alternately, consider the experiment of flipping a single coin. Here there are only two members in the sample space and these are usually designated as 'heads' and 'tails' (the possibility of the coin landing on its edge is excluded). In other words the discrete sample space in this case is:

$$\check{S} = \{\text{heads, tails}\} \ .$$

A continuous sample space can be illustrated by considering a manufacturing process that produces ball-bearings that have a nominal diameter of 1 cm. Because of inherent imprecision at the various manufacturing stages, the actual diameters of the ball-bearings have random variations. A basic screening device ensures that all bearings that are not in the range [0.9, 1.1] cm are rejected. Before packaging the bearings in the 'accepted stream', the diameter of each one is measured as part of a quality control procedure. The sample space for this experiment (the result obtained from the diameter measurements) is, at least in principle, the interval [0.9, 1.1]. (The qualifier 'at least in principle' is intended to acknowledge the finite precision of any measuring device).

The notions of probability theory are, of course, equally applicable to random experiments with either discrete or continuous sample spaces. However, acquiring an appreciation for the ideas involved is considerably simpler within the framework of discrete sample spaces. Consequently the presentation that follows first explores the fundamentals exclusively in the simpler framework of discrete sample spaces. Extension of these notions to the case of continuous sample spaces is presented later.

A1.3 Discrete Sample Spaces

A1.3.1 Events

An *event* is any subset of a sample space. The individual elements that populate a discrete sample space are therefore valid events but they are somewhat special and hence are usually called *simple events*. A total of 2^n events can be associated with a sample space having n elements. Included here is the entire sample space as well as the empty set.

Within the context of a random experiment, an event $\tilde{E}$ is said to occur if the outcome of the experiment is a member of $\tilde{E}$. For example, in the rolling dice experiment, we might designate an event E as

$$\tilde{E} =. \{(1,6), (2,5), (3,4), (4,3), (5,2), (6,1)\} \ .$$

This set has been constructed by selecting all the possible experiment outcomes for which the sum of the number of dots displayed on the two dice is 7. Thus, this event $\tilde{E}$ occurs if a 7 is rolled.

As we have pointed out, a large number of events can be associated with a random experiment. It is of some interest to observe that predefined events can be combined to yield 'new' events. There are two basic ways of carrying out such a combination, namely, intersection and union.

a) The intersection of two events E_1 and E_2 is denoted $E_1 \cap E_2$ and represents the event that corresponds to those experiment outcomes (simple events) that are common to both E_1 and E_2. Clearly, $E_1 \cap E_2 = E_2 \cap E_1$. Note that it is possible that $E_1 \cap E_2$ is the empty set φ.
b) The union of two events E_1 and E_2 is denoted $E_1 \cup E_2$ and represents the event that corresponds to those experiment outcomes (simple events) that are either in E_1 or in E_2 or in both. Clearly, $E_1 \cup E_2 = E_2 \cup E_1$.

A1.3.2 Assigning Probabilities

A fundamental concern in probability theory is with the characterisation of the likelihood of the occurrence of events where the notion of an 'event' is interpreted in the special sense outlined above. However a prerequisite for any such exploration is the assignment of a 'valid' set of probabilities to the simple events in the sample space under consideration. Each assigned value is intended to reflect the likelihood that that simple event will be the outcome of the random experiment.

Consider then a sample space S with n members (i.e., n simple events):

$$S = \{s_1, s_2, s_3, \ldots, s_n\} \ .$$

The probability assigned to the simple event s_k is a nonnegative real number denoted $p(s_k)$. An assignment of probabilities to the members of S is valid if:

$$\sum_{k=1}^{n} p(s_k) = 1 . \tag{A1.2}$$

Notice that because $p(s_k) \geq 0$ for each k, the validity condition of Equation (A1.2) naturally implies the constraint that $p(s_k) \leq 1$ for each k.

The assignment of a valid set of probabilities to the simple events in a sample space has no fundamental and/or formal basis. However, the chosen values generally do have a logically or psychologically satisfying basis. Often they reflect some idealised circumstance (such as an ideal coin or a pair of unloaded dice). In any case, once a valid set of probabilities has been assigned, it is often referred to as a 'probability model' for the random phenomenon under consideration.

To illustrate, let's return to the rolling dice example. A valid assignment of probabilities to each of the 36 possible outcomes is 1/36. This particular assignment corresponds to an assumption that both the red and green dice are unbiased and consequently each of the 36 possible outcomes is equally likely. This is a particular case of the general result that if a random experiment has N possible outcomes and each is equally likely, then assigning a probability of $1/N$ to each of the possible outcomes (i.e., to each of the simple events in the sample space) is both a valid and a logically satisfying assignment of probabilities.

Suppose E is an event associated with some random experiment. The notation $P[E]$ represents the probability that the outcome of the experiment will be the event E. In particular, note that for any simple event s_k, $P[s_k] = p(s_k)$. Furthermore, if E_a is the event:

$$E_a = \{s_1, s_2, \quad --- s_m\}$$

where each s_j is a simple event, then

$$P[E_a] = \sum_{i=1}^{m} p(s_j) . \tag{A1.3}$$

Suppose in the rolling dice example, we consider the event:

$$E = \{(1,6), (2,5), (3,4), (4,3), (5,2), (6,1)\} .$$

As pointed out earlier, this event E occurs if and only if a 7 is rolled. From Equation (A1.3) it follows that $P[E] = 6/36 = 1/6$; in other words, the probability of rolling a 7 is 1/6.

A fundamental result that relates the union and intersection of two events E_1 and E_2 is the following.

$$P[E_1 \cup E_2] = P[E_1] + P[E_2] - P[E_1 \cap E_2] . \tag{A1.4}$$

A1.3.3 Conditional Probability and Independent Events

Recall again that a large number of events can be associated with a random experiment. The notion of 'conditional probability' is concerned with the following question: what impact can knowledge that an event, say E_2, has occurred have on the probability that another event, say E_1, has occurred? As an extreme case (in the context of the rolling dice example), suppose E_1 is the event 'the sum of the dots showing is odd' and E_2 is the event 'the sum of the dots showing is seven.' Clearly if it is known that E_2 has occurred, then the probability of E_1 is 1 (i.e., certainty). Without the knowledge that E_2 has occurred, $P[E_1] = 0.5$ (this can be easily verified by examining the sample space $\hat{S}$ given in Equation (A1.1)).

In general, if E_1 and E_2 are two events associated with a random experiment then the probability that the event E_1 has occurred given that event E_2 has occurred (hence $P[E_2] \neq 0$) is called 'the conditional probability of E_1 given E_2.' Such a conditional probability is written as $P[E_1 \mid E_2]$. The following is a fundamental result relating to conditional probabilities:

$$P[E_1 \mid E_2] = P[E_1 \cap E_2] \,/\, P[E_2] \ . \qquad \text{(A1.5)}$$

Notice that if the events E_1 and E_2 have no simple events in common (i.e., $E_1 \cap E_2 = \phi$ (the empty set)), then $P[E_1 \mid E_2] = 0$ because $P[\phi] = 0$. This result is entirely to be expected because if E_2 has occurred and E_1 shares no simple events with E_2, then the occurrence of E_1 is not possible. Notice also that if E_1 and E_2 are mutually exclusive ($E_1 \cap E_2 = \phi$) then there is a suggestion of a 'lack of independence' (i.e., dependence) in the sense that if one occurs then the simultaneous occurrence of the other is excluded. This notion of independence is explored further in the discussion that follows.

From Equation (A1.5) it necessarily follows that the conditional probability of E_2 given E_1 is:

$$P[E_2 \mid E_1] = P[E_1 \cap E_2] \,/\, P[E_1] \ ,$$

where we have incorporated the fact that $E_2 \cap E_1 = E_1 \cap E_2$. Combining this result with Equation (A1.5) gives:

$$P[E_1 \cap E_2] = P[E_1] * P[E_2 \mid E_1] = P[E_2] * P[E_1 \mid E_2] \ . \qquad \text{(A1.6)}$$

Consider again our earlier example where E_1 and E_2 are two specific events within the context of the rolling dice example (E_1 is the event 'the sum of the dots showing is odd' and E_2 is the event 'the sum of the dots showing is seven'). We already know $P[E_1] = 0.5$ and $P[E_2] = 1/6$. Also $E_1 \cap E_2 = E_2$ (because E_2 is a subset of E_1) hence $P[E_1 \cap E_2] = P[E_2]$. Thus it follows from Equation (A1.5) that

$$P[E_1 \mid E_2] = P[E_2] \,/\, P[E_2] = 1$$

which confirms the result from our earlier intuitive arguments.

Consider, on the other hand, the conditional probability $P[E_2|E_1]$, namely, the probability of E_2 given E_1. From Equation (A1.5) we have:

$$P[E_2 \mid E_1] = P[E_2 \cap E_1]/P[E_1] = (1/6)/\, 0.5 = 1/3 \;\; .$$

In other words the probability that the outcome of the roll is 7 is doubled if it is already known that the outcome of the roll has an odd sum.

To illustrate another possibility, consider the case where we alter the definition of E_2 to be the event that the red dice show an even number. Now

$$E_1 \cap E_2 = \{(2,1), (2,3)\; (2,5),(4,1),(4,3),(4,5),(6,1),(6,3),(6,5)\}$$

and $P[E_1 \cap E_2] = 1/4$. Then from Equation (A1.5) we have:

$$P[E_1 \mid E_2] = P[E_1 \cap E_2] \,/\, P[E_2] = 0.25/0.5 = 0.5 \;\; .$$

Here the information that event E_2 has occurred (red dice show an even number) does not alter the probability of the occurrence of E_1 (the sum of the dots is odd).

A particularly important relationship that can exist between two events is that of independence. The event E_1 is said to be independent of the event E_2 if

$$P[E_1 \mid E_2] = P[E_1] \;\; . \qquad \text{(A1.7)}$$

In other words, knowledge that event E_2 has occurred does not change the probability of event E_1. This is the case that is illustrated in the previous example.

From Equation (A1.6), it follows that $P[E_1]$ in Equation (A1.7) can be replaced with $P[E_2] * P[E_1 \mid E_2] \,/\, P[E_2 \mid E_1]$ which then gives the result that

$$P[E_2 \mid E_1] = P[E_2].$$

In other words if event E_1 is independent of event E_2 then E_2 is independent of E_1.

Note now that if the independence condition $P[E_1 \mid E_2] = P[E_1]$ is substituted in Equation (A1.5) then we get the following equivalent condition for independence,

$$P[E_1 \cap E_2] = P[E_1] * P[E_2] \;\; , \qquad \text{(A1.8)}$$

which, in particular, implies that $P[E_1 \cap E_2] \neq 0$ is a necessary condition for the events E_1 and E_2 to be independent.

A1.3.4 Random Variables

Suppose that S is a sample space. A *random variable* X is a function which assigns a real value to each $s \in S$. (This is sometimes equivalently expressed by saying that X is a mapping from the sample space S to some particular set of the real numbers.) In more formal terms, a random variable is a function whose domain is a sample space and whose range is a set of real numbers. If Ψ represents the range then this assertion, in conventional mathematical notation, is written as

$$X: S \rightarrow \Psi \ . \tag{A1.9}$$

The feature that needs to be emphasised here is that a random variable is not, in fact, a variable but rather is a function! In spite of this apparently inappropriate terminology, the usage has become standard. It is often useful to review this definition when related concepts become cloudy.

It is interesting to observe that if $X: S \rightarrow \Psi$ is a random variable and g is a real-valued function whose domain includes the range of X, that is,

$$g: \Phi \rightarrow R \ ,$$

where $\Psi \subseteq \Phi$, then $X^* = g(X)$ is also a random variable because for each $s \in S$, $X^*(s)$ has the numerical value given by $g(X(s))$. In other words, X^* is likewise a mapping from the sample space S to the real numbers.

The notation of Equation (A1.9) conveys the important information that X maps each member of S onto a value in Ψ. However, it does not reveal the rule that governs how the assignment takes place. More specifically, if $s \in S$ we need to know the rule for determining the value in Ψ that corresponds to s (this value is denoted by $X(s)$).

To illustrate, let's continue with the earlier example of the pair of rolled dice and introduce a random variable Y whose value is the sum of the dots showing on the two dice. In this case, the domain set is $\hat{S}$ as given in Equation (A1.1) and the range set Ψ_y is:

$$\Psi_y = \{2, 3, 4, 5, \ldots, 11, 12\} \ .$$

The definition for Y is:

$$Y(s) = n_1 + n_2 \quad \text{when } s = (n_1, n_2) \ .$$

The random variable Y is one of many possible random variables that can be identified for this random experiment of throwing two (distinct) dice. Consider, for example, Z, whose value is the larger of the two numbers that show on the pair of dice. The range set in this case is:

$$\Psi_z = \{1, 2, \ldots, 6\}$$

and the definition for Z is:

$$Z(s) = \max(n_1, n_2) \quad \text{when } s = (n_1, n_2) \ .$$

Both random variables Y and Z have the same domain (namely, $\hat{S}$ as given in Equation (A1.1)) but different range sets. Note also that in both of these cases the size of the range set is smaller than the size of the domain set, implying that several values in the sample space get mapped onto the same value in the range set (in other words, both Y and Z are 'many-to-one' mappings). This is frequently the case but certainly is not essential. In the discussion that follows, we frequently make reference to the 'possible values' which a random variable can have; this refers to the range set for that random variable.

Consider the random variable X: $S \rightarrow \Psi$. Suppose we use # to represent any one of the relational operators $\{=, <, >, \leq, \geq\}$. Observe now that for each $x \in \Psi$, the set

$$\{s \in S: X(s) \# x\}$$

is, in fact, an event . The conventional simplified notation for this event is $(X \# x)$ and the probability of its occurrence is $P[X \# x]$. With any specific choice for the operator #, $P[X \# x]$ can be viewed as a function of x. Two particular choices have special relevance:

1. Choose # to be '='. This yields $f(x) = P[X = x]$ which is generally called the probability mass function (or simply the probability function) for the random variable X.
2. Choose # to be '≤'. This yields $F(x) = \mathrm{P}[X \leq x]$ which is generally called the cumulative distribution function (or simply the distribution function) for the random variable X.

The cumulative distribution function F, for the random variable X, has several important properties that are intuitively apparent:

- $0 \leq F(x) \leq 1, -\infty < x < \infty$.
- If $x_1 < x_2$ the $F(x_1) \leq F(x_2)$; that is, $F(x)$ is nondecreasing.
- $\lim_{x \to -\infty} F(x) = 0$ and $\lim_{x \to \infty} F(x) = 1$.

Earlier we introduced the random variable Y: $\hat{S} \rightarrow \Psi_y$ to represent the sum of the dots showing on the dice in our dice rolling example. The probability mass function for Y is shown in Figure A1.1a and the cumulative distribution function is shown in Figure A1.1b.

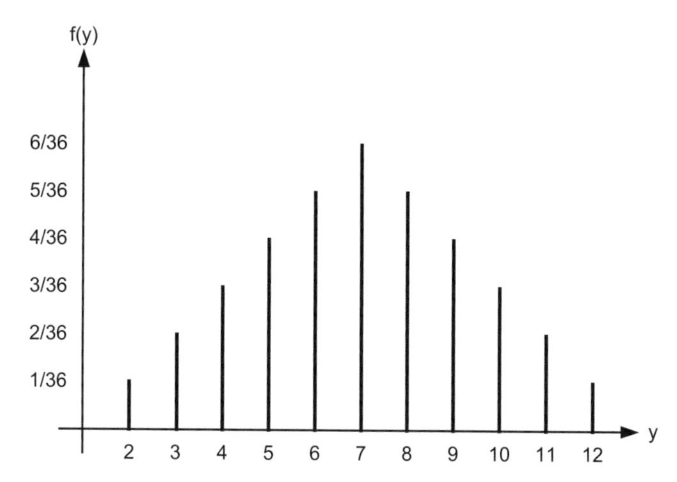

FIGURE A1.1a. Probability mass function for the random variable Y.

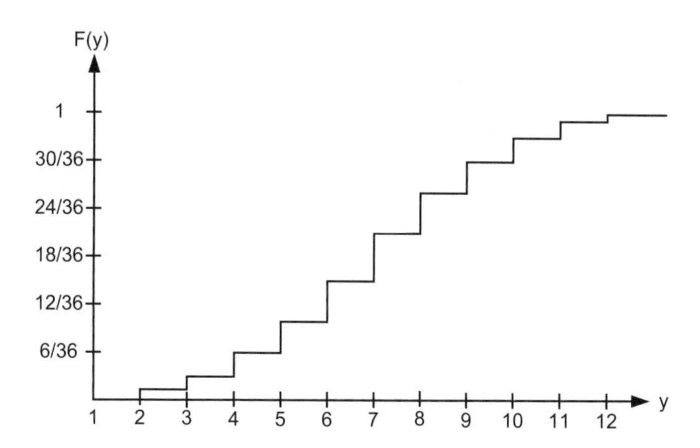

FIGURE A1.1b. Cumulative distribution function for the random variable Y.

A1.3.5 Expected Value, Variance, and Covariance

Let X be a random variable with possible values $x_1, x_2, \ldots, x_n$ and probability mass function $f(x)$. Thus the possible values occur with probabilities $f(x_1), f(x_2), \ldots, f(x_n)$, respectively. The *expected value* (or *mean*) of X is denoted $E[X]$ and has the value:

$$E[X] = \sum_{k=1}^{n} x_k f(x_k) \quad . \tag{A1.10}$$

The expected value of X is the weighted sum of its possible values where each value is weighted by the probability of its occurrence.

Consider again the random variable Y that represents the sum of the dots showing on the pair of rolled dice. From the information presented in Figure A1.1a it can be easily verified that $E[Y] = 7$.

Suppose X is a random variable with possible values $x_1, x_2, \ldots, x_n$ and probability mass function $f(x)$. Let the random variable Y be specified as $Y = g(X)$, then:

$$E[Y] = E[g(X)] = \sum_{k=1}^{n} g(x_k) f(x_k) \ . \tag{A1.11}$$

Using (A1.344) it can be easily demonstrated that:

$$E[aX + b] = aE[X] + b \ . \tag{A1.12}$$

The expected value of a random variable X is one of the parameters of its probability mass function and it serves to reflect where it is 'located' or 'centred' in the sense that it provides a rough indication of the middle of the likely values that X can assume. Another property (or parameter) that is generally of interest is a measure of the spread or dispersion of the probability mass function. For this purpose we might consider evaluating the expected value of the deviation of X from its mean, namely $E[X - \mu_x]$ where we have introduced μ_x as an alternate symbol for $E[X]$ to simplify notation. However, from (A1.12) it follows (taking $a = 1$ and $b = -E[X]$) that

$$E[(X - \mu_x)] = E[X] - E[X] = 0 \ ,$$

which shows that this particular measure is not useful.

The measure of dispersion that is generally used is called the *variance* of X which is denoted Var[X]. Its definition is:

$$\text{Var}[X] = E[(X - \mu_x)^2] = \sum_{k=1}^{n} (x_k - \mu_x)^2 f(x_k) , \tag{A1.13}$$

where, as earlier, we assume that $x_1, x_2, \ldots, x_n$ are the possible values for the random variable X and its probability mass function is $f(x)$. A related notion is the *standard deviation* of X which is normally denoted σ_x and defined as

$$\sigma_x = \sqrt{\text{Var}[X]}$$

(hence Var[X] = σ_x^2).

Two properties relating to the variance of X are especially noteworthy and can be easily demonstrated:

- $\text{Var}[X] = E[X^2] - \mu_x^2$.
- $\text{Var}[aX + b] = a^2 \text{Var}[X]$ where a and b are arbitrary constants.

The second of these (with $a = 1$) reinforces the dispersion feature of the variance notion inasmuch as it demonstrates (in a formal mathematical sense) that the addition of a constant to a random variable does not alter the dispersion of its distribution, a result which is entirely consistent with intuitive expectation.

It is sometimes convenient to 'normalise' a random variable X by creating an associated variable which we denote Z, where

$$Z = (X - \mu_x)/\sigma_x \ .$$

The random variable Z is called the *standardised* random variable corresponding to X. It can be demonstrated that $E[Z] = 0$ and $\text{Var}[Z] = 1$.

Covariance is a notion that relates to a pair of random variables, X and Y, and provides a measure of their mutual dependence. The covariance of the random variables X and Y is denoted $\text{Cov}[X,Y]$ and, by definition,

$$\text{Cov}[X,Y] = E[(X - \mu_x)\,(Y - \mu_y)] \ ,$$

where $\mu_x = E[X]$ and $\mu_y = E[Y]$. An alternate expression can be obtained by expanding the right-hand side and using the distributive property of the expected value operator. This yields:

$$\text{Cov}[X,Y] = E[XY] - E[X]E[Y] \ . \qquad \text{(A1.14)}$$

The covariance relationship is especially useful because it makes possible a convenient expression for the variance of the sum of two random variables. We have that:

$$\begin{aligned}\text{Var}[X + Y] &= E[(X + Y - \mu_x - \mu_y)^2] \\ &= E[(X - \mu_x)^2 + (Y - \mu_y)^2 + 2\,(X - \mu_x)\,(Y - \mu_y)] \qquad \text{(A1.15)} \\ &= \text{Var}[X] + \text{Var}[Y] + 2\,\text{Cov}[X,Y] \ .\end{aligned}$$

On the basis of Equations (A1.14) and (A1.15) it is now possible to show that if the random variables X and Y are independent then the variance of their sum is equal to the sum of their variances where independence means that

$$P[X = x_j, Y = y_k] = P[X = x_j]\, P[Y = y_k] \qquad \text{(A1.16)}$$

for all relevant x_j and y_k. .

Observe that the desired result follows directly from Equation (A1.15) if $\text{Cov}[X,Y] = 0$ when X and Y are independent. To show that this is indeed the case we only need to show (from Equation (A1.14)) that $E[XY] = E[X]E[Y]$ when X and Y are independent. where the above

$$\begin{aligned}
E[XY] &= \sum_{j=1}^{n}\sum_{k=1}^{m} x_j y_k P[X = x_j, Y = y_k] \\
&= \sum_{j=1}^{n}\sum_{k=1}^{m} x_j y_k P[X = x_j] P[Y = y_k] \\
&= \sum_{j=1}^{n} x_j P[X = x_j] \sum_{k=1}^{m} y_k P[Y = y_k] \\
&= E[X] E[Y]
\end{aligned} \tag{A1.17}$$

development has used Equation (A1.16). Thus we have demonstrated that if the random variables X and Y are independent then:

$$\mathrm{Var}[XY] = \mathrm{Var}[X]\,\mathrm{Var}[Y] \ . \tag{A1.18}$$

In general the 'strength' of the relationship between the random variables X and Y is measured by a dimensionless quantity $\rho[X, Y]$ called the correlation coefficient where

$$\rho[X,Y] = \frac{\mathrm{Cov}[X,Y]}{\sqrt{\mathrm{Var}[X]\,\mathrm{Var}[Y]}} \ . \tag{A1.19}$$

When X and Y are independent then $\rho[X, Y] = 0$. It can be shown that in the general case, the value of $\rho[X, Y]$ lies in the interval $[-1, +1]$. When the correlation coefficient $\rho[X,Y]$ is positive, a positive correlation exists between X and Y which implies that if a large value is observed for one of the variables then the observed value for the other will likely also be large. Negative correlation indicates that when an observation of one of the variables is large, the observation for the other will tend to be small. The value of the correlation coefficient reflects the level of this dependency. Values close to +1 or –1 indicate high levels of correlation.

A1.3.6 Some Discrete Distribution Functions

A1.3.6.1 Bernoulli Distribution

A random variable X has a Bernoulli distribution with parameter p if it has the two possible values 1 and 0 and $X = 1$ with probability p and $X = 0$ with probability (necessarily) of $(1 - p)$. X could, for example, characterise a random experiment that has two possible outcomes; such as 'success' or 'failure' (such binary-valued experiments are often called Bernoulli trials). The probability mass function for X is:

$$f(1) = P[X = 1] = p$$

$$f(0) = P[X = 0] = (1 - p) \ .$$

The mean value of X is $E[X] = f(1) + 0\, f(0) = p$ and the variance of X is $\mathrm{Var}[X] = p(1-p)^2 + (1-p)\,(0-p)^2 = p(1-p)$.

A1.3.6.2 Binomial Distribution

The binomial distribution can be viewed as a generalisation of the Bernoulli distribution. More specifically, if X is a random variable that represents the number of 'successes' in n consecutive Bernoulli trials, then X has a binomial distribution. In other words, the binomial distribution relates to the probability of obtaining x successes in n independent trials of an experiment for which p is the probability of success in a single trial. If the random variable X has a binomial distribution then its probability mass function is:

$$f(x) = \frac{n!}{x!(n-x)!} \quad p^x (1-p)^{n-x} \ .$$

The mean value of X is $E[X] = n\, p$ and the variance of X is $\mathrm{Var}[X] = n\, p\, (1-p)$.

A1.3.6.3 Poisson Distribution

The Poisson distribution is a limiting case of a binomial distribution. It results when p is set to μ/n and $n \to \infty$. The probability mass function is:

$$f(x) = \frac{\mu^x e^{-\mu}}{x!} \ .$$

If X has a Poisson distribution then the mean and variance of X are equal and $E[X] = \mu = \mathrm{Var}[X]$. The mean value μ is sometimes called the rate parameter of the Poisson distribution.

The Poisson distribution is commonly used as a model for arrival rates, for example, messages into a communication network or vehicles into a service station. Suppose, for example, that the random variable X represents the number of vehicles arriving at a service station per half-hour. If X has a Poisson distribution with mean of 12 (i.e., the rate parameter $\mu = 12$ implying a mean arrival rate of 12 cars per half-hour) then the probability of 6 cars arriving during any particular half-hour interval is:

$$f(6) = \frac{\mu^6 e^{-\mu}}{6!} = \frac{12^6 e^{-12}}{6!} = 0.0255$$

The probability mass function for this example is shown in Figure A1.2.

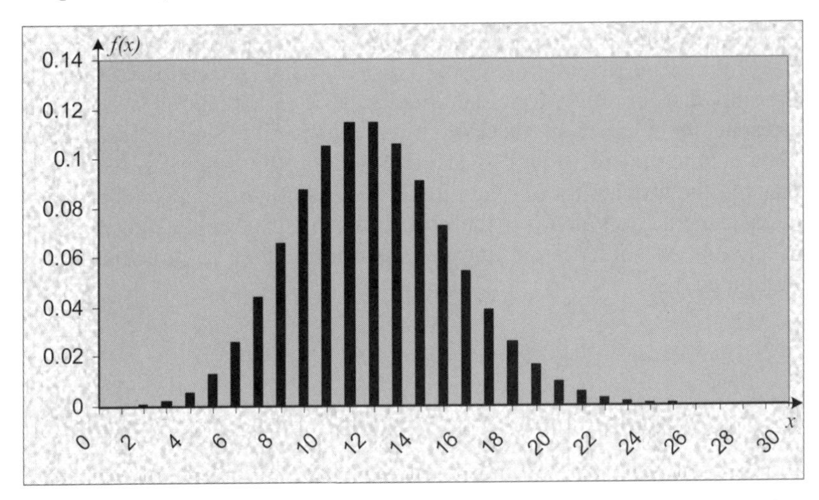

FIGURE A1.2. Example of a probability mass function for the Poisson distribution (μ =12).

A1.4 Continuous Sample Spaces

A1.4.1 Background

Continuous sample spaces are typically associated with experiments that involve the measurement of physical attributes, for example, weight, length, temperature, and time (where the latter may arguably not fall in the category of a physical attribute). In reality, however, no such measurement can be carried out with unbounded precision hence the notion of a continuous sample space is one of theoretical rather than practical relevance. Nevertheless, it is a concept that has credible modelling power in many practical situations.

As an example, consider a circumstance where the manufacturer of a particular consumer product needs to characterise the 'time-to-failure' of the product in order to determine a realistic warranty to offer purchasers.

The sample space S associated with experiments carried out to measure time-to-failure is the positive real line (i.e., $S = R^+$) and a random variable X introduced to represent this property would have the specification X: $S \rightarrow R^+$. Because the domain of X is a segment of the real line, X is a continuous random variable. Note also that $X(s) = s$ for all $s \in S$. In other words, the random variable in this case is the identity function. Such situations are frequently encountered with continuous random variables.

A1.4.2 Continuous Random Variables

The distinguishing feature of a continuous random variable is that it can assume an uncountably infinite number of values. Because of this it is not possible to enumerate the probabilities of these values as is done in the case of discrete random variables. In fact, we are logically obliged to conclude that the probability that X will assume any particular value that lies within its domain is zero. But, on the other hand, the probability that it will assume some value in its domain is 1 (certainty).

As a consequence of this it is preferable to begin the characterisation of a continuous random variable, X: $S \rightarrow R$ (where R is the real line and S is a collection of disjoint intervals from R), by first noting the properties of its cumulative distribution function $F(x) = P[X \leq x]$. These are:

- $F(x)$ is continuous.
- The derivative of $F(x)$, that is, $F'(x)$, exists except possibly at a finite set of points.
- $F'(x)$ is piecewise continuous.

Recall that a discrete random variable has a probability mass function. Its analogue in the case of a continuous random variable X is a *probability density function* $f(x)$ which has the following properties.

- $f(x) \geq 0$.
- $f(x) = 0$ if x is not a member of the range set of X.
- $\int_{-\infty}^{\infty} f(x)dx = 1$.
- $F(x) = \int_{-\infty}^{x} f(t)dt$.

A1.4.3 Expected Value and Variance

As in the case of a discrete random variable, a continuous random variable X likewise has an expected value and a variance and these reflect characteristics of the probability density function for X. Not surprisingly, the specifications for these parallel those given in Equations (A1.10) and (A1.13) but with the summation operation replaced with an integration operation:

$$E[X] = \int_{-\infty}^{\infty} x\, f(x)dx$$

$$\mathrm{Var}[X] = \int_{-\infty}^{\infty} (x - E(X))^2\, dx \ .$$

As in the discrete case, $\sigma_x = \sqrt{\mathrm{Var}[X]}$ is called the standard deviation of X and μ_x is used as an alternate symbol for $E[X]$. It can be demonstrated that:

$$\mathrm{Var}[X] = \int_{-\infty}^{\infty} x^2 f(x)dx - \mu_x^{\,2} = E[X^2] - \mu_x^{\,2} \ . \tag{A1.20}$$

If g is a real-valued continuous function whose domain includes the range of X and the continuous random variable Y is specified as $Y = g(X)$, then

$$E[Y] = \int_{-\infty}^{\infty} g(x) f(x)dx$$

from which it follows that

$$E[aX + b] = aE[X] + b \ .$$

It can likewise be shown that

$$\mathrm{Var}[aX + b] = a^2\, \mathrm{Var}[X].$$

A1.4.4 Some Continuous Distribution Functions

A1.4.4.1 The Uniform Distribution

A random variable X is uniformly distributed if its probability density function (PDF) has the form:

$$f(x) = \begin{cases} \dfrac{1}{(b-a)} & \text{for } a \le x \le b \\ 0 & \text{otherwise} \end{cases} \ .$$

It then follows that the cumulative distribution function is:

$$F(x) = \begin{cases} 0 & for \quad x < a \\ \dfrac{x-a}{b-a} & for \quad a \le x \le b \\ 1 & for \quad x > b \end{cases} .$$

The probability distribution function and the cumulative distribution function for the uniform distribution are shown in Figure A1.3a and b, respectively.

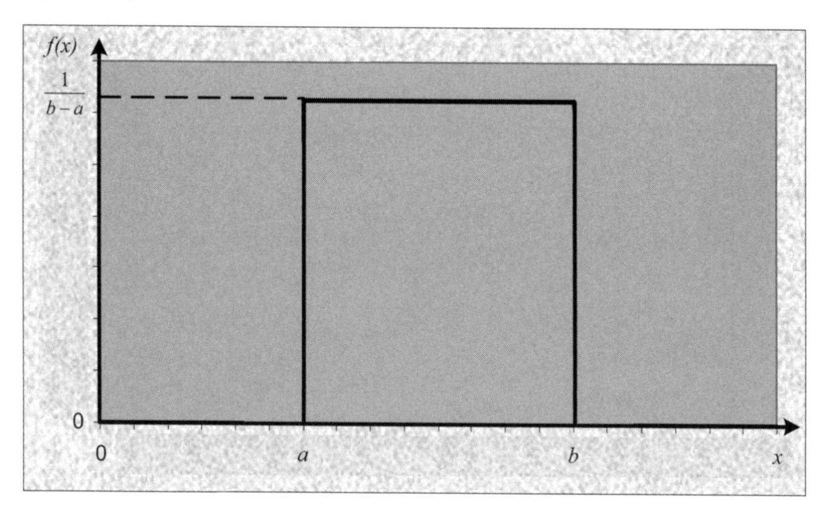

FIGURE A1.3a. Probability density function for the uniform distribution.

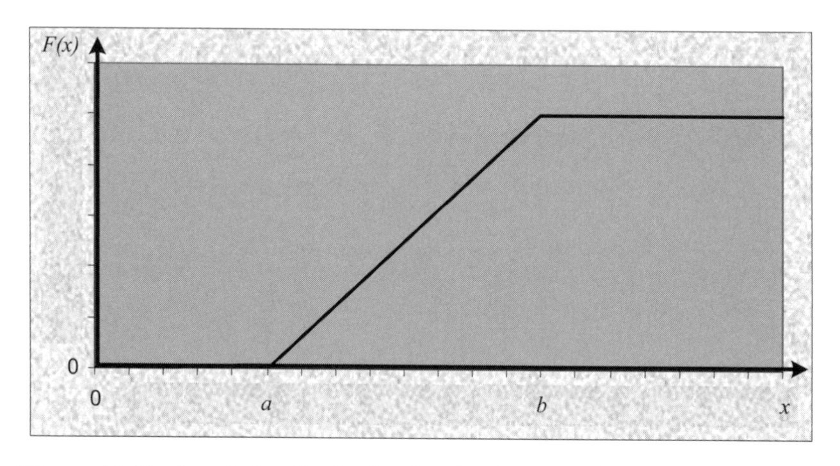

FIGURE A1.3b. Cumulative distribution function for the uniform distribution.

It can be readily shown that the mean and variance of the uniformly distributed random variable X are:

$$\mathrm{E}[X] = \frac{a+b}{2}$$

$$\mathrm{Var}[X] = \frac{(b-a)^2}{12}$$

A1.4.4.2 The Triangular Distribution

The triangular distribution is especially useful for modelling random phenomena in circumstances where very little information is available. If the random variable X has a triangular distribution then the probability density function (PDF) for X has the following form.

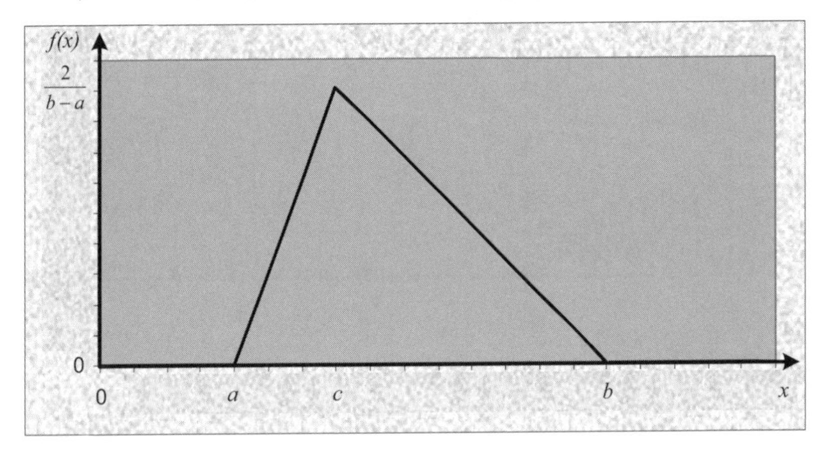

FIGURE A1.4. Probability density function for the triangular distribution.

The specifications for the PDF given in Figure A1.4 are as follows.

$$f(x) = \begin{cases} 0 & \mathit{for} \quad x < a \\ \dfrac{2(x-a)}{(b-a)(c-a)} & \mathit{for} \quad a \le x \le c \\ \dfrac{2(b-x)}{(b-a)(b-c)} & \mathit{for} \quad c < x \le b \\ 0 & \mathit{for} \quad x > b \end{cases}.$$

The corresponding cumulative distribution function is:

$$F(x)=\begin{cases}0 & for \quad x<a \\ \dfrac{(x-a)^2}{(b-a)(c-a)} & for \quad a \le x \le c \\ 1-\dfrac{(b-x)^2}{(b-a)(b-c)} & for \quad c<x\le b \\ 1 & for \quad x>b\end{cases}.$$

The mean and the variance of X are:

$$\mathrm{E}[X]=\frac{a+b+c}{3}$$

$$\mathrm{Var}[X]=\frac{a^2+b^2+c^2-ab-ac-bc}{18}$$

A statistical parameter that is especially relevant to the triangular distribution is the *mode*. For X a continuous distribution, the mode is the value of x where the maximum value of the PDF occurs. When X has a triangular distribution as shown above, the mode equals c.

A1.4.4.3 The Exponential Distribution

A random variable X is exponentially distributed if its probability density function has the form:

$$f(x)=\begin{cases}\lambda\, e^{-\lambda x} & for\ x\ge 0 \\ 0 & otherwise\end{cases}. \tag{A1.21a}$$

It then follows that the cumulative distribution function is:

$$F(x)=\begin{cases}\int_{-\infty}^{x} f(t)\,dt=\lambda\int_{0}^{x} e^{-\lambda t}\,dt=1-e^{-\lambda x} & for\ x\ge 0 \\ 0 & otherwise\end{cases}. \tag{A1.21b}$$

The probability density function and the cumulative density function for the exponential distribution are shown in Figures A1.5a and A1.5b, respectively, for the cases where $\lambda = 0.5$, 1.0, and 2.0.

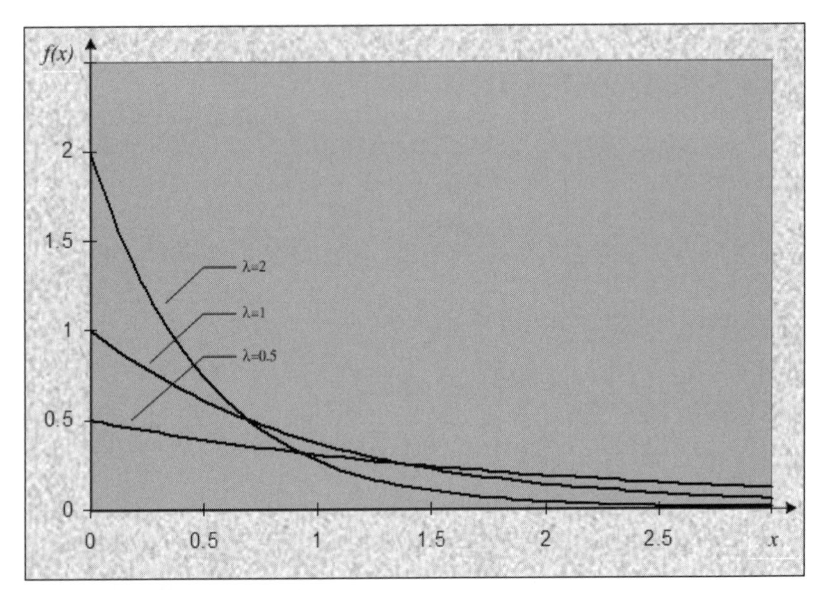

FIGURE A1.5a. Some probability density functions for the exponential distribution.

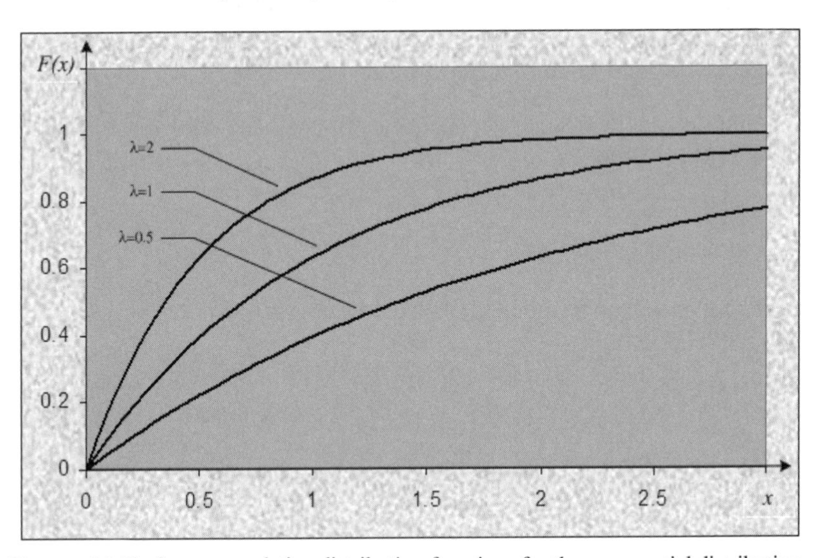

FIGURE A1.5b. Some cumulative distribution functions for the exponential distribution.

It can be readily shown that the mean and variance of the exponentially distributed random variable X are:

$$E[X] = 1/\lambda$$

$$\mathrm{Var}[X] = (1/\lambda)^2 = (E[X])^2 \ .$$

One of the most significant properties of the exponential distribution is that it is 'memoryless'. (It is the only distribution with this property.) This notion corresponds to the following relationship,

$$P[X > s + t \mid X > s] = P[X > t] \ . \qquad \text{(A1.22)}$$

The implication here is well illustrated in terms of the life of many electronic components, Suppose, for example, that the random variable X represents length of time that a memory chip in a cell phone will function before failing. If X has an exponential distribution then (from Equation (A1.22) the probability of it being functional at time $(s + t)$ given that it is functional at time s is the same as the probability of it being functional at time t. In other words, the chip does not 'remember' that it has already been in use for s time units. If, for example, X has a mean of 4 (years), that is, $\lambda = 0.25$, then the probability of the cell phone remaining functional for 2 years is:

$$\begin{aligned} P[\text{remaining life} > 2 \text{ years}] &= 1 - F(2) \\ &= 1 - (1 - e^{-0..5}) \\ &= e^{-0..5} \\ &= 0.604 \end{aligned}$$

and this is independent of how many years the phone has already been in use.

The following important relationship exists between the Poisson distribution and the exponential distribution:

> Suppose that arrivals have a Poisson distribution with mean μ; then the corresponding interarrival times have an exponential distribution with mean $1/\mu$.

A1.4.4.4 The Gamma Distribution

A random variable X has a gamma distribution if its probability density function has the form

$$f(x) = \begin{cases} \dfrac{\lambda^\alpha \, x^{\alpha - 1} \, exp(-\lambda \, x)}{\Gamma(\alpha)} & for \ x \geq 0 \\ 0 & otherwise \end{cases} ,$$

where α and λ are positive parameters. The function $\Gamma(\alpha)$ is called the *gamma function* and has the following definition:

$$\Gamma(\alpha) = \int_0^{\infty} x^{\alpha-1} exp(-x)\, dx \quad .$$

A general analytic expression for the cumulative distribution $F(x)$ cannot be formulated. Values can only be obtained by numerical integration of $f(x)$.

If the random variable X has a gamma distribution with parameters α and λ, then it can be shown that

$$E[X] = \frac{\alpha}{\lambda}$$
$$\mathrm{Var}[X] = \frac{\alpha}{\lambda^2} = \frac{(E[X])^2}{\alpha} \quad . \tag{A1.23}$$

Notice that if the mean is held constant and α increases then the variance approaches zero.

The probability density function and the cumulative density function for the gamma distribution are shown in Figures A1.6a and A1.6b, respectively, for the cases where α = λ = 1, α = λ = 2, and α = λ = 3.

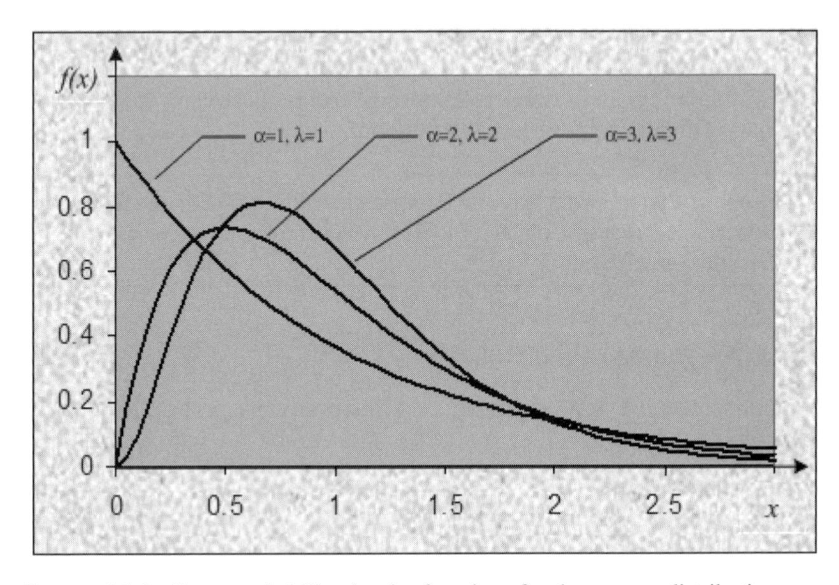

FIGURE A1.6a. Some probability density functions for the gamma distribution.

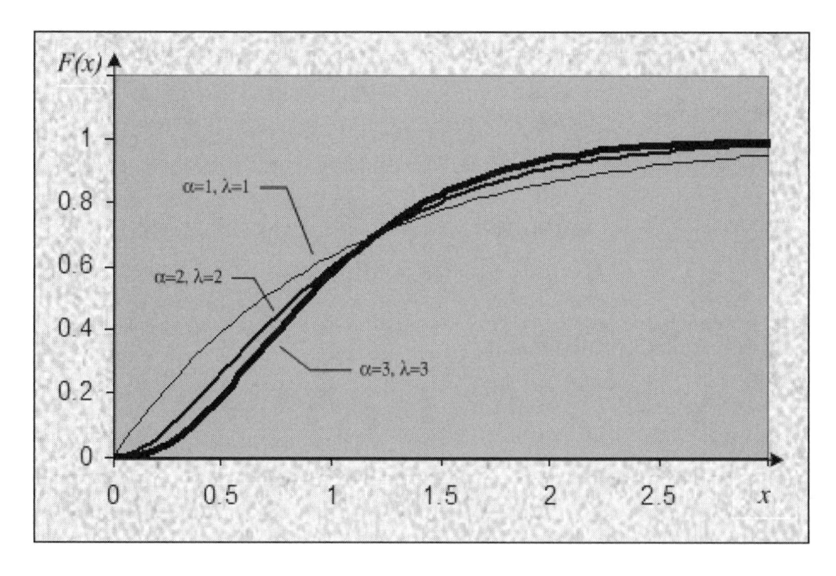

FIGURE A1.6b. Some cumulative distribution functions for the gamma distribution.

A1.4.4.5 The Erlang Distribution

The Erlang distribution is a specialisation of the gamma distribution which corresponds to the case where the parameter α is restricted to (positive) integer values; i.e., $\alpha = n > 0$. In this case it can be shown that:

$$\Gamma(n) = (n-1)! = (n-1)\,\Gamma(n-1)$$

To facilitate examples of the application of this distribution, it's convenient to replace the parameter λ of the gamma distribution with $n\omega$. and to rewrite the distribution in the form:

$$f(x) = \begin{cases} \dfrac{n\omega\,(n\omega\,x)^{n-1}\,exp(-n\,\omega\,x)}{(n-1)!} & for\ x > 0 \\ 0 & otherwise \end{cases}.$$

The density function of the Erlang distributon as given above can be explicitly integrated and the cumulative distribution $F(x)$ can be written as

$$F(x) = \begin{cases} 1 - \sum_{i=0}^{n} \dfrac{(n\,\omega\,x)^i\,exp(-n\,\omega\,x)}{i!} & for\ x > 0 \\ 0 & otherwise \end{cases}.$$

If the random variable X has a Erlang distribution with parameters n and ω then:

$$E[X] = \frac{1}{\omega}$$

$$\text{Var}[X] = \frac{1}{n\,\omega^2} = \frac{(E[X])^2}{n} \quad .$$

It can be demonstrated that if

$$X = \sum_{i=1}^{n} X_i \quad ,$$

where the random variables X_i are independent and exponentially distributed with mean $(1/n\omega)$, then X has an Erlang distribution with parameters n and ω.

A1.4.4.6 The Chi-Square Distribution

The chi-square distribution is another special case of the gamma distribution. It can be obtained by setting $\alpha = (m/2)$ and $\lambda = 0.5$. This results in:

$$f(x) = \begin{cases} \dfrac{(0.5\,x)^{\frac{m}{2}-1}\,exp(-0.5\,x)}{2\,\Gamma(\frac{m}{2})} & for\ x > 0 \\ 0 & otherwise \end{cases} \quad .$$

The parameter m is called the 'degrees of freedom' of the distribution. It follows directly from Equation (A1.23) that if X has a chi-square distribution, then:

$$E[X] = m$$

$$\text{Var}[X] = 2\,m \quad .$$

It can be demonstrated that if

$$X = \sum_{i=1}^{m} X_i^2 \quad ,$$

where the random variables X_i are independent and have a standard normal distribution (see below) then X has a chi-square distribution with m degrees of freedom.

A1.4.4.7 The Normal Distribution

A random variable X has a normal distribution with mean μ and variance σ^2 if its probability density function is given by:

$$f(x) = \frac{1}{\sqrt{2\pi}\ \sigma} exp(\frac{-(x-\mu)^2}{2\sigma^2}); \quad -\infty < x < \infty \ .$$

This function has the classic bell-shape which is symmetric about the mean μ. The maximum value of f occurs at $x = \mu$ and $f(\mu)$ $\frac{1}{\sqrt{2\pi}\,\sigma}$. It is sometimes convenient to use the notation $X \sim N[\mu, \sigma^2]$ to indicate that X is a normally distributed random variable with mean μ and variance σ^2.

The probability density function and the cumulative density function for the normal distribution are shown in Figures A1.7a and A1.7b, respectively, for the cases where $\mu = 0$ and $\sigma = 1$, 2, and 4. Note, however, that a closed-form analytic expression for the cumulative distribution function is not possible and consequently the data for the construction of Figure A1.7b was necessarily obtained using numerical approximation for the required integration operation.

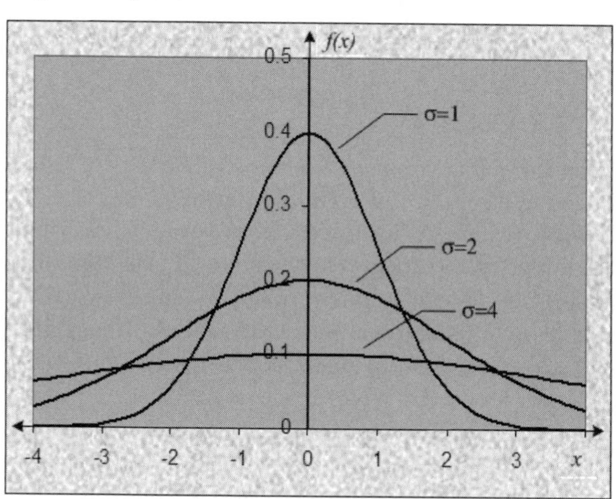

FIGURE A1.7a. Some probability density functions for the normal distribution.

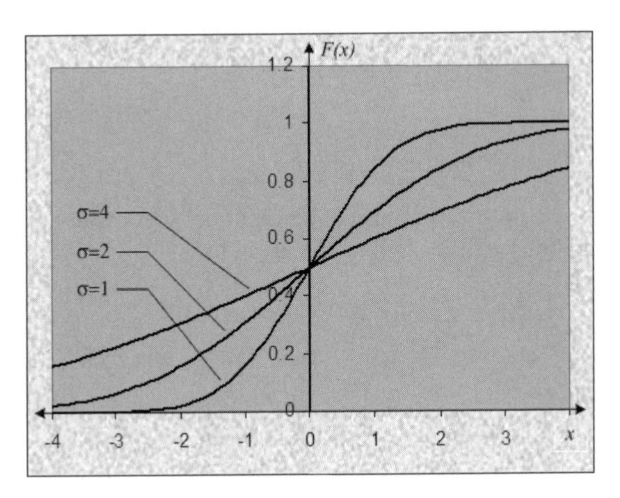

FIGURE A1.7b. Some cumulative distribution functions for the normal distribution.

Suppose Y is a random variable that is linearly related to the normally distributed random variable, $X \sim N[\mu, \sigma^2]$; for example, $Y = aX + b$. where a and b are constants. Then Y likewise has a normal distribution with $E[Y] = (a\mu + b)$ and $\text{Var}[Y] = a^2\sigma^2$; that is, $Y \sim N[a\mu + b, a^2 \sigma^2]$.

The above feature provides a convenient way of 'normalising' the normal distribution. In particular, by choosing $a = (1/\sigma)$ and $b = -\mu/\sigma$ then the normally distributed random variable

$$Z = aX - b = \frac{X - \mu}{\sigma}$$

has zero mean and unit variance; that is, $Z \sim N[0,1]$. A random variable such as Z is said to have a *standard normal distribution.*

The following feature of a collection of normally distributed random variables has important practical relevance. Let X be a normally distributed random variable with mean μ and variance σ^2 and let $X_1, X_2, \ldots, X_n$ be random variables that correspond to a sequence of n independent samples of X (hence the X_k are identically distributed). The random variable:

$$\overline{X}(n) = \frac{1}{n}\sum_{k=1}^{n} X_k$$

which is called the sample mean, is normally distributed with mean μ and variance σ^2/n.; that is, $\overline{X}(n) = N[\mu, \sigma^2/n.]$. This result can be viewed as a special case of the central limit theorem that is given in the following section.

A1.4.4.8 The Beta Distribution

A random variable X has a beta distribution with parameters α_1 and α_2 if its probability density function is given by:

$$f(x) = \begin{cases} \dfrac{x^{\alpha_1 - 1}(1-x)^{\alpha_2 - 1}}{B(\alpha_1, \alpha_2)} & \text{for } 0 < x < 1 \\ 0 & \text{otherwise} \end{cases},$$

where $\alpha_1 > 0$, $\alpha_2 > 0$, and

$$B(\alpha_1, \alpha_2) = \frac{\Gamma(\alpha_1)\Gamma(\alpha_2)}{\Gamma(\alpha_1 + \alpha_2)} .$$

The corresponding cumulative distribution does not exist in analytic form. The probability density function and the cumulative distribution function for the beta distribution are shown for several values of the parameters α_l and α_2 in Figures A1.8a and A1.8b. These Figures illustrate, one of the interesting features of the beta PDF; namely the wide range of 'shapes' which it can assume as its parameters are varied. This can be useful in dealing with the problem of selecting appropriate data models for random phenomena as discussed in Chapter 3. Some general properties are as follows:

a) If $\alpha_1 = \alpha_2$ then the beta PDF is symmetric about 0.5.
b) If $\alpha_1 > 1$ and $\alpha_2 \leq 1$ or $\alpha_1 = 1$ and $\alpha_2 < 1$ then the beta PDF is strictly increasing.
c) If $\alpha_1 < 1$ and $\alpha_2 \geq 1$ or $\alpha_1 = 1$ and $\alpha_2 > 1$ then the beta PDF is strictly decreasing.

Several specific shapes that can be achieved are:

a) Constant (i.e., the uniform distribution) when: $\alpha_1 = \alpha_2 = 1$
b) Uniformly rising to the right (special case of the triangular distribution where $a = 0$, $b = c = 1$) when $\alpha_1 = 2$ and $\alpha_2 = 1$
c) Uniformly falling to the right (special case of the triangular distribution where $a = c = 0$ and $b = 1$) when $\alpha_1 = 1$ and $\alpha_2 = 2$
d) Parabolic when $\alpha_1 = 2 = \alpha_2$
e) U-shaped when $\alpha_1 < 1$ and $\alpha_2 < 1$ (e.g., $\alpha_1 = 0.5 = \alpha_2$)

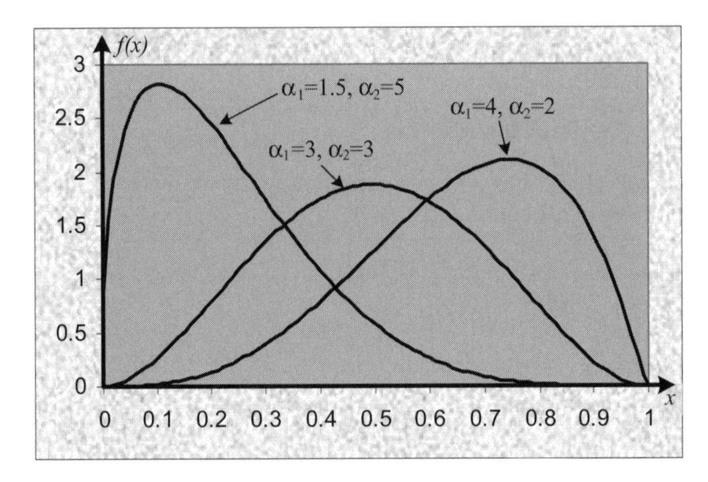

FIGURE A1.8a. Some probability density functions for the beta distribution.

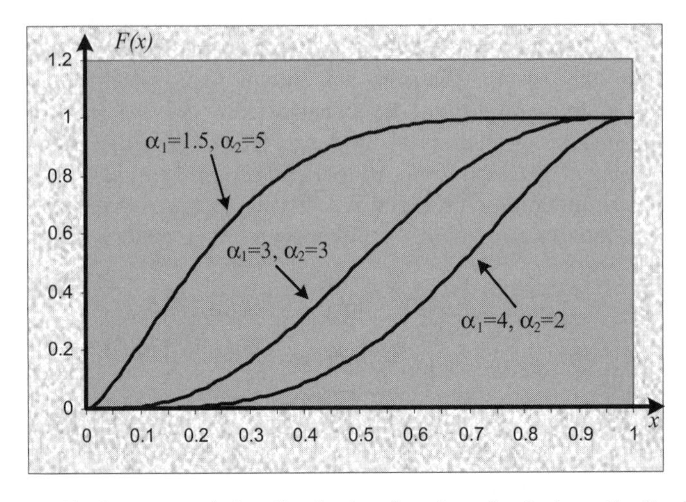

FIGURE A1.8b. Some cumulative distribution functions for the beta distribution.

An 'extension' of the beta distribution as given above can be obtained by letting $Y = a + (b - a)X$. The distribution of random variable Y maintains the shape of the distribution of X but its domain extends over the interval (a, b) (rather than $(0, 1)$). The mean and variance of Y are:

$$E[Y] = a + (b-a)\frac{\alpha_1}{\alpha_1 + \alpha_2}$$

$$Var[Y] = (b-a)^2 \frac{\alpha_1 \alpha_2}{(\alpha_1 + \alpha_2)^2 (\alpha_1 + \alpha_2 + 1)}$$

A1.5 Some Theorems

Two results in probability theory stand out as being especially significant and valuable. These are the strong law of large numbers and the central limit theorem and they are stated below (without proof). Because of their wide relevance they are frequently referenced in the discussions in this textbook.

The results in question are all formulated in terms of a sequence, $X_1, X_2, \ldots, X_n$ of random variables that are independent and identically distributed.[1] Suppose, for definiteness, that for each k, $E[X_k] = \mu$ and $Var[X_k] = \sigma^2$, where we assume that both μ and σ are finite. The random variable

$$\overline{X}(n) = \frac{1}{n}\sum_{k=1}^{n} X_k \tag{A1.24}$$

is called the sample mean.

Theorem (Strong Law of Large Numbers)

With probability 1,

$$\lim_{n\to\infty} \overline{X}(n) = \mu \quad .$$

Various intuitive interpretations are possible. One is simply that the average of sequence of independent, identically distributed random variables converges to the common mean as the sequence becomes larger.

[1] Typically, within the context of our consideration in this textbook, this collection of random variables corresponds to a sequence of n independent samples of a population.

Theorem (Central Limit Theorem)

The sample mean random variable $\overline{X}(n)$ is normally distributed with mean μ and variance σ^2/n *as* $n \to \infty$. Consequently, the random variable

$$Z = \frac{\overline{X}(n) - \mu}{(\sigma / \sqrt{n})}$$

has a distribution that approaches the standard normal distribution as $n \to \infty$.

Typically the random variables X_k are samples from an underlying population X. The central limit theorem states that regardless of what distribution X has, the distribution of the sample mean $\overline{X}(n)$ will tend towards a normal distribution as $n \to \infty$. Furthermore, $E[\,\overline{X}(n)\,] \to \mu$ and $\mathrm{Var}[\,\overline{X}(n)\,] \to \sigma^2 / n$.

Although the central limit theorem has exceptional theoretical importance it does give rise to the very natural practical question of how large n needs to be in order to achieve a satisfactory approximation. No specific result is available, however, it is generally true that if the parent distribution is symmetric and short-tailed then approximate normality is reached with smaller sample sizes than if the parent distribution were skewed and/or long-tailed. The value of 30 is a widely recommended as a satisfactory value for n. There is convincing experimental evidence that when $n > 50$ the actual shape of the parent distribution has minimal impact.

A1.6 The Sample Mean as an Estimator

Let $X_1, X_2, \ldots, X_n$ be random variables that correspond to a sequence of n independent samples of a population whose mean (μ) and variance (σ^2) are unknown. Note that the X_k are therefore identically distributed and have the mean and variance of the underlying population, namely, μ and σ^2, respectively. The random variable

$$\overline{X}(n) = \frac{1}{n} \sum_{k=1}^{n} X_k$$

is called a sample mean. Notice that:

$$E[\overline{X}(n)] = E[\frac{1}{n}\sum_{k=1}^{n} X_k] = \frac{1}{n}\sum_{k=1}^{n} \mathrm{E}[X_k] = \mu \ . \tag{A1.25}$$

Because the expected value of $\overline{X}(n)$ yields the value of the distribution parameter μ of the underlying distribution, $\overline{X}(n)$ is said to be an (unbiased) estimator of μ. (It is unbiased because E[$\overline{X}(n)$] is not displaced from μ by a nonzero constant.) The interesting and important observation here is that the value of a parameter of a distribution (μ in this case) can be obtained in a somewhat indirect fashion.

The variance of $\overline{X}(n)$ provides insight into its 'quality' as an estimator of μ because Var[$\overline{X}(n)$] is a measure of deviation of $\overline{X}(n)$ from the mean μ. In other words, the estimator's quality is inversely proportional to the value of $E[(\overline{X}(n) - \mu)^2]$ = Var[$\overline{X}(n)$]. The determination of Var[$\overline{X}(n)$] is straightforward.

$$\begin{aligned} \mathrm{Var}[\overline{X}(n)] &= \mathrm{Var}[\frac{1}{n}\sum_{k=1}^{n} X_k] \\ &= \frac{1}{n^2}\sum_{k=1}^{n} \mathrm{Var}[X_k] \\ &= \frac{\sigma^2}{n} \end{aligned} \tag{A1.26}$$

where we have used the relation

$$\mathrm{Var}\left[\sum_{k=1}^{n} c_k X_k\right] = \sum_{k=1}^{n} c_k^2 \ \mathrm{Var}[X_k]$$

whose validity is restricted to the case where the X_k are independent. Note that the implication of Equation (A1.26) is that the quality of our estimator $\overline{X}(n)$ improves with increasing n.

From a practical point of view the result of Equation (A1.26) is not particularly useful because, in general, σ^2, like μ, is not known. To some extent this can be remedied by introducing the random variable:

$$\overline{S}^2(n) = \frac{\sum_{k=1}^{n} (X_k - \overline{X}(n))^2}{n-1}$$

which is called the sample variance. It can now be demonstrated that:

$$E[\overline{S}^2(n)] = \sigma^2 \tag{A1.27}$$

In other words, $E[\overline{S}^2(n)]$ is an (unbiased) estimator of σ^2. This demonstration begins by first noting that

$$\begin{aligned}(n-1)E[\overline{S}^2(n)] &= E\left[\sum_{k=1}^{n} X_k^2\right] - nE[\overline{X}^2(n)] \\ &= n\,(E[X_1^2] - E[\overline{X}^2(n)])\end{aligned} \tag{A1.28}$$

because

$$\sum_{k=1}^{n}(X_k - \overline{X}(n))^2 = \sum_{k=1}^{n} X_k^2 - n\overline{X}^2(n)$$

and because the X_k all have the same distribution. From the general result of Equation (A1.20) we can write:

$$E[X_1^2] = \text{Var}[X_1] + (E[X_1])^2$$

and similarly that:

$$E[\overline{X}^2(n)] = \text{Var}[\overline{X}(n)] + (E[\overline{X}(n)])^2$$

Substitution into Equation (A1.28) together with the observations that $E[X_l] = \mu$, $\text{Var}[X_l] = \sigma^2$, $E[\overline{X}(n)] = \mu$, and $\text{Var}[\overline{X}(n)] = \sigma^2/n$ yields the desired result of Equation (A1.27).

A1.7 Interval Estimation

In Section A1.6 we explored the problem of obtaining a 'point' (single value) estimate of the mean of an unknown distribution. In this section our goal is to identify an interval in which the unknown mean is likely to fall together with an indication of the 'confidence' that can be associated with the result. We begin the discussion in the context of the standard normal distribution whose density function $\phi(z)$ is described in Section A1.4.4.

Choose a value a in the range (0,1). Then there necessarily exists a value of z (which we denote by z_a) such that the area under $\phi(z)$ that is to the right of z_a is equal to a. (see Figure A1.9).

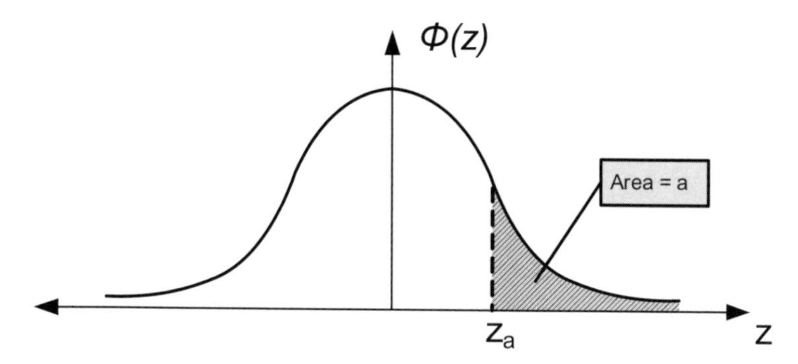

FIGURE A1.9. A standard normal density function (perspective 1).

Observe furthermore that this area corresponds to $P[Z > z_a] = 1 - P[Z \leq z_a] = 1 - \Phi(z_a)$, where $\Phi(z)$ is the cumulative distribution function of the standard normal distribution. In other words we have:

$$a = \int_{z_a}^{\infty} \phi(z)dz = 1 - \Phi(z_a) = P[Z > z_a] \ .$$

Using tabulated values for $\Phi(z)$, a tabular summary of the form shown in Table A1.1 can be developed

TABLE A1.1. Standard normal distribution data.

a = 1 – Φ(z_a) = P[Z>z_a]	***Φ(z_a)***	z_a
0.05	0.95	1.64
0.025	0.975	1.96
0.01	0.99	2.33

Thus it follows that $z_{0.05} = 1.64$, $z_{0.025} = 1.96$, and $z_{0.01} = 2.33$.

We consider now a slight variation of the above discussion. Again we choose a value of a in the range (0, 1). As earlier, there necessarily is a value of z (which we again denote by z_a) such that (see Figure A1.10)

$$C = P[-z_a \leq Z \leq z_a] = (1 - 2a) \ . \qquad \text{(A1.29)}$$

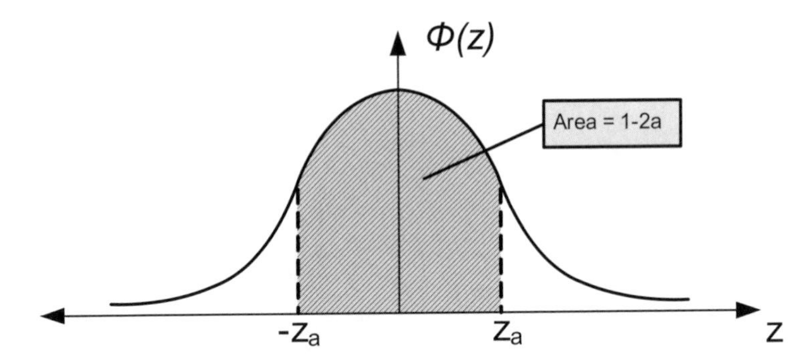

FIGURE A1.10. Standard normal density function (perspective 2).

Our interest here is to find a value for z_a given a value for C. However we know that $a = (1 - C)/2$ and that:

$$P[Z > z_a] = a = 1 - \Phi(z_a) \ .$$

Consequently Table A1.2 can be derived.

TABLE A1.2. Standard normal distribution data (extended).

C	$a = \frac{1-C}{2}$	$\Phi(z_a)$	z_a
0.90	0.05	0.95	1.64
0.95	0.025	0.975	1.96
0.98	0.01	0.99	2.33

It then follows that:

$$\begin{aligned} C &= P[-z_{0.05} \leq Z \leq z_{0.05}] = 0.90, \quad \text{with } z_{0.05} = 1.64 \\ C &= P[-z_{0.025} \leq Z \leq z_{0.025}] = 0.95, \quad \text{with } z_{0.025} = 1.96 \\ C &= P[-z_{0.01} \leq Z \leq z_{0.01}] = 0.98, \quad \text{with } z_{0.01} = 2.33. \end{aligned} \tag{A1.30}$$

The illustrative results given above have an important practical interpretation. For example, the first can be interpreted to imply that a random sample from a standard normal distribution will fall, with 90% confidence (100C%), in the range [–1.64, 1.64]. (The 0.9 probability is interpreted here as an indicator of confidence level.)

We know from the central limit theorem, that the distribution of the random variable

$$Z(n) = \frac{\overline{X}(n) - \mu}{(\sigma / \sqrt{n})} \tag{A1.31}$$

approaches the standard normal distribution as n becomes large. Here μ and σ^2 are the mean and the variance of a parent distribution and our task is to formulate an interval estimate for μ. To the extent that we can assume that n is sufficiently large, $Z(n)$ could be substituted in Equation (A1.30). Instead, however, we substitute the right-hand side of Equation (A1.31) and, with some straightforward manipulation, the first result of Equation (A1.30) becomes:

$$C = P\left[\overline{X}(n) - z_{0.05}\frac{\sigma}{\sqrt{n}} \le \mu \le \overline{X}(n) + z_{0.05}\frac{\sigma}{\sqrt{n}}\right] = 0.9 \tag{A1.32}$$

with $z_{0.05} = 1.64$.

Unfortunately Equation (A1.32) has little practical value because the bounds on the unknown mean μ depend on σ which likewise is unknown. Recall however, that earlier it was demonstrated that $\mathrm{E}[\overline{S}^2(n)] = \sigma^2$ (see Equation (A1.27)). The implication here is that for large n the random variable $\overline{S}^2(n)$ can be used as an estimator for σ^2. This in effect, provides the basis for replacing σ in Equation (A1.32) with a random variable for which an observation (i.e., value) can be established. In effect, then, we can restate our earlier illustrative conclusion of Equation (A1.32) in the following way:

$$C = P\left[\overline{X}(n) - z_{0.05}\frac{\overline{S}(n)}{\sqrt{n}} \le \mu \le \overline{X}(n) + z_{0.05}\frac{\overline{S}(n)}{\sqrt{n}}\right] = 0.9 \tag{A1.33}$$

with $z_{0.05} = 1.64$.

Suppose now that observed values of $\bar{x}$ and $\bar{s}$ have been obtained for $\overline{X}(n)$ and $\overline{S}(n)$, respectively, (with the use of suitably large values for n). A practical interpretation for the result of Equation (A1.33) is that with approximately 90% confidence, we can assume that the unknown mean μ lies in the interval:

$$[\bar{x} - z_{0.05}\frac{\bar{s}}{\sqrt{n}},\ \bar{x} + z_{0.05}\frac{\bar{s}}{\sqrt{n}}] \tag{A1.34}$$

with $z_{0.05} = 1.64$.

Or alternately, the above can be interpreted to mean that if a large number of independent observations for $\bar{x}$ and $\bar{s}$ are obtained then, in approximately 90% of the cases, the mean will fall in the interval given by Equation (A1.34). The qualifier 'approximately' is required for two reasons: the use of the estimator $\overline{S}(n)$ and the uncertainty about the adequacy of n.

The approximation of the interval estimate given above that arises because of the uncertainty introduced by reliance on n being sufficiently large can be remedied in one special case. This is the special case where the random variables X_k that are the constituents of $\overline{X}(n)$ are themselves normally distributed. Earlier we pointed out (see Section A1.4.4.7) that under such circumstances $\overline{X}(n) = N[\mu, \sigma^2/n.]$; that is, for any n, the sample mean is normally distributed with mean equal to the mean of the underlying distribution and with variance that is reduced by a factor of n from the variance of the underlying distribution. Under these circumstances the analysis can be recast in the context of the Student t-distribution. This distribution is similar in shape to the standard normal distribution but is less peaked and has a longer tail (see Figure A1.11).

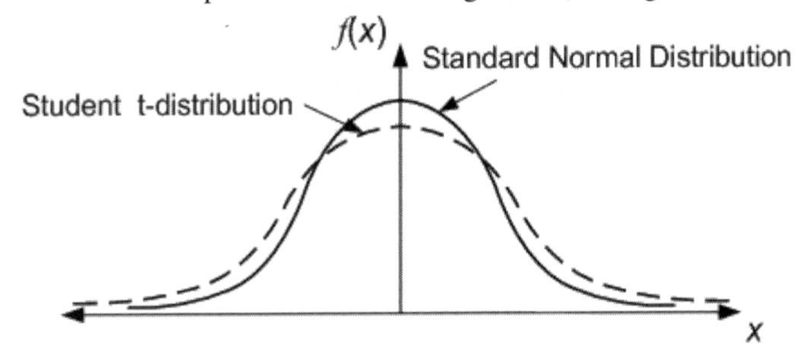

FIGURE A1.11. Illustration of the relationship between the Student t-distribution and the standard normal distribution.

More specifically, the random variable,

$$T(n) = \frac{\overline{X}(n) - \mu}{(\overline{S}(n)/\sqrt{n})} \tag{A1.35}$$

has a Student t-distribution with $(n - 1)$ *degrees of freedom*. The feature that is especially significant here is that this distribution accommodates the approximation introduced by replacing σ by $\overline{S}(n)$. The distribution is, in fact, a family of distributions that are indexed by the value of n. Consequently the transition from our earlier analysis to the framework of the Student t-distribution requires that the anchor point z_a acquire a second index to identify the degrees of freedom that relate to the circumstances under consideration. To emphasise this transition to the

Student t-distribution, we also re-label the anchor point to get $t_{n\text{-}1,a}$. Equation (A1.33) can be now be re-written as

$$C = P\left[\overline{X}(n) - t_{n-1,0.05}\frac{\overline{S}(n)}{\sqrt{n}} \le \mu \le \overline{X}(n) + t_{n-1,0.05}\frac{\overline{S}(n)}{\sqrt{n}}\right] = 0.9 \ . \qquad \text{(A1.36)}$$

The value of $t_{n-1,0.05}$ is determined from tabulated data relating to the Student t-distribution but a prerequisite is the specification of a value for n. Some representative values for $t_{n-1,0.05}$ are give in Table A1.3.

TABLE A1.3. Some representative values for $t_{n-1,0.05}$ ($C = 0.9$).

n	$t_{n-1,0.05}$
10	1.83
20	1.73
30	1.70
∞	1.64

It should be observed that as n becomes large the value of $t_{n-1,a}$ approaches the value z_a for any value of a (see last row of Table A1.3). A more complete section of the Student t-distribution is given in Table A1.4 where the limiting case is again given in the last row.

A practical implication of Equation (A1.36) (which parallels that of Equation (A1.34)) is that if a large number of independent observations for $\overline{x}$ and $\overline{s}$ are obtained then, in 90% of the cases, the unknown mean μ will fall in the interval given by Equation (A1.37).

$$[\overline{x} - t_{n-1,0.05}\frac{\overline{s}}{\sqrt{n}}, \ \overline{x} + t_{n-1,0.05}\frac{\overline{s}}{\sqrt{n}}] \qquad \text{(A1.37)}$$

with $t_{n\text{-}1,\,0.05} = 1.83$ when $n = 10$.

Notice furthermore that there is no need here for the qualifier 'approximately' because of our assumption that the unknown distribution is normal.

It needs to be emphasised that the discussion above with respect to the application of the Student t-distribution was founded on the assumption of a normal distribution for the underlying population being investigated. This is clearly an idealised case. Nevertheless the Student t-distribution is the tool generally used to establish interval estimates for population means even when the normality assumption is not applicable. This naturally means that the results obtained are approximate and this should be clearly indicated.

TABLE A1.4. A section of the Student *t*-distribution.

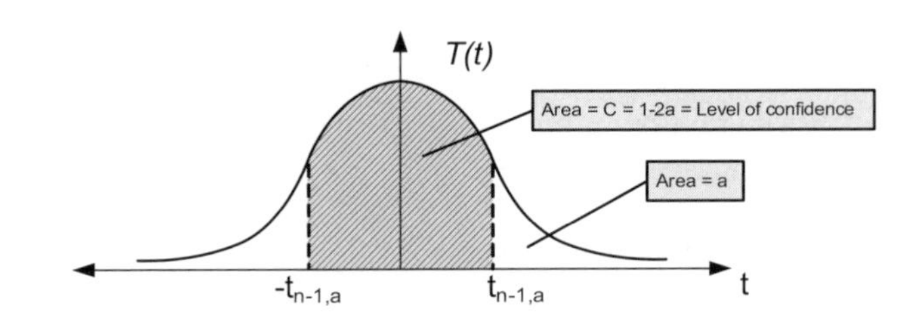

	a						
	0.2	**0.15**	**0.1**	**0.05**	**0.025**	**0.005**	**0.0025**
				C			
(n-1)	**0.6**	**0.7**	**0.8**	**0.9**	**0.95**	**0.99**	**0.995**
1	1.376382	1.962611	3.077684	6.313752	12.7062	63.65674	127.3213
2	1.06066	1.386207	1.885618	2.919986	4.302653	9.924843	14.08905
3	0.978472	1.249778	1.637744	2.353363	3.182446	5.840909	7.453319
4	0.940965	1.189567	1.533206	2.131847	2.776445	4.604095	5.597568
5	0.919544	1.155767	1.475884	2.015048	2.570582	4.032143	4.773341
6	0.905703	1.134157	1.439756	1.94318	2.446912	3.707428	4.316827
7	0.89603	1.119159	1.414924	1.894579	2.364624	3.499483	4.029337
8	0.88889	1.108145	1.396815	1.859548	2.306004	3.355387	3.832519
9	0.883404	1.099716	1.383029	1.833113	2.262157	3.249836	3.689662
10	0.879058	1.093058	1.372184	1.812461	2.228139	3.169273	3.581406
11	0.87553	1.087666	1.36343	1.795885	2.200985	3.105807	3.496614
12	0.872609	1.083211	1.356217	1.782288	2.178813	3.05454	3.428444
13	0.870152	1.079469	1.350171	1.770933	2.160369	3.012276	3.372468
14	0.868055	1.07628	1.34503	1.76131	2.144787	2.976843	3.325696
15	0.866245	1.073531	1.340606	1.75305	2.13145	2.946713	3.286039
16	0.864667	1.071137	1.336757	1.745884	2.119905	2.920782	3.251993
17	0.863279	1.069033	1.333379	1.739607	2.109816	2.898231	3.22245
18	0.862049	1.06717	1.330391	1.734064	2.100922	2.87844	3.196574
19	0.860951	1.065507	1.327728	1.729133	2.093024	2.860935	3.173725
20	0.859964	1.064016	1.325341	1.724718	2.085963	2.84534	3.153401
21	0.859074	1.06267	1.323188	1.720743	2.079614	2.83136	3.135206
22	0.858266	1.061449	1.321237	1.717144	2.073873	2.818756	3.118824
23	0.85753	1.060337	1.31946	1.713872	2.068658	2.807336	3.103997
24	0.856855	1.059319	1.317836	1.710882	2.063899	2.796939	3.090514
25	0.856236	1.058384	1.316345	1.708141	2.059539	2.787436	3.078199
26	0.855665	1.057523	1.314972	1.705618	2.055529	2.778715	3.066909
27	0.855137	1.056727	1.313703	1.703288	2.05183	2.770683	3.05652
28	0.854647	1.055989	1.312527	1.701131	2.048407	2.763262	3.046929
29	0.854192	1.055302	1.311434	1.699127	2.04523	2.756386	3.038047
30	0.853767	1.054662	1.310415	1.697261	2.042272	2.749996	3.029798
40	0.8507	1.050046	1.303077	1.683851	2.021075	2.704459	2.971171
50	0.848869	1.047295	1.298714	1.675905	2.008559	2.677793	2.936964
75	0.84644	1.043649	1.292941	1.665425	1.992102	2.642983	2.89245
100	0.84523	1.041836	1.290075	1.660234	1.983971	2.625891	2.870652
∞	0.841621	1.036433	1.281552	1.644854	1.959964	2.575829	2.807034

Generated using Microsoft Excel functions TINV and NORMINV

A1.8 Stochastic Processes

A discrete stochastic process is a finite sequence of random variables defined on the same population; for example, $X = (X_1\ X_2,\ X_3\ \ldots, X_n)$. Often the indices have a correspondence with a set of discrete points in time; that is, index i corresponds to t_i and $t_i \le t_j$ for $i < j$). In general, each of the X_i's may have its own distinct distribution. Consider, for example, the customer waiting times in a queue in front of a store cashier. A random variable W_i can be used to represent the waiting time of the customer whose service begins at time t_i for $i = 1, 2, \ldots, n$. Thus we can associate the discrete stochastic process $(W_1, W_2, W_3, \ldots, W_n)$ with this activity.

The following properties of a discrete stochastic process $X = (X_1, X_2, X_3, \ldots, X_n)$ are of particular relevance:

$$\text{mean}: \mu_i = E[X_i] \quad (\text{for } i = 1,2,---- n)$$
$$\text{variance}: \sigma_i^2 = E[(X_i - \mu_i)^2] \quad (\text{for } i = 1,2,---- n)$$
$$\text{covariance}: Cov[X_i, X_j] = E[(X_i - \mu_i)(X_j - \mu_j)]$$
$$= E[X_i X_j] - \mu_i \mu_j \quad (\text{for } i, j = 1,2,---- n)$$
$$\text{correlation}: \rho_{i,j} = \frac{Cov[X_i, X_j]}{\sigma_i \sigma_j} \quad (\text{for } i, j = 1,2,---- n)$$

Note that $\text{Cov}[X_i, X_i] = \sigma^2_i$ and $\rho_{i,i} = 1$ for any i.

A discrete stochastic process is stationary when the following invariance properties hold:

$$\mu_i = \mu \quad \textit{for all } i$$
$$\sigma_i^2 = \sigma^2 \quad \textit{for all } i$$
$$Cov[X_i, X_{i+k}] = \gamma_k \quad \textit{for all } i$$
$$\rho_{i,i+k} = \frac{\gamma_k}{\sigma^2} \quad \textit{for all } i$$

Note that when a discrete stochastic process is stationary, covariance (and hence correlation) between two members of the sequence is dependent only on their separation (or lag) k and is independent of the specific indices involved.

We define a *homogeneous stochastic process* to be a stationary discrete stochastic process for which the constituent random variables are independent and identically distributed (in other words, the stochastic process becomes a sequence of independent, identically distributed random variables defined on the same population). Recall (see Equation (A1.17)) that independence implies that $E[X_i X_j] = \mu_i \mu_j$ for $i \neq j$ and consequently $\text{Cov}[X_i, X_j] = 0$ for $i \neq j$. The properties of special interest become:

$$\mu_i = \mu \quad \textit{for all } i$$

$$\sigma_i^2 = \sigma^2 \quad \textit{for all } i$$

$$Cov[X_i, X_j] = \begin{cases} \sigma^2 & \textit{for } i = j \\ 0 & \textit{for } i \neq j \end{cases}$$

$$\rho_{i,j} = \begin{cases} 1 & \textit{for } i = j \\ 0 & \textit{for } i \neq j \end{cases}$$

An observation of a discrete stochastic process, $X = (X_1, X_2, \ldots, X_n)$, yields a sequence of values $x = (x_1, x_2, \ldots, x_n)$ where each x_i is obtained as an observed value of X_i. Such a sequence of values is often called a *time series*. The possibility of a recurring relationship between values in this sequence (e.g., every seventh value is consistently almost zero) is an important statistical consideration. The notion of serial correlation addresses this issue and a measure called autocorrelation is used as a means of quantification.

The underlying idea is to determine a set of sample autocovariance values where each such value corresponds to a different 'alignment' of the values in x; that is,

$$c_k = \frac{1}{(n-k)} \sum_{i=1}^{n-k} (x_i - \overline{x}(n))(x_{i+k} - \overline{x}(n)) \quad ,$$

where $\overline{x}(n) = \sum_{i=1}^{n} x_i$ is the sample mean. The measure c_k is called the sample autocovariance for lag k. In principle c_k has a meaningful definition for $k = 1, 2, \ldots, (n - 1)$ but in practice, the largest k should be significantly smaller than n to ensure that a reasonable number of terms appear in the summation. An associated measure is the sample autocorrelation for lag k which is defined as

$$\hat{\rho}(k) = \frac{c_k}{s^2(n)} \quad ,$$

where $s^2(n) = \frac{1}{(n-1)} \sum_{i=1}^{n} (x_i - \overline{x}(n))^2$ is an estimator for the sample variance. A nonzero value for $\hat{\rho}(k)$ implies that the stochastic process X cannot be regarded as homogeneous because there is dependence among the constituent random variables.

We describe a discrete stochastic process that is stationary as having steady-state behaviour. Generally real-life stochastic processes are nonstationary. Often however, a nonstationary stochastic process may move from one steady-state behaviour to another steady-state behaviour where the underlying distribution parameters have changed. In other words the nonstationary process does have *piecewise stationarity*. During the transition, the process has transient behaviour. Note that a discrete stochastic process can also start with transient behaviour that eventually leads to steady-state behaviour.

Our discussion thus far has been restricted to discrete stochastic processes. Not surprisingly there exists as well an alternate category of *continuous* stochastic processes. Each member of this category is a function whose domain is an interval of the real line. Because of the continuous nature of this domain, the stochastic process in question could be viewed as an uncountably infinite sequence of random variables.

A special class of continuous stochastic process evolves when the function in question is piecewise constant (see Figure A1.12). The function $L(t)$ in Figure A1.12 could, for example, represent the length of a queue. Although $L(t)$ has a value for every t in the observation interval, there are only a finite number of relevant time points in the domain set, namely, those at which L changes value. Because the domain set for this special class of continuous stochastic process can be viewed as being finite, members of this class can be represented as a finite collection of ordered random variables. In this regard they share an important common feature with discrete stochastic processes. This similarity, however, does not extend to the manner in which statistical parameters are evaluated.

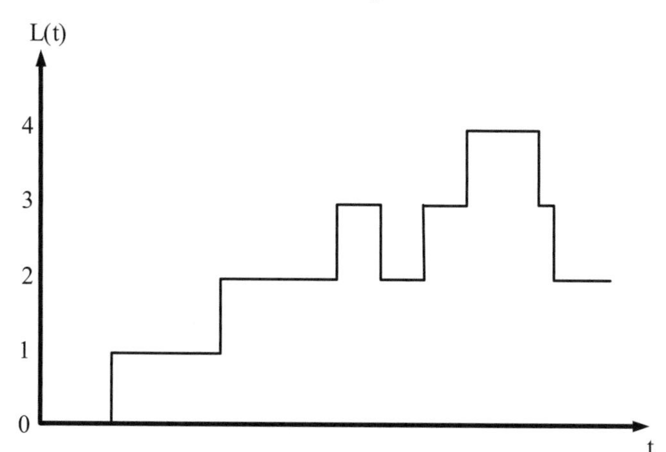

FIGURE A1.12. Piecewise constant stochastic process.

A1.9 References

A1.1. Bethea, R.M. and Rhinehart, R.R., (1991), *Applied Engineering Statistics*, Marcel Dekker, New York.

A1.2. Hogg, R.V. and Craig, A.T., (1995), *Introduction to Mathematical Statistics*, 5th edn., Prentice Hall, Upper Saddle River, NJ.

A1.3. Kleijnen, J.P.C., (1987), *Statistical Tools for Simulation Practitioners*, Marcel Dekker, New York.

A1.4. Ross, S.M., (2000), *Introduction to Probability and Statistics for Engineers and Scientists*, 2nd edn., Academic Press, San Diego.

Annex 2 GPSS Primer

A2.1 Introduction

This primer provides an introduction to GPSS (general purpose simulation system) and thus provides background material for Chapter 5 (Section 5.5) for readers unfamiliar with this software. GPSS is a general process-oriented simulation software environment. A student version of GPSS World, a current GPSS system developed by Minute Man Software, is freely available at http://www.minutemansoftware.com/product.htm. The reader is encouraged to obtain a copy of this version and use it with this primer. A number of examples are presented and problems suggested allowing the reader to gain practical experience with creating process-oriented simulation models.

GPSS World is a Win32 application designed to run on the various Windows operating systems (Win98, Win2000, XP, etc.). The student version of the software includes a complete reference manual. As well a 20-lesson tutorial developed by Minute Man is available from their Web site.

The objective of this primer is to describe how GPSS functions and to highlight several important subtleties. It is by no means a replacement for the *GPSS World Reference Manual* made available by Minute Man. The authors hope that this primer will enable the student to understand the basic operation of GPSS which can then be extended with the details provided by Minute Man documentation.

All GPSS program elements are referred to as entities. Note that in this text we adopt the convention of capitalising the first letter of GPSS entity terms. For example, the GPSS Transaction entity will be referred to as a GPSS **T**ransaction, or simply as a **T**ransaction.

GPSS carries out simulation tasks by processing GPSS Transaction entities (a data structure that consists of a list of attributes called Transaction parameters). These entities are viewed as flowing through a simulation model. The programmer's view of this movement consists of the Transaction flowing through a sequence of *Block* entities. In reality the Transaction actually moves through a number of underlying data structures (that are classified as structural entities). Each GPSS Block in fact represents a GPSS function triggered by Transactions that traverse, or try to traverse, the Block. A GPSS Block corresponds to a GPSS programming statement. Figure A2.2 in the next section provides a simple example.

Consider the following example of a Transaction moving through the SEIZE Block (shown in Figure A2.2). The SEIZE Block is used to gain the ownership of a Facility structural entity (more on the Facility entity later). When a Transaction tries to enter the SEIZE Block, it triggers a function that determines if the Transaction is allowed ownership. If the Transaction is allowed ownership (i.e., the Facility is not currently owned by another Transaction and the Facility is available), it is allowed into the SEIZE block. Otherwise, it cannot enter the Block and stays in the current Block until the Facility is freed and/or made available (actually the Transaction is placed on a list in the Facility). The relationship between the movement through the SEIZE Block and gaining ownership of the Facility is clarified in the next few sections.

A GPSS segment is composed of a sequence of Blocks. It provides a specification of how Transaction entities flow through the simulation model beginning with their arrival in the simulation model until their exit from the simulation model (note that is it permissible for Transactions to never exit the model). A GPSS program generally contains multiple segments that can interact with each other as shown later. A segment generally corresponds to a process.

This primer introduces GPSS by exploring how Transactions flow among the GPSS data structures or more specifically among lists that are called chains in GPSS. When a GPSS program is compiled and executed, its statements (Blocks) will reference structural entities that contain these chains. The structural entities are often places of interaction between processes where Transactions from different processes compete with each other.

The flow of Transactions from Block to Block is the most common perspective taken in describing GPSS. In reality, however, GPSS is a list processor that moves Transactions from chain to chain. This list processing is, of course, automatic. The simulation modeller typically focuses on the functions provided by Blocks but will benefit considerably from an understanding of the background list processing.

A2.1.1 GPSS Transaction Flow

As suggested above, the most fundamental entities in a GPSS program are the Transaction entities and Block entities. Recall that Transactions flow through Blocks to invoke their functions. These functions include list processing that moves the Transactions between chains that are part of structural entities. For example, one of the chains associated with a Facility entity is a chain called the delay chain. When a Transaction is denied entry into the SEIZE block, the Transaction is moved into the delay chain, creating a queue of Transactions waiting to gain ownership of the Facility. Let's introduce a simple example to explore how GPSS list processing is related to the flow of Transactions within a GPSS simulation model.

We consider a simple manufacturing process where a workstation WK0 works on parts one at a time. These parts can be of two different types (called A and B). Figure A2.1 illustrates the system. Transactions can be used to represent parts that are serviced by the workstation. The workstation and queue on the other hand would typically be represented as a GPSS Facility entity. Only one Transaction can 'own' the workstation Facility because only one part can be serviced by the workstation at any moment in time. All other Transactions will be placed on the Facility delay chain thereby forming a queue of components waiting for servicing.

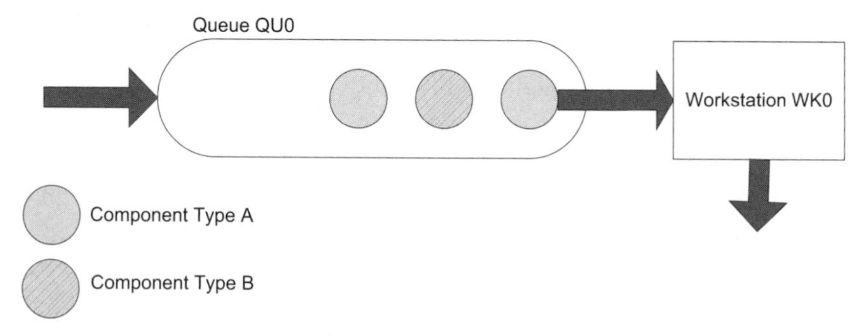

FIGURE A2.1. Manufacturing system.

Figure A2.2 shows a corresponding GPSS simulation model. Note that it is organised into two code groups, each composed of a sequence of Blocks and called a Block segment. Block segments generally start with a GENERATE Block that inserts Transactions into the simulation model and end with a TERMINATE block that removes Transactions from the simulation model. Such a Block segment specifies a process, that is, a lifecycle, for Transactions (in some very special cases more than one segment is required to represent a GPSS process). Two segments (processes) are included in the simulation model for the manufacturing system, one for each component type. For the moment, we focus on the first Block segment that represents the servicing of type A components.

```
*****Component A Process (single segment)
CompA     GENERATE 6,3
          SEIZE WK0
          ADVANCE 10
          RELEASE WK0
          TERMINATE

*****Component B Process (single segment)
CompB     GENERATE 7,4
          SEIZE WK0
          ADVANCE 5
          RELEASE WK0
          TERMINATE
```

FIGURE A2.2. GPSS simulation model for the manufacturing system.

Each GPSS statement in Figure A2.2 corresponds to a GPSS Block entity. Section A2.6.1 provides details about the use of these blocks. Here we simply present a brief summary of each Block.

- GENERATE: This Block inserts a Transaction into the simulation model according to its arguments. For the component A case, Transactions enter the simulation model every 6 ± 3 time units (uniformly distributed).
- SEIZE: A Transaction enters this Block when it gains ownership of the WK0 Facility.
- ADVANCE: Used to represent a service time duration. A Transaction remains in this Block for some duration of time as defined by the statement argument(s). In the case of component A, the duration is 10 time units.
- RELEASE: A Transaction releases ownership of the named Facility when it enters this block.
- TERMINATE: Transactions leave the simulation model when they enter this Block.

The functions invoked by the Blocks generally update the status of the simulation model (variables, data structures, etc.). In fact many Blocks do not affect the Transaction that entered the Block but instead update some unrelated part of the simulation model such as a structural entity. For example, the statement 'SEIZE WK0' which references a Facility entity named WK0 (the Facility entity is one of the GPSS structural entities). In general, any GPSS simulation model can be viewed as some set of functions that act on the program's structural entities such as the Facility entity.

The Transaction itself is a data structure consisting of a data-tuple that passes from chain to chain (i.e., list to list) in the simulation model. The data values in the data-tuple represent values for *parameters* (attributes). Standard parameters exist for each Transaction and are essential for processing the Transaction. The most common standard parameters include:

- Number: An integer that uniquely identifies each Transaction currently within the simulation model.
- Current Block: Indicates the Block entity at which the Transaction is located.
- Next Block: Indicates the Block entity the Transaction will next enter.
- Priority: Defines the priority[1] of the Transaction which is used to order Transactions within GPSS chains.
- Mark time: Initially set to the (simulation) time the Transaction entered the simulation model.
- BDT (Block Departure Time): Contains a time value which indicates when the Transaction can be processed. This value is present when the Transaction is part of the Future Events Chain, a list used to advance time. It is created when the Transaction enters an ADVANCE Block or when a Transaction is created by a GENERATE Block. The Future Events Chain is described below.

Additional parameters can be defined and created for Transactions in a GPSS program (by using the ASSIGN Block). These additional parameters are specified by the simulation modeller and reflect requirements flowing from the conceptual model.

The passage of a Transaction through a sequence of Blocks is managed with the Current Block and Next Block parameters of that Transaction. These parameters contain numerical identifiers that GPSS assigns to each Block in the simulation model. The GPSS processor consults the value of the Next Block parameter to identify the Block (and the operation) when processing a Transaction. When a Transaction 'enters' a Block, the Block's numeric identifier is saved in the Current Block parameter and the Next Block parameter is updated with the numeric identifier of the subsequent Block. If a Transaction is refused entry, then no changes are made to the Block parameters and the Transaction is typically placed on some chain, such as the Facility Delay chain. Some Blocks (see the Transfer Block) will update the Next Block parameter of the entering Transaction.

As previously indicated, Transactions are moved between chains that are usually priority-based first-in first-out lists. When a Transaction is

[1] In GPSS World, larger values indicate a higher priority. Negative values are valid as priority values.

placed on a chain, it is placed ahead of any lower priority Transactions and behind any Transaction with equal or higher priority. Let's examine now the main GPSS chains used to process Transactions.

The Current Events Chain (CEC) and the Future Events Chain (FEC) are the main chains used in the processing of Transactions and in moving simulation time forward. Note that there exists only one CEC and one FEC within any GPSS simulation program. Other chains that are part of GPSS structural entities such as the Facility, are nevertheless very important in managing the execution of a GPSS program.

The Transaction at the head of the CEC is the 'active' Transaction. GPSS moves the active Transaction using the Next Block parameter. More specifically, the Block identified by the Next Block parameter provides the functions invoked by the active Transaction. Returning to our SEIZE Block example, an active Transaction whose Next Block parameter refers to a SEIZE Block undergoes the following processing:

- If it can gain ownership of the Facility, it enters the SEIZE Block and
  - Its Current Block parameter is updated to the Block number of the SEIZE Block,
  - The Next Block parameter is updated to the Block number of the Block that would logically follow the SEIZE Block,
  - The Facility is tagged with the Transaction identifier to identify it as the owner,
- OR, if it cannot gain ownership of the Facility it is refused entry into the SEIZE Block and
  - The active Transaction is removed from the CEC and placed on the Delay Chain of the Facility. The Delay Chain corresponds to the queue of Transactions waiting to access the Facility.

In the first case (the Transaction is successful in gaining ownership of the Facility), GPSS then moves the Transaction to the Block following the SEIZE Block, namely, the ADVANCE Block in our example. In the second case, the Transaction is removed from the CEC and placed on the Facility delay chain where it can move no further. Now the next Transaction at the head of the CEC becomes the active Transaction and it will be processed by GPSS using its Next Block parameter.

Consider now the case where all Transactions on the CEC have been processed; that is, the list is empty. The chain must somehow be replenished. The FEC is the source for replenishing the CEC in this situation (as discussed later, the CEC can also be replenished from other sources). Note that the FEC is populated with Transactions flowing from two sources, namely, ADVANCE Blocks and GENERATE Blocks.

Consider the first case. When a Transaction enters an ADVANCE Block, the invoked function will calculate the time when the delay will terminate and assigns this value to the Transaction's BDT parameter. The Transaction is then placed on the FEC according to the value of the BDT parameter, with due regard for the priority of Transactions already on the FEC. Thus the FEC contains a list of Transactions that are time-ordered.

In the second case, the GENERATE Block places the Transaction on the FEC for entry into the simulation model with the BDT set to the time when the Transaction is scheduled to arrive in the simulation model. The current Block parameter is set to 0 (nonexisting Block) and the Next Block parameter is set to the Block number of the GENERATE Block. When a Transaction is moved from the FEC to the CEC and it enters the GENERATE Block, a successor Transaction will be placed on the FEC (ready to enter the GENERATE Block) by the GENERATE Block. This bootstrapping technique ensures a continuous flow of Transactions into the simulation model.

To advance time, GPSS moves the Transaction at the head of the FEC to the CEC and updates the time according to that Transaction's BDT parameter. If more than one Transaction on the FEC has the same BDT value, they are all moved to the CEC. Transactions with the same priority will be moved to the CEC in random order, otherwise priority conditions are enforced.

The CEC can also be replenished from other chains during the processing of the active Transaction. To see how the CEC could be replenished from the Facility delay chain (see Figure A2.3), consider the case when the GPSS processor finds a reference to a RELEASE Block in the Next Block parameter of the active Transaction. This Transaction enters the RELEASE Block, relinquishes ownership of the Facility, and the Transaction at the head of the delay chain becomes the new owner of the Facility. The new owner is also moved into the SEIZE block by updating its Current Block parameter and is moved from the delay chain to the CEC. This illustrates how the CEC can be replenished from chains other than the FEC.

Figure A2.3 shows the state of the Facility when Transaction A1 is the owner of the Facility and five Transactions are on the delay chain, namely, B1, A2, A3, B2, and A4. Assuming that Transactions A1 to A4 represent components of type A and B1 and B2 represent components of type B, it is clear that Transactions from different processes (traversing different Block segments) can be found in the delay chain (more on this in the next section). Note that the owner A1 is part of the CEC and in fact is the active Transaction. Assuming that A1 just entered the SEIZE Block (see Figure A2.4), it will be moved to the FEC when it enters the ADVANCE Block; that is, the ADVANCE Block function will place the Transaction on the FEC. As a consequence of the bootstrapping process of the two GENERATE Blocks in Figure A2.2, the FEC will always contain two

Transactions (one of type A and one of type B) that are scheduled to enter the simulation model at some future time.

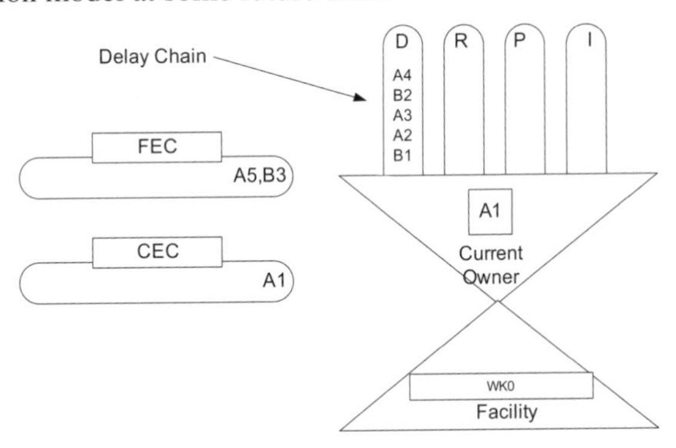

FIGURE A2.3. Structure diagram for the manufacturing problem.

The above demonstrates that although GPSS simulation models are typically constructed from the perspective of Transactions flowing through some designated sequence of Blocks, the underlying reality is that Transactions are moved among the chains that are implicit in a GPSS simulation model. By keeping this fact in mind, the program developer should be better able to understand the interaction between different processes. In fact it is this background processing that allows interaction between processes. The advantage of GPSS, and other process-oriented simulation environments, is that the simulation software takes care of managing the interactions among processes. There is no need for the program developer to be concerned with this detail. But structure diagrams are useful for keeping the program developer mindful of this underlying processing when creating Block segments. The next section outlines how the GPSS Blocks are used to implement processes.

A2.1.2 Programmer's View of Transaction Flow

A GPSS program specifies how Transactions flow through a simulation model, that is, through Block entities (and implicitly through the structural entities) of a simulation model. In the manufacturing example in the previous section, Transactions are used to represent parts that are serviced by the workstation WK0. The workstation together with the Transaction queue waiting for the workstation becomes a GPSS Facility entity. Transactions flow through this Facility, one at a time.

Recall that the GPSS Block segment (a sequence of Blocks that are traversed by Transactions having a common interpretation) is generally a specification of a process. A GPSS simulation program specifies one or more interacting processes formulated using multiple Block segments. This specification includes, but is not limited to:

- How Transactions enter into the model
- How they flow among structural entities
- How their service times are established
- The decisions for selecting different paths through the Block segment
- How their existence in the model is terminated

A GPSS Transaction can represent a consumer entity instance (cei) as presented in Section 4.2.2. Examples of such cei's are messages flowing in a communications network or customers in a department store. But it is also possible to use Transactions to represent resources that move within a model. Consider a tugboat that moves in a harbour, a waiter or waitress in a restaurant, or repair personnel in a manufacturing plant.

A GPSS Block segment can be graphically represented thereby providing a pictorial representation of a process. Figure A2.4 illustrates the Blocks used to specify the process of the type A components for the simple manufacturing system described in the previous section. The arrows indicate paths for the flow of Transactions from Block to Block triggering the functions associated to the Blocks. We consider now the manufacturing problem in more detail in order to examine these concepts more carefully.

The GENERATE Block defines how Transactions typically enter the model, that is, when they first appear on the FEC. In our example a type A Transaction is generated randomly every 6 ± 3 time units using the uniform distribution (i.e., randomly between 3 and 9 time units). Other distributions are available; for example, GPSS World provides over 20 different probability distributions for this purpose. As pointed out earlier, the operation of the GENERATE Block is based on a bootstrapping approach.

The Transaction exiting the GENERATE Block moves to the SEIZE Block that will let the Transaction enter it if the associated Facility WK0 is free (a Facility is free when it is not already owned by some Transaction). Otherwise, the Transaction remains in the GENERATE Block. Entering the SEIZE block triggers the function that causes the Transaction to become the owner of the Facility. Subsequent Transactions that try to enter the SEIZE Block will remain in the GENERATE Block until the owner Transaction releases the Facility (by entering the RELEASE Block).

The description in the previous paragraph is a process view of the Transaction flow presented in terms of updating the Block location parameters. As we saw in the previous section, the Transactions that cannot gain access to the Facility are placed on the Facility's delay chain.

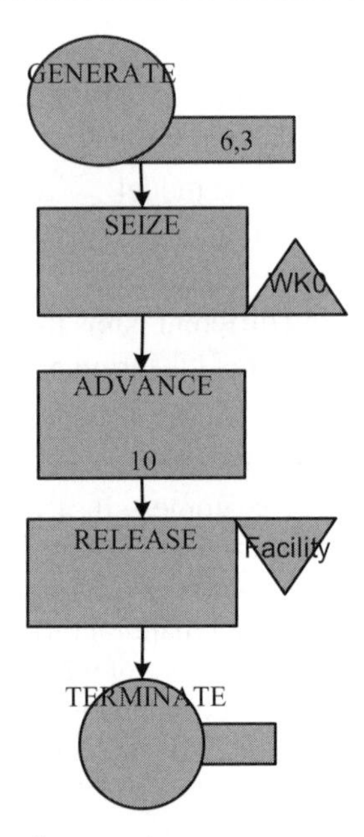

FIGURE A2.4. GPSS process for type A components.

The owner Transaction then proceeds from the SEIZE Block to the ADVANCE Block where it remains for a specified length of time (10 time units in this case). This duration represents the service time the workstation WK0 requires to service the part (i.e., the 'type A Transaction').

Once the Transaction exits the ADVANCE Block, it will pass through the RELEASE Block, triggering the function that relinquishes the ownership of the WK0 Facility. The Transaction then exits the model by entering the TERMINATE Block. When the Facility ownership is released, a Transaction waiting to enter the SEIZE Block will be enabled to do so. When the Facility is released, the Transaction at the head of the Facility delay chain will become owner of the Facility and allowed to move into the SEIZE Block. Note that the GPSS processor continues to move an active Transaction until it cannot be moved any further (i.e., it comes to rest in a chain when refused entry into a Block or it exits the model). Only then does the processor consider other Transactions first on the CEC otherwise on the FEC.

Figure A2.4 shows a single Block-segment that specifies the processing of type A components. Figure A2.5, on the other hand, shows two Block segments (hence two processes), one for each of the two component types, A and B. Recall that the workstation WK0 can service either component type. This is reflected by the SEIZE Block in each Block segment that references the same Facility WK0. This demonstrates that a process represented by one Block segment can affect the operation of another process represented by another Block segment. If a Transaction from either Block segment has seized the WK0 workstation, no other Transaction from either Block segment can seize it. The SEIZE and RELEASE Blocks are points of interaction between each Block segment, or in other words, between each process. This illustrates how the processes for each component type interact with each other.

The Block segments provide a program developer's view of the Transaction flow. They are used to specify the behaviour of the system. As was seen in the previous section, when a GPSS program is compiled, each Block is numbered and the data structures associated with GPSS structural entities such as the Facility referenced by the Block entities are created. During execution, Block functions will change the state of these data structures. Each Transaction is tagged both with its current location in the Block segment and the Next Block it should enter.

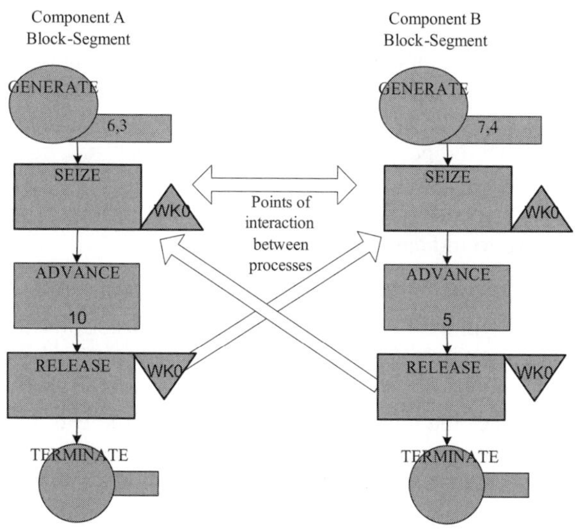

FIGURE A2.5. GPSS simulation model for the simple manufacturing problem (block segments for processing both types of components).

The remainder of this annex presents system numeric attribute (SNA), GPSS commands, details about the various GPSS Transaction chains, and an overview of available GPSS entities. The SNA plays an important role in accessing the state of entities in the simulation model. Commands are used to declare and initialise various entities and to control the execution of simulations.

A2.2 System Numeric Attributes

The system numeric attributes provide the means for accessing a whole range of entity data in a simulation model, for example, Transaction parameters, Facility utilisation (time Facility is occupied), Facility available (indicates if it is free), system clock, active transaction number, number of Transactions in a Block, etc.

The SNAs fall into categories. Each category is aligned with a specific entity type. Any particular SNA has relevance only to the entities of the designated entity type. An SNA has two parts. The first is the type of data to be delivered and the second identifies the specific entity (or sometimes parameter) of the relevant type from which the data are to be acquired. The entity identifier references the specific entity of interest. Consider the SNA used to provide the number of Transactions currently in a Block entity labelled *CompA*. The first part of the necessary SNA is 'W' (the number of Transactions in a Block entity) and the entity identifier is *CompA*, hence the SNA is 'W$CompA' (the $ serves to indicate a symbolic reference for the entity identifier).

The entity identifier can be explicitly given as a numerical value or a symbolic representation, or a reference to a parameter of the active Transaction whose value is the numeric identifier of the entity. This latter mechanism for referencing the entity identifier is called indirect addressing.

The forms of specific SNA class W are as follows (note that the characters * serve to indicate indirect addressing).

- **W22**: References the number of Transactions in Block 22.
- **W$CompA**: References the number of Transactions in the Block labelled CompA (the GENERATE block in the example of the previous section).
- **W*5**: References the number of Transactions in the Block identified by the value of the active Transaction's parameter 5.
- **W*BLKNUM (or W*$BLKNUM)**: References the number of Transactions in the Block identified by the value of the active Transaction's parameter named BLKNUM.

Other examples of SNAs include:

- **P$Name**: To reference the parameter labelled *Name* of the active Transaction.
- **F$WK0**: Indicates if a Facility labelled WK0 is busy (i.e., owned by a Transaction).

- **FC$WK0:** Facility capture count that gives number of times the Facility labelled WK0 has been SEIZEed.
- **FR$WK0**: Facility utilisation that indicate the percentage of time the Facility labelled WK0 was busy.
- **N$CompA:** Block entry count that gives the number of Transactions that have entered the Block labelled CompA.

Note that all the symbolic names (together with the $) in the above examples can be replaced with a numeric value that identifies the entity in question.

The following 'atomic' SNAs do not include an entity identifier.

- Parameters of the active Transaction:
  - A1: Assembly Set.
  - M1: Transit Time which is initially set to AC1 when the transaction enters the simulation (often called the mark time).
  - PR: Priority.
- XN1: Transaction number.
- AC1: Absolute value of the system clock (measured with respect to the start of the simulation run).
- C1: Elapsed simulation time since the last RESET Command.
- TG1: Remaining termination count. See the TERMINATE Block.

The example in Figure A2.6 illustrates the use of an SNA with the TEST Block to check the availability of the Facility called 'Machine'. The SNA FV$Machine is 1 when the Facility is available in which case the TEST Block allows the Transaction to move to the Block labelled 'Continue'. Otherwise the SNA is 0 and the TEST Block halts further progress of the Transaction (i.e., it will be placed on the retry chain of the Machine Facility).

```
            TEST E 1,FV$Machine
Continue    .
            .
            .
            SEIZE Machine
```

FIGURE A2.6. Using SNAs.

To demonstrate the use of indirect addressing, we revise the example in Figure A2.6 by replacing the explicit reference to the Machine Facility with an indirect address as shown in Figure A2.7.

```
            TEST E 1,FV*Station
            .
            .
            .
            SEIZE  P$Station
```

FIGURE A2.7. Using indirect addressing.

We assume here that the Transaction parameter named ‘Station’ contains the number of the Machine Facility. The TEST Block uses an SNA with indirect addressing to reference the Facility using the Station parameter of the active Transaction. It will only allow Transactions to pass if the SNA is equal (E) to 1; that is, the Facility is available. The SEIZE statement references the Facility using the parameter value reference given by the SNA ‘P$Station’.

When a Transaction parameter is used to identify entities, a generic block segment (process) specification can be defined. The Block segment is generic in the sense that explicit references to the entities are not included in the Block segment but rather are provided by the parameters of Transactions traversing the Block segment. This approach can simplify GPSS programs by collapsing a family of identical Block segments into a single (parameterised) Block segment. This is particularly useful, for example, when there are a large number of Facilities (consider modelling a system with many tens or hundreds of Facilities).

A2.3 GPSS Standard Output

Figure A2.8 shows part of the GPSS reports generated at the end of a simulation run for the simple manufacturing model. For Facility entities, GPSS provides the following values at the end of a simulation run.

1. ENTRIES – Total number of Transactions that entered the Facility entity.
2. UTIL. – The utilisation of the Facility, that is, the percentage of time that the Facility was busy.
3. AVE. TIME – The average time a Transaction owns the Facility entity.
4. AVAIL. – Indicates the availability of the Facility entity, that is, equals 1 when the Facility is available, 0 otherwise.
5. OWNER – Transaction identifier of the current owner.
6. PEND. – Number of Transactions on the pending chain of the Facility entity.
7. INTER. – Number of Transactions on the interrupt chain of the Facility entity.

8. RETRY – Number of Transactions on the retry chain of the Facility entity.
9. DELAY – Number of Transactions on the delay chain of the Facility entity.

Note that the ENTRIES, UTIL., and AVE. TIME can be used as DSOV values. Consult the *GPSS World Reference Manual* for details on available output values for other GPSS entities.

The first part of Figure A2.8 provides information on the Block entities including the Block Number (LOC), the type of Block (BLOCK TYPE), the number of Transactions that have traversed the Block (ENTRY COUNT), the number of Transactions currently in the Block (CURRENT COUNT), and the number of Transactions on the Block's delay chain (DELAY). The last part of the figure shows the contents of the FEC and provides for each Transaction its Transaction number (XN), its priority (PRI), its block departure time (BDT), its assembly set identifier (ASSEM), its current block parameter (CURRENT), its next block parameter (NEXT), and nonstandard parameter values (PARAMETER and VALUE).

A2.4 GPSS Commands

The existence of certain entities needs to be declared to the GPSS processor and also initialised. One of the purposes of GPSS commands is to carry out these tasks and the following commands are relevant:

- BVARIABLE: Used to declare Boolean Variable entities that implement Boolean expressions. The expression is evaluated during a simulation run using the 'BV' SNA.
- VARIABLE or FVARIABLE: Used to declare Variable entities that implement arithmetic expressions. The expression is evaluated where the 'V' SNA is specified.
- FUNCTION: Used to declare a Function entity. The Function is invoked where the 'FN' SNA is specified.
- MATRIX: Used to declare a Matrix entity.
- QTABLE: Used to declare a Table entity for collecting statistics from a Queue Entity.
- TABLE: Used to declare a Table entity for collecting statistics from other simulation data items.
- STORAGE: Used to declare and configure a Storage entity.
- EQU: Used to assign a value to a label, for example, for assigning labels to constants.

```
LABEL              LOC  BLOCK TYPE     ENTRY COUNT CURRENT COUNT RETRY
COMPA               1    GENERATE           2417             0     0
                    2    SEIZE              2417             0     0
                    3    ADVANCE            2417             1     0
                    4    RELEASE            2416             0     0
                    5    TERMINATE          2416             0     0
COMPB               6    GENERATE           2053             0     0
                    7    SEIZE              2053             0     0
                    8    ADVANCE            2053             0     0
                    9    RELEASE            2053             0     0
                   10    TERMINATE          2053             0     0
                   11    GENERATE              1             0     0
                   12    TERMINATE             1             0     0


FACILITY         ENTRIES  UTIL.   AVE. TIME AVAIL. OWNER PEND INTER RETRY DELAY
 WK0               4470    0.906       2.918  1     4470    0    0     0     0


FEC XN   PRI        BDT      ASSEM  CURRENT  NEXT  PARAMETER    VALUE
  4470    0      14401.728   4470      3       4
  4472    0      14402.357   4472      0       6
  4473    0      14403.421   4473      0       1
  4474    0      28800.000   4474      0      11
```

FIGURE A2.8. GPSS output for a simulation run of the manufacturing model.

- INITIAL: Used to initialise a Matrix entity, a SaveValue entity, or a LogicSwitch entity.
- RMULT: To set the seeds for the random number generators.

Another role for GPSS commands is the control of the simulation execution.

- CLEAR: Completely reinitialises the simulation including, but not limited to, the removal of all Transactions, clearing all data values, resetting the system clock, and priming the GENERATE blocks with their first Transactions. (See the GPSS World documentation for details.)
- RESET: Resets the statistical accumulators. This command is useful for clearing collected data after a transient warm-up period.
- START: To start a simulation run.
- STOP: To set or remove a stop condition (break point) at a Block.
- HALT: To halt the execution of the simulation run. (The GPSS World hot key Cntrl-Alt-H can be used to invoke HALT.)
- STEP: To execute a specified number of Blocks (useful for debugging).
- CONTINUE: Used to continue the simulation run following a stop condition.
- SHOW: To display the state of the simulation model. Can be used to examine SNAs or calculate expressions.

Other miscellaneous Commands include:

- REPORT: At the end of a simulation run a standard report is generated. This standard report can be generated before the end of the simulation run with this command.
- INCLUDE: To include contents from another file into a simulation model. This feature is frequently needed in complex models.

GPSS World provides a GUI interface that can invoke many of the commands to control the simulation run. Details about this interface can be found in the *GPSS World Reference Manual*

A2.5 Transaction Chains

Different types of Transaction chains exist in a GPSS simulation model The two main Transactions chains used for processing Transactions have already been introduced in Section A2.1.1. These are:

- Current Events Chain: Transactions waiting for processing.
- Future Events Chain: Transactions waiting for time to advance.

Other chains used by the GPSS processor include: User chains, delay chains, retry chains, pending chains, and interrupt chains. Note that at any moment in time a Transaction will be on only one of the following chains.

- Current Events Chain
- Future Events Chain
- Delay Chain
- Pending Chain
- User Chain

Various other rules about occupying a chain also exist. For example, a Transaction on an Interrupt chain cannot be on the Future Events Chain. A Transaction may be on any number of Interrupt Chains or any number of Retry Chains at the same time.

a) **Retry Chains:** Blocks such as GATE or TEST perform tests using some logical expression. These expressions typically contain SNA's related to one or more entities. When a Transaction tries to enter such a Block and the test fails the Transaction is placed on all Retry chains of the referenced entities. When the underlying data structure one of entities changes, the Transactions on the entity's retry chain are moved back onto the CEC. Thus the GPSS processor will re-evaluate the test to see if the change in the entity has caused the logical expression to become TRUE. If the test is successful, then the Transaction will enter the Block and be removed from all Retry chains.
b) **Delay Chains:** Delay chains are used to queue Transactions that have been refused access to a Facility or a Storage entity. (See the discussion on the Facility and the Storage entities for details.) These chains use a priority FIFO queueing discipline. User chains must be used for implementing more complex queueing disciplines.
c) **Pending Chain:** Each Facility entity contains a Pending chain used to queue Transactions that are waiting to pre-empt a Facility via an interrupt mode PREEMPT Block, that is, were refused entry into the PREEMPT Block. (See the discussion of the Facility entity for further details.)
d) **Interrupt Chain:** Each Facility entity contains an Interrupt chain used to queue Transactions that have been pre-empted. (See the discussion of the Facility entity for further details.)

e) **User Chains:** The User chain entity is simply a general Transaction chain. Such chains can be used to implement different queuing disciplines. The UNLINK block provides many features for selecting a Transaction from a User Chain.

A2.6 GPSS Entities

We first divide all available GPSS entities into four categories and examine each category in turn. An overview of each entity is presented that includes related Commands, Blocks, and SNAs where applicable. Further details about the usage of these various entities can be found in the *GPSS World Reference Manual.*

A2.6.1 GPSS Entity Categories

We separate the GPSS Entities into the following categories.

1 Basic Category: Transaction entities and Block entities are the two members of this category. They represent the principle building blocks for GPSS simulation models.

2 Structural Category: This category is comprised of entities used to represent the structural aspect of the simulation model. In Section A2.1, the Facility entity is used to represent a workstation together with the queue leading to it. This category has the following members:

   a) Facility: Represents a resource together with a queue leading to it. (There are also two other queues defined for the Facility that are used in the implementation of pre-emption).
   b) Storage: Represents a group together with a queue leading to it. A Transaction occupies space in a Storage entity. The amount of space available and the amount of space a Transaction occupies in the Storage entity are assignable by the program developer.
   c) User Chain: A queue that is a priority ordered list of Transactions.
   d) Logic Switch: A simple entity with two states: Set or Reset.
   e) Numeric Group: A set of numeric values that can be useful for recording events or representing the state of a process.

   f) Transaction Group: A set of Transactions. This is different from an the Storage entity. A Transaction can be member of one or more Transaction Groups. This membership serves as a means of classifying Transactions.

3 Data and Procedures Category: This category is comprised of entities that can be used to represent data or be used to develop expressions and functions.

   a) Matrix: A multi-dimension array of data values. Matrices are global in scope and can be accessed (including modification) by both the PLUS procedures and GPSS Blocks. Within GPSS statements, the "MX" SNA is used to reference a matrix, while PLUS procedures use the Matrix label.
   b) SaveValue: A variable that can take on any value.
   c) Variable: The variable entity is an arithmetic or logical expression (of arbitrary complexity) that can be evaluated at any time using an SNA reference. There are three types of Variable entities: arithmetic, floating point and Boolean.
   d) Function: Used to create data functions. For example, this entity is useful in creating discrete probability mass functions.

4 Output Category: This category contains entities that can be used to collect data during a simulation run and carry out subsequent statistical analysis.

   a) Queue: This entity is often used to collect data and provide average values (such as the average resident time) relative to the delay chains for Facility or Storage entities. SNAs for Queue entities include average queue size, maximum queue size, average queue resident time (with and without entries with zero time in queue), and number of entries with zero time in the queue. The Queue entity can also be used to determine other times such as the time that a Transaction spends in an area in the system (such as the time waiting for and being serviced by some resource, e.g. a Facility) or the total (simulation) time in the model. Note that the Queue entity is not, in fact, a genuine queue (as is a User chain).
   b) Table: This entity provides a means of creating a histogram for data obtained from the simulation run. Tables can be created for Queue entities or for any other data values in the system, such as parameter values.

A2.6.2 Basic Category

A2.6.2.1 Transaction Entity

Recall that Transactions are data-tuples that 'move' through a simulation model. Although the programmer views movement of Transactions from Block to Block, GPSS actually moves Transactions from chain to chain (and invokes list processing operations).

The values in the data-tuple that represents a Transaction are values for a collection of Transaction parameters. Parameter values can be referenced using system numeric attributes (see section A2.2). Some standard parameters are:

- Transaction Number: This parameter uniquely identifies the Transaction in the simulation model. It is set when the Transaction enters the simulation model.
- Priority: The priority affects the selection of Transactions from the various GPSS chains. Higher priority Transactions will be selected first.
- Mark Time: A time stamp that indicates the time that a Transaction entered the simulation model. This value can by altered by a MARK Block.
- Assembly Set: It is possible to split a Transaction into a number of 'parallel' Transactions that form a set called an assembly set. The SPLIT Block carries out this split operation. Each of the resulting Transactions is given its own Transaction number. All Transactions resulting from such an operation have the same value for the Assembly Set parameter (most other parameters are also duplicated from the parent). This makes it possible to reassemble the split Transactions with the ASSEMBLE Block. When a Transaction enters a simulation model, its Assembly Set parameter is set to its Transaction number.
- Current Block: This parameter contains the number of the Block that the Transaction currently occupies.
- Next Block: This parameter contains the number of the Block that the Transaction wishes to enter.
- Trace Indicator: A flag that enables printing of tracing messages to the console each time a Transaction enters a Block. This flag is set and reset by the TRACE and UNTRACE Blocks, respectively.

The state of a Transaction is a reference to the chain on which it is located. The possible states of a Transaction are as follows.

- Active: The Transaction has reached the head of the CEC and is the one that is currently being processed.
- Suspended: The Transaction is either on the FEC or the CEC and is waiting for processing.

- Passive: The Transaction is resting in a Delay Chain, Pending Chain, or User Chain.

A Transaction can also be in the pre-empted state which is not mutually exclusive with the states listed above. A Transaction is in this state when located in an interrupt Chain. The Transaction is in the Terminated state when it has left the simulation model.

Related Blocks

As Transactions move into Blocks they invoke some processing that affects the simulation model's status. Some Blocks, however, have a direct impact on the Transactions themselves. A list of these Blocks is given below.
TRANSFER SBR – The TRANSFER Block, in subroutine mode, assigns the Block number of the TRANFER Block to one of the Transaction's parameters. This allows the Transaction to return to this point thereby providing a subroutine function.
SELECT – The SELECT Block provides a variety of features to select an entity from a set specified by lower and upper entity numbers, such as selecting from a set of Facility entities. The number of the selected entity is saved in a parameter of the active Transaction (which can be used with indirect addressing).
SPLIT – This Block splits a Transaction into a number of 'replicated' Transactions. These replicated Transactions have the same value for their respective assembly set parameter.
ASSEMBLE – This Block reassembles the Transactions created by the SPLIT Block. When a specified number of Transactions in the assembly set have been collected, the first Transaction will exit the ASSEMBLE Block and the rest destroyed.
GATHER – This Block collects Transactions created by the SPLIT Block (i.e., from the same assembly set). When a specified number of Transactions from the same assembly set have been collected, they are allowed to exit the Block. Note that this Block is similar to the Assemble Block but does not destroy Transactions.
MATCH – This Block works in pairs. That is, two MATCH Blocks are placed in a simulation model and will allow Transactions through only when both Blocks see Transactions from the same assembly set.
COUNT – This Block carries a counting operation on a set of entities (identified by lower and upper entity numbers) and places the count in one of the active Transaction's parameters.
INDEX – INDEX has two operands. This Block adds the value of a parameter of the active Transaction identified by the first operand to the value of the second operand and saves the result into parameter 1 of the

active Transaction. For example, INDEX 4,23.4 will add 23.4 to the value in parameter 4 and save the results in parameter 1.
LOOP – This Block decrements the parameter in the active Transaction that is identified by a LOOP operand. If the resulting parameter value is not 0, then the Transaction will be sent to a Block identified by another LOOP operand. Otherwise, the Transaction proceeds to the Block following the LOOP Block.
TRACE and UNTRACE – These Blocks turn the Transaction's trace indicator on and off.
ADOPT – This Block will change the assembly set value of the active Transaction.
GENERATE – This Block generates Transactions.
TERMINATE – This Block removes Transactions from the simulation model.

Related SNAs

The following SNAs provide information about the Active Transaction.

- A1: References the value of the assembly set number of the Active Transaction.
- MB: This SNA can be used to determine if another Transaction from the same assembly set is located at a specific Block with number *EntNum* using the expression MB*EntNum*. The SNA returns a 1 if such a Transaction exists and 0 otherwise.
- MP: This SNA can be used to compute a transit time. It provides a value equal to the current simulation time minus the value in a parameter referenced by the parameter identifier in the SNA (e.g., MP5 yields the value of the current time minus the value found in the 5th parameter of the current Transaction).
- M1: This SNA returns a value equal to the current simulation time minus the Mark Time.
- P: This SNA can be used to reference the value of a parameter (e.g., P*Param*).
- PR: This SNA is a reference to the value of the priority parameter of the active Transaction.
- XN1: This SNA is a reference to the value of the Transaction's number (i.e., its unique identifier within the program as assigned by the GPSS processor).

A2.6.2.2 Block Entity

The Block entities are the basic programming elements of the GPSS environment. The Block segments (a sequence of Blocks) are used to specify the various processes within a simulation model. The Blocks and the Transactions function together to generate changes in the various structural entities in the simulation model and these, in turn, give rise to the model's behaviour. Furthermore, it is through these structural entities that processes interact with each other (see the Structural entity category). As was already seen in Section A2.1.2 is possible to represent Block segments using a graphical representation of the Blocks.

Related SNAs

- N: The total number of Transactions that have entered a Block
- W: The current number of Transactions occupying a Block

Several Block entity types will be examined in the following sub-sections in terms of their relationship with various other GPSS entities. First, however, we list some Blocks that have a variety of specialised roles.

Miscellaneous Blocks

- ADVANCE: This block provides the fundamental mechanism for representing the passage of time (typically some service time). This is achieved by placing the Transaction entering the Block on the FEC with its BDT parameter set as described in Section A2.1.1. Many options exist for defining the time delay.
- BUFFER: This Block allows a Transaction to interrupt its processing and allows other Transactions on the CEC to be processed first. In effect, the active Transaction on the CEC is moved behind other Transactions of equal or greater priority.
- EXECUTE: The EXECUTE operand identifies a Block number. When a Transaction enters this Block, the action associated with the Block identified in the operand is executed.
- PLUS: This Block evaluates an expression that may include PLUS procedures. If a second operand is provided, the result of the evaluation is saved in the Active Transaction's parameter as specified by the second operand.

Test and Displacement Blocks:

- DISPLACE: This Block allows a Transaction to displace any other Transaction in the simulation model. The displaced Transaction is the one identified by the Transaction number in the first operand of the DISPLACE Block.
- GATE: Changes the flow of a Transaction based on testing the state of some entity in the simulation model. It operates in two modes. If the test is successful the Transaction traverses the Block. Otherwise, in the Refuse mode, the GATE will block Transactions until the test becomes

successful and in the Alternate Exit mode, the GATE will send the Transaction to an alternate Block.

- TEST: Changes the flow of a Transaction based on a test that compares values, often using expressions with SNAs. As with the GATE Block, it can also operate in either Refuse mode or an Alternate Exit mode.
- TRANSFER: This Block results in the entering Transaction jumping to a new location. The Block basically sets the value for the Transaction's Next Block parameter by evaluating its operands. Many options are available with this Block, such as unconditional transfer, picking a location randomly, selecting among a number of Blocks that are nonblocking, using a Function to compute the Block location, and so on.

Data Stream Blocks

- OPEN: Opens a 'data stream'. For example, OPEN ("Data.Txt"),2,ERROR will open the file *Data.Txt* as stream 2. If an error occurs, then the active Transaction is sent to the Block labelled *ERROR*.
- READ: Retrieves a line of text from a data stream identified by an operand.
- WRITE: Writes a line of text to a data stream identified by an operand.
- SEEK: Sets the line position of a data stream identified by an operand.
- CLOSE: Used to close a data stream identified by an operand.

A2.6.3 Structural Category

A2.6.3.1 Facility Entity

The Facility entity represents a resource that can be owned by a single Transaction. Our representation for the Facility as shown in Figure A2.9 is intended as a snapshot of the Facility during the simulation run. A Facility contains a number of Transaction chains. These support various queuing requirements of waiting Transactions, as well as pre-emption and interruption. GPSS maintains statistical information for Facilities. The Facility is always in one of two states: namely 'busy' when owned by a Transaction or 'idle' otherwise.

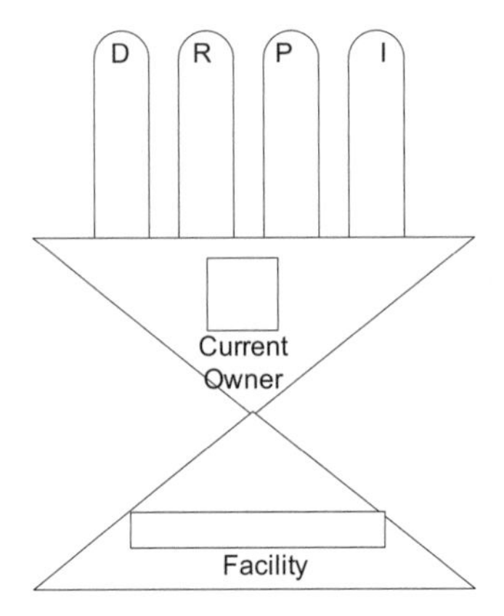

FIGURE A2.9. A representation of the GPSS facility.

In Figure A2.9, the box labelled Facility is the name assigned to the Facility and the box labelled Current Owner identifies the Transaction that currently owns the Facility. The following Transaction chains are associated with a Facility.

- Delay Chain (D)
- Retry Chain (R)
- Pending Chain (P)
- Interrupt Chain (I)

The Retry chain contains Transactions waiting for a change in the Facility's state (e.g., from busy to idle). The Delay, Pending, and Interrupt chains are used to support pre-emption of the Facility. Two modes of pre-emption are available:

1. Interrupt mode: Allows a Transaction to pre-empt the Facility if the current owner of the Facility has not already pre-empted the Transaction. Transaction priority is not relevant in this mode.
2. Priority mode: A Transaction will pre-empt the Facility if its priority is higher than the priority of the current owner.

When a Transaction tries to pre-empt the Facility using the Interrupt mode of the PREEMPT Block, it is refused ownership when the Facility is owned by a Transaction that acquired the Facility through pre-emption. In this case it is placed on the pending chain.

When a Transaction tries to pre-empt the Facility using the Priority mode of the PREEMPT Block it is refused ownership when the Facility is owned by a Transaction of equal or higher priority. In this case it is placed on the delay chain.

When the Transaction that owns the Facility is pre-empted, it is placed on the Interrupt chain. Transactions may be allowed to move through Blocks while on an interrupt chain, but movement is restricted. For example, such Transactions can never be placed on the FEC (e.g., enter an ADVANCE Block).

A Transaction that owns a Facility must return ownership using the RETURN or RELEASE Blocks; for example, a Transaction that owns a Facility must not directly exit the simulation model. This applies equally to Transactions that are on a Delay chain, a Pending chain, or an Interrupt chain. Pre-empted Transactions, on the other hand, are allowed to exit the simulation model. Certain Blocks such as the PREEMPT and FUNAVAIL Blocks have options that can remove Transactions that are on Delay chains, Pending chains, or Interrupt chains (such Transactions are said to be 'in contention').

When a Transaction releases ownership of a Facility, ownership is given to a Transaction from one of the Facility's chains in the following order.

- Pending chain, the Transaction enters the PREEMPT Block and is placed on the CEC.
- Interrupt chain.
- Delay chain, the Transaction enters the SEIZE or PREEMPT Block and is placed on the CEC.

SNAs related to a Facility's state are:

- F: Returns 1 if the Facility is busy and 0 otherwise.
- FI: Returns 1 if the Facility is currently pre-empted using Interrupt Mode pre-emption and 0 otherwise.
- FV: Returns 1 if the Facility is available and 0 otherwise.

SNAs related to data collection are:

- FC: Returns the number of Transactions that have gained ownership of the Facility (SEIZEd or PREEMPTed).
- FR: Returns the fraction of time the Facility has been busy. It is expressed as parts per thousand using a value between 0 and 1000.
- FT: Returns the average time that a Transaction owns the Facility.

Related Block Entities

- SEIZE and RELEASE: The SEIZE Block is used by a Transaction to gain ownership of a Facility and RELEASE is then used to release the ownership.

- PREEMPT and RETURN: The PREEMPT Block is used to pre-empt a busy Facility. PREEMPT allows both interrupt mode and priority mode pre-emption. A pre-empting Transaction must enter a RETURN Block to release the ownership of the Facility.
- FAVAIL and FUNAVAIL: The FUNAVAIL Block will place the Facility in the unavailable state whereas the FAVAIL Block places the Facility in the available state. The FUNAVAIL Block offers many options to remove Transactions from contention.

A2.6.3.2 Storage Entity

The Storage entity provides the means for allowing a Transaction to occupy 'space'. Such an entity must be declared using the STORAGE command to define its size, that is, how much space is available in the entity. By default, a Transaction occupies a single unit of space when entering the Storage entity, but this default can be overridden with an ENTER Block operand. There are two chains associated with a Storage entity as shown in our representation in Figure A2.10, namely, a Delay chain (D) and a Retry (R) chain. The Members box can be used to record the Transactions that currently occupy space in the Storage Entity. The Max box provides the maximum storage units available and the Storage box gives the name of the Storage entity.

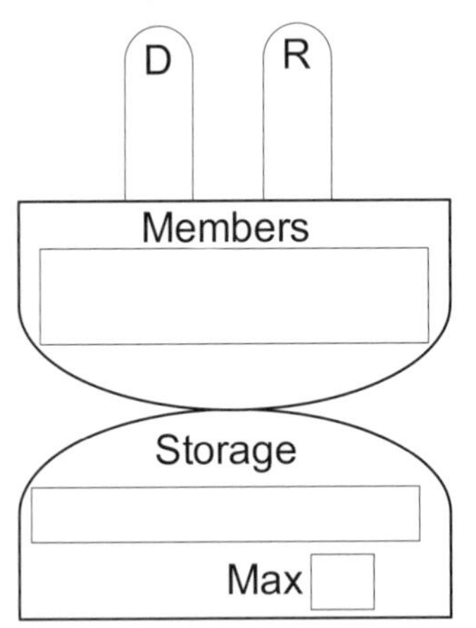

FIGURE A2.10. A representation of the storage entity.

When there is insufficient space to accommodate a storage request from a Transaction, the Transaction is placed on the Storage entity's Delay chain in order of priority. When space is freed by the departure of a member Transaction, the Delay chain is scanned and the first Transaction that can be accommodated becomes a member in the Storage entity. The Retry chain contains Transactions waiting for a change in the underlying data structure of the Storage entity.

SNAs related to the Storage Entity's state are:

- R: Provides the number of storage units available within the Storage entity.
- S: Provides the number of storage units occupied within the Storage entity.
- SE: Returns 1 if the Storage entity is empty and 0 otherwise.
- SF: Returns 1 if the Storage entity is full and 0 otherwise.
- SV: Returns 1 if the Storage entity is available and 0 otherwise.

SNAs related to data collection are:

- SA: Returns the time-weighted average of storage capacity in use.
- SC: Returns the total number of storage units that have been occupied.
- SR: Returns utilisation as the fraction of total possible utilisation, expressed in parts per thousand, that is, using a value between 0 and 1000.
- SM: Returns the maximum storage value used within the entity, that is, the high-water mark.
- ST: Returns the average holding time per unit storage.

Related Block Entities

- ENTER and LEAVE: A Transaction uses the ENTER Block to enter a Storage Entity and the LEAVE Block to exit the Storage entity.
- SAVAIL and SUNAVAIL: The SAVAIL makes the Storage entity available and the SUNAVAIL makes the Storage entity unavailable. Transactions that are already members of the Storage entity are not affected by the SUNAVAIL Block.

Related Commands

- STORAGE: Must be used to declare the capacity of the Storage entity.

A2.6.3.3 User Chain

A user chain provides a means to queue Transactions. Our representation of the User chain is shown in Figure A2.11. It offers a priority FIFO queuing discipline. It is possible to alter such a queuing discipline by using the UNLINK Block that offers many options for selecting Transactions for removal based on expressions and Transaction parameter values.

The User chain also contains a link indicator, a sort of 'red' and 'green' light that turns to red when a Transaction encounters an empty User chain with a green light indicator. When the Transaction subsequently enters an UNLINK block, the red light is turned green if the User Chain is empty. If the User chain contains Transactions, the one at the head of the chain is removed and the light is left red. This link indicator feature is useful when placing a User chain in front of a Facility (in essence replacing the Facility's Delay chain).

FIGURE A2.11. A representation of the user chain entity.

SNAs related to the User chain entity's state are:

- CH: Returns the number of Transactions currently in the User chain.

SNAs related to data collection are:

- CA: Returns the average number of Transactions that have been in the User chain.
- CC: Returns the total number of Transactions that have been in the User chain.
- CM: Returns the maximum number of Transactions that have occupied the User chain, that is, the high-water mark.
- CT: Returns the average time that Transactions resided in the User chain.

Related Block Entities

- LINK and UNLINK: When a Transaction enters the LINK Block it is placed on the User chain (unless the link indicator option is used). A Transaction is removed from a User chain by some other Transaction that enters the UNLINK Block. The UNLINK Block offers many options for selection of Transactions based on expressions and Transaction parameters.

A2.6.3.4 Logic Switch

A Logic Switch (see our representation in Figure A2.12) simply assumes one of two states, Set or Reset.

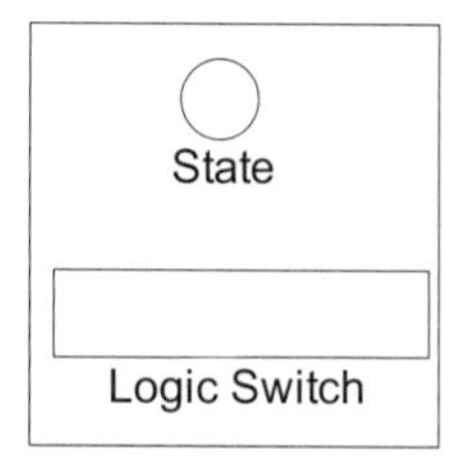

FIGURE A2.12. A representation of the logic switch entity.

There is one SNA related to the Logic Switch entity; namely,

- LS: Returns 1 if when the Logic Switch is Set and 0 for Reset.

Related Blocks

- LOGIC: This Block can be used to Set, Reset, or Invert the Logic Switch state.

Related Commands

- INITIAL: Used to initialise a Logic Switch.

A2.6.3.5 Numeric Group

The Numeric Group entity can be useful for describing the state of a process or recording events in a simulation run. Such entities contain a set of integer or real values.

There is one SNA related to the Numeric Group; namely,

- GN: Returns the number of members in the Numeric Group

Related Blocks

- JOIN and REMOVE: The JOIN Block adds a value to a Numeric Group and the REMOVE Block removes a value from a Numeric Group.
- EXAMINE: This block tests for the existence of specified values in a Numeric Group.

A2.6.3.6 Transaction Group

A Transaction Group is a set of Transactions. Transactions can be members of any number of Transaction Groups. These entities are useful for classifying Transactions.

There is one SNA related to the Transaction Group; namely,

- GT: Returns the number of Transactions in a Transaction Group.

Related Blocks

- JOIN and REMOVE: The JOIN Block adds a Transaction to a Transaction Group and the REMOVE Block removes a Transaction from a Transaction Group.
- EXAMINE: This Block tests for the presence of the active Transaction in a Transaction Group.
- SCAN and ALTER: These Blocks can test and/or modify the members of a Transaction Group.

A2.6.4 Data and Procedure Category

A2.6.4.1 Matrix Entity

GPSS World supports matrices with up to six dimensions. All six dimensions are available in PLUS procedures, but only two dimensions are accessible when used with SNAs, that is, within Block arguments.

There is only one SNA related to the Matrix entity; namely,

- MX: The expression MX*matA*(m,n) returns the value in row *m* and column *n* of the matrix *matA*.

Related Blocks

- MSAVEVALUE: This Block can be used to assign a value to an element in the matrix. Only two dimensions of the matrix can be accessed by this Block.

Related Commands

- INITIAL: The command can be used to initialise a matrix element.
- MATRIX: This command declares the existence of a Matrix entity, from two to six dimensions (in GPSS World). This command must be used before a Matrix entity can be referenced.

A2.6.4.2 SaveValue Entity

SaveValue entities are variables that can assume any value.

There is only one SNA related to the SaveValue entity; namely,

- X: Returns the value of the SaveValue entity, for example, X$VarB.

Related Block

- SAVEVALUE: This Block assigns or modifies the value of a SaveValue entity.

Related Command

- INITIAL: Can be used to initialise the value of a SaveValue entity.

A2.6.4.3 Variable Entity

The Variable entity is associated with an expression. These expressions can contain constants, SNAs, procedure calls, and include logic and arithmetic operators.

SNA's related to the Variable entity are:

- BV: Returns the result of evaluating a Boolean Variable Entity.
- V: Returns the results of evaluating an arithmetic or floating point Variable entity.

Related Commands

- VARIABLE: creates an arithmetic Variable entity.
- FVARIABLE: creates a floating-point Variable entity.
- BVARIABLE: creates a Boolean Variable entity.

A2.6.4.4 Function Entity

The Function entity provides a number of options for computing a value. These options include the use of random numbers. For example, it is possible to return a random value according to some discrete probability mass function.

Many times it is more convenient to use a PLUS procedure rather than using a Function entity. Functions are particularly useful for creating list base functions and empirical probability distributions.

There is only one SNA related to the Function entity; namely,

- FN: Returns the result of evaluating the Function entity.

Related Commands

- FUNCTION: Used to define a Function.

A2.6.5 Output Category

A2.6.5.1 Queue Entity

Queue entities are used for collecting data. It is important not to confuse Queue entities with chains. Transaction queues are formed in GPSS chains, not in Queue entities. As in the case of Transaction Groups, Transactions can become members of one or more Queue entities. GPSS will then compute queuing information based on this membership.

Typical use of Queue entities is to compute statistics related to Facility Delay chains (a Transaction becomes a member of the Queue entity just before trying to SEIZE a Facility and releases the membership just after acquiring the ownership of the Facility). This can be done by sandwiching the SEIZE Block between the QUEUE and DEPART Blocks.

SNAs related to the Queue entity's state are:

- Q: Returns the current number of Transactions that are members of the entity.

SNAs related to data collection are:

- QA: The average number of members in the Queue entity.
- QC: The total number of Transactions that entered the entity.
- QM: The maximum number of Transactions that occupied the entity.
- QT: The average time a Transaction spent in the entity.
- QX: The average time a Transaction spent in the entity without including Transactions that spent 0 time in the entity.
- QZ: The number of Transactions that spent 0 time in the entity.

Related Block

- QUEUE and DEPART: When a Transaction enters a QUEUE Block, it becomes member of the associated Queue entity. When a Transaction enters the DEPART Block it releases its membership from the associated Queue entity.

Related Command

- QTABLE: Declares a Table Entity that can be used to create a histogram of the times that Transactions spend in a Queue entity.

A2.6.5.2 Table Entity

The Table entity is used to accumulate data for creating a histogram.

SNA's related to the Table entity are:

- TB: Average of the entries in the Table entity.
- TC: Number of entries in the entity.
- TD: Standard deviation of the entries in the entity.

Related Blocks

- TABULATE: Used by a Transaction to add entries to a Table entity.

Related Commands

- TABLE: Used to declare a Table entity.
- QTABLE: Declares a Table Entity that can be used to create a histogram of the time that Transactions spend in a Queue entity.

Annex 3 Open Desire Primer

A3.1 Introduction

Open Desire is a software tool designed to support modelling and simulation projects within the realm of continuous-time dynamic systems. This support has two main facets; one relates to the numerical solution of the differential equations embedded in the conceptual model and the other relates to the execution of the experiments that need to be carried out in order to achieve the project goals.

Open Desire has been developed by Professor Granino A. Korn, Professor Emeritus at the University of Arizona. It has been written in C with an orientation to the Linux/UNIX environment. However, the source code can be recompiled to run under CYGWIN (see www.cygwin.com), a UNIX/Linux-like environment that runs under the Windows operating system. Both Open Desire and CYGWIN are freely available. Relevant information relating to Open Desire can be found at http://www.members.aol.com/gatmkorn.

There is considerable flexibility allowed in organising an Open Desire program. However, for pedagogical reasons, we superimpose in our presentation a structure that, for the most part, is not intrinsic to the Open Desire programming protocol. Likewise some of the functionality provided by Open Desire is not presented, e.g. features relating to neural network , Fourier transforms, and the solution of partial differential equations. The interested reader is urged to explore more comprehensive presentations of this simulation environment for CTDS models, for example [A3.1] and [A3.2].

A3.2 Programming Perspectives

An Open Desire program can be viewed from two perspectives which, to some extent, do overlap. One is from a functional perspective. This perspective has three facets:

a) Solution of the differential equations that generally comprise the major part of the conceptual model

b) Execution of the experiments that need to be carried out in order to achieve the project goals
c) Generation of the output that embraces the information implicitly or explicitly needed for achieving the project goals

Within the context of the program development approach we advocate, an Open Desire program can be likewise be viewed from a structural perspective. The structural perspective has four segments each beginning with a keyword.[1] These are as follows.

Keyword: ---CONSTANTS

This segment contains the assignment of values for physical constants (e.g., the acceleration due to gravity) and for other constants that are embedded within the conceptual model, provided they are not subject to change as a consequence of the experimentation activity, such as a friction coefficient or the mass of various system components. Likewise the initial values for the state variables and the boundaries of the observation interval are included here if these values are not subject to change as a consequence of the experimentation activity. Constants and initial values that are subject to change as part of the experimentation (and hence are parameters) are best assigned in the EXPERIMENT segment.

Keyword: ---EXPERIMENT

We include here values for various 'system parameters', that is, parameters used, for example, to specify features of the numerical solution procedure to be used in solving the underlying differential equations or for the characterisation of desired output. This segment typically contains program code that specifies some particular series of solutions to the equations of the conceptual model that will yield the results needed for achieving the goals of the project.

Keyword: ---OUTPUT

There generally is particular output that is needed from the experiments to satisfy the requirements of the project goals. The specification of this output takes place in this segment. (Note: often this segment is somewhat idealised inasmuch as it is not always constituted from contiguous code.)

Keyword: DYNAMIC

This segment contains the specifications of the differential equations that define the dynamic behaviour of the model being investigated. This is

---

[1] Of the four keywords discussed, only DYNAMIC is intrinsic to Open Desire. The others are simply 'comments' that are part of our recommended structuring (the three leading dashes are the required signal for a comment).

provided according to a specific protocol that is described in the following section.

A3.3 Solving the Differential Equations

The implicit assumption in Open Desire is that differential equations that need to be studied have been written in the form of a set of first-order ordinary differential equations; for example,

$$\begin{aligned} x_1'(t) &= f_1(x_1(t), x_2(t), \ldots, x_N(t), t) \\ x_2'(t) &= f_2(x_1(t), x_2(t), \ldots, x_N(t), t) \\ &\ \vdots \\ x_N'(t) &= f_N(x_1(t), x_2(t), \ldots, x_N(t), t) \ . \end{aligned} \tag{A3.1}$$

Here the N variables $x_1(t), x_2(t), \ldots, x_N(t)$ are the state variables for the model. Within the Open Desire protocol, each of these equations gives rise to a program statement of the form:[2]

d/dt xk = [*exp*] ,

where *exp* (i.e., "expression") is the programming equivalent to the derivative function $f_k(x_1(t), x_2(t), \ldots, x_N(t), t)$. For example, the program code for the equation:

$$P'(t) = \alpha\, P(t)\,(1 - K\,P(t))$$

could have the form:

d/dt P = alpha**P**(1 – *K***P*) .

Each first-order differential equation gives rise to an Open Desire program statement which begins with *d/dt* and is followed by the name of a state variable. These statements (which we collectively refer to as the *derivative specifications*) must all be included in the DYNAMIC segment of the program, that is, must follow the keyword DYNAMIC. The ordering of the statements within the derivative specification is arbitrary. Note that each state variable must have an initial value and these are specified either in the CONSTANTS or in the EXPERIMENT segment of the program.

Often, in the interests of clarity and/or efficiency some collection of *auxiliary variables* is sometimes introduced in the formulation of the

---

[2] We use square brackets to enclose a generic item.

derivative specifications. Consider, for example, the following CTDS model,

$$x_1'(t) = (x_1^2(t) + x_2^2(t))^{0.5}\, x_2(t)$$
$$x_2'(t) = [(x_1^2(t) + x_2^2(t))^{0.5} - |\, x_2(t)|\,]\, x_1(t) \;.$$

The corresponding derivative specification could be formulated as:

$A = \text{sqrt}(x1^{\wedge 2} + x2^{\wedge 2})$
$B = \text{abs}(x2)$
$C = A - B$
$d/dt\, x1 = A * x2$
$d/dt\, x2 = C * x1$.

Here A, B, and C are auxiliary variables. The assignment of their values must precede the derivative specifications and furthermore their ordering must respect the normal programming requirement that values on the right of the assignment must be defined. Hence the sequence

$C = A - B$
$A = \text{sqrt}(x1^{\wedge 2} + x2^{\wedge 2})$
$B = \text{abs}(x2)$

would not be meaningful.

Open Desire provides a variety of numerical solution methods for solving the differential equations of the model. Method selection is via the value assigned to the system parameter **irule** (this assignment should take place in the EXPERIMENT segment of the program). Values in the range 0 through 14 are permissible (although the value 0 has a special meaning; (see [1, page 132]). Some of the most commonly used solution procedures and their corresponding **irule** codes are listed below:

irule = 1: A fixed step-size second-order Runge–Kutta–Heun process

irule = 2: A fixed step-size, first-order Euler process

irule = 3: A fixed step-size, fourth-order Runge–Kutta process

irule = 4: A variable step-size, fourth-order Runge–Kutta process

The Open Desire parameter **DT** holds the value of the step size used in the solution process when a fixed step-size method has been selected. When a variable step-size procedure is specified (e.g., **irule** = 4) the assigned value of **DT** serves only to initiate the solution process and will generally be altered (by internal mechanisms) as the solution progresses. The value of **DT** should be assigned in the EXPERIMENT segment.

The Open Desire variable t is used to represent the independent variable for the model and typically t corresponds to time (the units of t have no particular relevance to the solution process but, in the interests of good programming practice, the units should be noted in a comment within the program code). The 'nominal' value for the right boundary of the observation interval for any particular simulation run is (t_0 + **TMAX**). Here **TMAX** is an Open Desire parameter whose value is assigned in the EXPERIMENT segment and t_0 is the initial value of t for the run in question. The initial value of t for the first simulation run is explicitly assigned in the EXPERIMENT segment. It should be emphasised that this need not be the initial value for t on subsequent runs. Nevertheless the observation interval for any simulation run is [t_0, t_0 + **TMAX**] where t_0 is the initial value of t for that run.

A simulation run (i.e., solution of the model equations) will never advance beyond t = (t_0 + **TMAX**). However, for a variety of reasons it may be necessary to terminate the solution prior to (t_0 + **TMAX**); for example, a projectile has struck its target or the pressure in a boiler has reached a level that necessitates shutting down a chemical process. Generally there are one or more events that can lead to such 'premature' termination. The **term** statement in Open Desire provides the means for dealing with such a situation. Its form is:

term [*exp*]

where *exp* is an expression (formulated in terms of program variables) whose algebraic sign changes from negative to positive when the termination event occurs. For example, if the program variable y represents the distance above the ground of a ball's trajectory and it is required to terminate the solution when the ball strikes the ground, then *exp* would be $-y$.

Multiple **term** statements are permitted and the solution terminates when the first is 'triggered'. All **term** statements must appear in the DYNAMIC segment of the program.

A3.4 Organising Experiments

Experimentation with a CTDS model generally involves the repeated solution of the underlying model equations during which behavioural features are observed/evaluated. This is typically associated with some organised set of changes in the model. Two essential requirements therefore emerge. The first is simply the need for a means to invoke a solution of the model equations (i.e., invoke a *simulation run*) and the second is a means to create a program structure that carries out the changes in the model that are essential for meeting the experimentation requirements.

The first requirement (invoking a simulation run) is achieved in Open Desire with the **drun** statement. The execution of this statement results in the solution of the equations specified in the DYNAMIC segment. A key issue relating to any simulation run is the initial value assigned to the state variables at the beginning of the simulation run and, as well, the nominal observation interval (because the right-hand boundary may be overridden by a **term** statement, the observation interval can only be regarded as 'nominal'). The rules are as follows.

The first simulation run (the first invocation of **drun** in the EXPERIMENT segment) begins with the values assigned to the state variables in the CONSTANTS and/or EXPERIMENT segment. The initial value of t is likewise the value explicitly assigned to it in the EXPERIMENT segment. If the **reset** statement is not invoked, the initial values assigned to the state variables for a subsequent run (resulting from invocation of **drun)** are the final values of the state variables from the preceding run; likewise the starting value of t is the final value of t from the preceding run.

If **drun** is followed by **reset** then, at the end of the simulation run, all state variables (and t as well) are reassigned the values they had at the beginning of the preceding run.

To illustrate, suppose $x1$ is one of the state variables (i.e., the DYNAMIC segment contains a statement of the form: *d/dt* $x1$ = . . .) and that the assignment $x1 = 5$ appears in the CONSTANTS segment and the assignments $t = 0$ and TMAX = 2 appear in the EXPERIMENT segment (to simplify the discussion, we assume there is no **term** statement). Consider now the following code fragment within the EXPERIMENT segment:

```
drun
write 'final value of x1 is: ', x1
drun
write 'final value of x1 is: ', x1
```

Two simulation runs will occur. The observation interval for the first is [0,2] and the first run will have an initial value of 5 for $x1$. The first write statement displays the value of $x1$ at the end of the observation interval, that is, at $t = 2$. Suppose this value is X. The second simulation run will have an initial value of X for $x1$. In fact, the initial value for all state variables will be the final value produced in the first run. Likewise the initial value of t for the second run will be the final value of t from the first run (namely 2) and the observation interval for the second run will be [2, 2 +**TMAX**] = [2, 4]. In other words the second simulation run is simply the continuation of the first and the value of $x1$ that is written by the second write statement is the solution value for $x1$ at time $t = 4$.

The **reset** statement provides the means to re-establish initial conditions on a subsequent simulation run. Consider the following replacement for the above code fragment.

```
drun
write 'final value of x1 is: ', x1
reset
x1=6
drun
write 'final value of x1 is: ', x1
```

Again two simulation runs will take place and the first will be identical to that produced by the original code fragment. The **reset** statement causes the initial values for all state variables to be reset to the initial values used in the first run; likewise t is reset to 0 and the observation interval for the second run will again be [0, 2]. However, because of the assignment $x1$ = 6, the state variable $x1$ will have an initial value of 6 instead of 5 for the second simulation run. Note also that the pair of statements: **drun** followed by **reset** can be replaced with the single statement: **drunr**

A3.4.1 Programming Constructs

A variety of programming constructs is provided in Open Desire to enable the specification of the experiments that need to be carried out. These are briefly reviewed below.

a) Decision
   if [*logexpr*] then [*expA*]
   else [*expB*]
   proceed

   Here the keyword **proceed** serves to terminate the if-construct and *logexp* is a logical expression with two permitted formats:

   (i) simple logical expression:
      [*exp1*] *rop* [*exp2*] where *exp1* and *exp2* are expressions and *rop* is one of the relational operators: <, >, =, <=, >=, <>; e.g.,

      a*b >= X1/5

   (ii) compound logical expression:
      [*slgexp1*] *lop* [*slgexp 2*] where *slgexp1* and *slgexp2* are simple logical expressions and *lop* is one of the logical operators *and* or *or*; e.g.,

(x-y) < 0 and (c <> 0)

(b**2 – 4*a*c) > 0 or (L < 5).

Note that if the "else clause" is empty, then **else proceed** must appear on the same program line.

b) Looping
Three looping mechanisms are available in Open Desire. Each is illustrated below (note that PI is a reserved Open Desire variable whose value is 3.14159≈ π). These code fragments each produce the same output.

(i)
```
PId10=PI/10
        for k=0 to 10 step 1
          TH=k*PId10
          write 'theta = '; TH . 'sin(theta) = '; sin(TH)
        next
```
(ii)
```
PId10=PI/10 | k=0
        while k<=10
          TH=k*PId10
          write 'theta = '; TH . 'sin(theta) = '; sin(TH)
          k=k+1
        end while
```
(iii)
```
PId10=PI/10 | k=0
        repeat
          TH=k*PId10
          write 'theta = '; TH . 'sin(theta) = '; sin(TH)
          k=k+1
        until k>10
```
Note that the logical expressions controlling the *while/end while* and the *repeat/until* constructs may incorporate a single '*or*' or '*and*' operator.

c) Functions
The Function constructs available in Open Desire fall into two categories; namely library Functions and user-defined Functions. We consider each in turn.

(i) Library Functions[3].
The library Functions listed below can be used anywhere in an Open Desire program. In all cases, both the value returned by the function and its arguments have "type REAL"

[3] Not all library Functions that are available in Open Desire are included in our presentation.

sin(x)	cos(x)	tan(x)	asin(x)
acos(x)	atan(x)		
sqrt(x)	abs(x)	ln(x)	log(x)

lim(x) = 0, x when $x < 0,\ x \geq 0$
swtch(x) = 0, 1 when $x < 0,\ x \geq 0$
sgn(x) = -1, 0, +1 when $x < 0, x = 0,\ x > 0$

sat(x) = -1, x, +1 when $x < -1,\ -1 \leq x \leq +1.\ x > +1$
deadz(x) = (x+1), 0, (x-1) when $x < -1,\ -1 \leq x \leq +1.\ x > +1$
deadc(x) = -1, 0 ,+1 when $x < -1,\ -1 \leq x \leq +1.\ x > +1$
rect(x) = 0, 1, 0 when $x < -1,\ -1 \leq x \leq +1.\ x > +1$

(ii) User-Defined Functions
A key restriction on the development of Open Desire user-defined Functions is that the specification must be constrained to a single line of program code. A user-defined Function generally has a parameter list and a further constraint is that the variable names used in that list cannot be used elsewhere in the program. Recursive application of Open Desire Functions is not permitted.

Some examples of user-defined Function's are given below:

```
FUNCTION sum(n1,n2) = n1+ n2
FUNCTION hypot(a,b) = sqrt(a^2 + b^2)
FUNCTION min(a,b) = a – lim(a-b)
FUNCTION max(a,b) = a – lim(b-a)
```

The first example simply returns the sum of its two arguments. The code fragment in Figure A3.1 illustrates the application of this Function. The execution of this fragment would print the first ten Fibonacci numbers (i.e., the sequence: $N_k = N_{k-1} + N_{k-2}$ with $N_1 = N_2 = 1$).

```
---CONSTANTS
INTEGER N1,N2,sum,Next,k
FUNCTION sum(n1,n2)=n1+n2
N1=1 | N2=1
write N1 | write N2
k=3
repeat
  Next=sum(N1,N2)
  write Next
  N1=N2 | N2=Next | k=k+1
until k>10
```

FIGURE A3.1. A code fragment illustrating a user-defined function.

The FUNCTION hypot() computes the hypotenuse of a right angle triangle. The third and fourth examples, respectively, provide a means for determining the larger and smaller of two numerical values. Each uses the library Function lim().

d) Procedures

The required definition format is:

PROCEDURE procname([*list1*];[*list2*])
  procedure body
end

Here *list1* and *list2* are parameter lists; their members typically play the role of inputs and outputs, respectively, of the procedure. The *procedure body* is the program code that specifies the operations carried out by the procedure. It is formulated from the parameters in *list1* and *list2* together with other variables within the Open Desire program. An admissible statement in a Procedure is **exit** which causes an immediate return from the Procedure. The names used for the parameters within *list1* and *list2* may not be used elsewhere in the program.

A defined Procedure may be subsequently invoked with the **call** statement provided allowable substitutions are made for the parameters. Allowable substitutions for the parameters in *list1* are program variables or expressions whereas only program variables are allowed substitutions for the parameters in *list2*. Procedures may call other Procedures but recursive calls are not permitted. Note also that user-defined Procedures are not permitted in the DYNAMIC segment of an Open Desire program.

We complete this section by noting some of the syntactic rules that must be respected when constructing an Open Desire program.

a) Variable names are restricted to a maximum length of eight characters; the first must be a letter and the remainder may be letters, digits, or the character '$'.
b) Program processing is case sensitive; for example, the variables xyz and xYz are different.
c) All variables have type REAL unless otherwise designated; for example, INTEGER k, Knt declares both k and Knt to have type INTEGER.
d) A program statement that begins with '---' (three dashes) is treated as a comment and is not processed.
e) A program line can contain multiple program statements separated by the character '|'; for example, the following is a valid program line.
   g=9.8 | c1=3 | ---g is the acceleration due to gravity (meters/sec*sec)
f) A program line can have up to 244 characters; when necessary, an automatic 'wrap' occurs on the screen.
g) The exponentiation operator is: '^'; for example, 2^3 = 8.

A3.5 An Example

In Chapter 2 (Section 2.2.5) we introduced an example problem in which a boy would like to throw a ball so that it bounces into a hole in the ice surface on which the boy is standing. Here, however, we restrict our consideration to the ball's trajectory up until it first strikes the ice surface; that is, we postpone for the moment the issue of dealing with the 'bounce'.

As previously outlined, the ball is released from the boy's hand with a velocity of V_0 where the velocity vector makes an angle of θ_0 with the horizontal (see Figure 2.4). At the moment of release, the ball has a vertical displacement of y_0 above the ice surface. The hole is located a horizontal distance of H from the boy whose horizontal position is at $x = 0$ (hence the hole is at $x = H$). In this simplified version of the problem our goal is simply to determine which of ten release angles: θ_0 = 15, 20, 25, . . . , 55, 60 (degrees), will place the ball closest to the hole when it first strikes the ice surface, given that the release velocity is $V_0 = 7$ meters/sec. In particular, we relax the originally stated objective of finding a value of θ_0 for which the ball drops into the hole (although there is nothing to preclude one of the test values having this consequence).

The ball's motion, up to the first encounter with the ice surface (i.e., the first bounce), can be described by the following CTDS model (i.e., set of differential equations):

$$\frac{dx_1(t)}{dt} = x_2(t)$$

$$\frac{dx_2(t)}{dt} = -W(t)/m$$

$$\frac{dy_1(t)}{dt} = y_2(t)$$

$$\frac{dy_2(t)}{dt} = -g$$

Here the state variables $x_1(t)$ and $x_2(t)$ represent the horizontal displacement and velocity of the ball and $y_1(t)$ and $y_2(t)$ represent its vertical displacement and velocity. The relevant initial conditions are: $x_1(0) = 0$, $x_2(0) = V_0*\cos(\theta_0)$, $y_1(0) = y_0$, $y_2(0) = V_0*\sin(\theta_0)$. We choose the following values for the problem parameters: $H = 5$ meters, $m = 0.25$ kg, $y_0 = 1.0$ meters and we take the wind force $W(t)$ to have a constant value of 0.15 Newton.

An initial Open Desire simulation program which yields the necessary information to solve the simplified bouncing ball problem as stated above, is provided in Figure A3.2. Some notes relating to this program are given below:

- A repeat/until loop in the EXPERIMENT segment re-solves the model equations for ten values of the initial release angle θ_0. At the end of each run, the distance from the hole is compared with the current 'least distance' and if the current result is superior, then values of several variables are recorded.
- The purpose of the variable **side** is to record whether the impact is to the left of (prior to) or to the right of (beyond) the hole. Its value is established using the library function sgn() introduced earlier.
- Upon completion of this sequence of ten experiments, the values of the variables chosen to characterise the best result uncovered are written.
- The screen output generated by the program of Figure A3.2 is shown in Figure A3.3.

```
----->>Bouncing Ball (one bounce; search over a range of release angles)
---CONSTANTS
 g=9.8 |  ---acceleration due to gravity (meters/sec^2)
 W=0.15 |  m=0.25 |  ---wind force (Newtons) and ball mass (kg)
 H=5 |  ---location of hole (meters)
inc=5 |  ---increment in initial release angle (degrees)
 V0=7.0 |  ---initial release velocity (meters/sec)
 x1=0 |  ---initial horizontal position of ball (meters)
 y1=1.0 |  ---initial vertical position of ball (meters)
 ------------------------------------------------------
 ---EXPERIMENT
 thetaD=15 |  ---initial release angle (degrees)
 x2=V0*cos(thetaD*(PI/180)) |  ---initial horizontal velocity (meters/sec)
 y2=V0*sin(thetaD*(PI/180)) |  ---initial vertical velocity (meters/sec)
 MinDist=H
 irule=3 |  ---fixed step size RK4
 t=0 |  DT=0.002 |  TMAX=2
 repeat
   drun
   dist=H-x1
   if abs(dist)<MinDist then bestTD=thetaD |  MinDist=abs(dist)
     side=sgn(dist) | timp=t
     else proceed
   reset
   thetaD=thetaD+inc
   x2=V0*cos(thetaD*(PI/180))
   y2=V0*sin(thetaD*(PI/180))
   until thetaD>60
 ------------------------------------------------------
 ---OUTPUT
 write 'Best release angle is',bestTD,'degrees'
 write 'Minimum distance is: ',MinDist,'meters, with impact to '
 if side>0 then write 'left of hole at time t =',timp
   else write 'right of hole at time t =',timp
   proceed
 ------------------------------------------------------
 DYNAMIC
 d/dt x1=x2
 d/dt x2=-W/m
 d/dt y1=y2
 d/dt y2=-g
term -y1
```

FIGURE A3.2. Open Desire simulation model: simplified bouncing ball, search over a range of release angles.

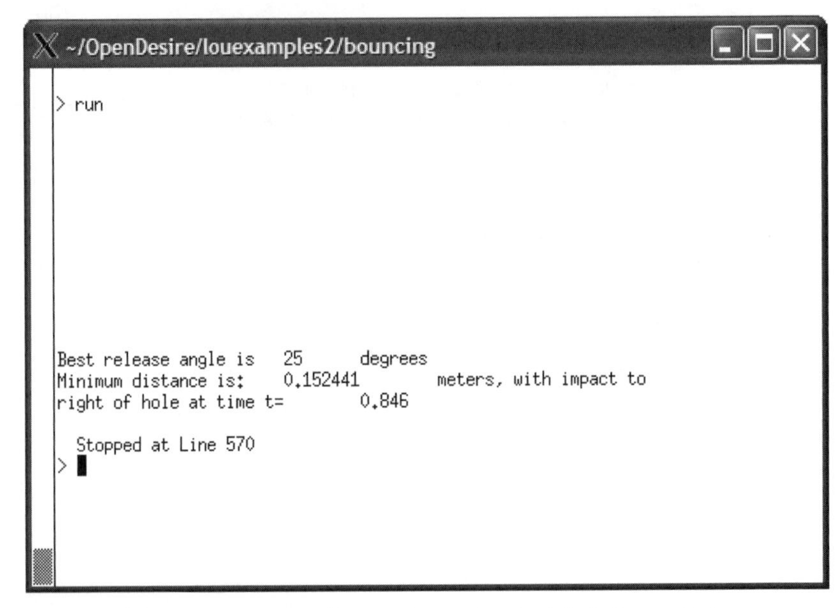

FIGURE A3.3. Output generated by the simulation model of Figure A3.3.

A3.6 Generating Output

The examples provided above have illustrated the most fundamental of the output mechanisms in Open Desire, namely the **write** statement. It provides the mechanism for displaying values within the INITIAL, EXPERIMENT, and OUTPUT segments of an Open Desire program; that is, its use is not permitted in the DYNAMIC segment. Our discussion in this section examines various mechanisms provided for displaying the time trajectories generated during the course of the solution of the model equations.

There are three categories of time trajectory output that are made available in Open Desire. In all cases the data for this output are provided at (or near) the *communication times* T_k, $k = 0, 1, 2, \ldots$, (**NN** – 1) where $T_k = (t_0 + k$ * **COMINT**) and:

$$\textbf{COMINT} = \max\{\textbf{DT}, \textbf{TMAX}/(\textbf{NN} - 1)\}$$

Here **DT**, **TMAX,** and **NN** are system parameters representing, respectively, the integration step size, the nominal length of the observation interval, and the nominal number of communication times; t_0 is the initial value of t for the simulation run in question (see Section A3.3). Notice that the actual number of communication times may be less than **NN** (namely, if **DT** > **TMAX**/(**NN** – 1)). In any event, there is a communication time at both ends of the observation interval [t_0, t_0 + **TMAX**], provided that a **term** condition does not terminate the run before $t = t_0$ + **TMAX**. The values for **DT** and **TMAX** are typically assigned in the EXPERIMENT section whereas NN is typically assigned in the OUTPUT section.

The three categories of time trajectory output are provided. These are summarised below.

a) A listing of increasing values of t (co-incident with the communication times), together with the corresponding values of a prescribed set of variables. The pertinent statement is:
   type [*list*]
   Here *list* is a list of up to five variables whose values will be tabulated.
b) A graph of one or more variables is plotted against time t. The pertinent statement is:
   dispt [*list*]
   Up to eight variables can appear in *list* and each will be plotted against time.
c) A 'cross-plot' of one or more variables plotted against some designated variable. The pertinent statement is:
   dispxy [*list*]
   The first variable in *list* provides the horizontal axis of the display and the remaining variables (up to seven) will be plotted against the first.

The **type**, **dispt,** and **dispxy** statements can only appear in the DYNAMIC segment of an Open Desire program. Furthermore only one of these statements may appear in any section of the DYNAMIC segment.[4]

Open Desire presents graphics in a fixed size *display window*. It is into this window that the graphical output from the **dispt** and **dispxy** statements is placed. The display window cannot be moved with the mouse. Its location can, however, be adjusted with the statement

display W p,q

where *p,q* are the co-ordinates of the upper left-hand corner of the display window. A display window can be erased or removed by striking the Enter key.

There are a number of program statements that control the display window and the presentation of the graphical output. These are summarised below.

a) *Background colour*. There are two options for the background colour for the display window, namely, black and white. The statement **display C15** produces white co-ordinate lines on a black background and **display C17** produces black co-ordinate lines on a white background.

[4] The notion of sections of the DYNAMIC segment is discussed in Section A3.9.

b) *Colour and intensity of the plotted graphs.* A display window may contain up to eight curves and these are presented in different colours. The sequence of colours associated with the sequence of variables listed in either **dispt** or **dispxy**is controlled with the statement **display Nn**. Here *n* is a colour designator for the first listed variable in **dispt** or **dispxy**. The numerical value for *n* can be in the range 1 through 15 and a good choice is 11 (red). Notation at the bottom of the display clarifies the connection between the colours and the graphed variables. The intensity of the lines used to display the graphs can be enhanced by including the statement **display R** in the INITIAL segment of the program (the enhancement is achieved through the use of larger 'dots').

c) *The virtualdisplay window.* It was pointed out above that two formats are available in Open Desire for graphical display of data, namely, time plots (one or more variables plotted against time *t* and cross-plots (one or more variables plotted against a designated variable). Recall furthermore that only one of the alternatives may be specified (in any section of the DYNAMIC segment). The display window serves as a template onto which graphical data (arising from the **dispt/dispxy** statements) are projected. This template has several attributes and these need to be understood in order to achieve the desired presentation.

We note first that although the display window has a fixed physical size, it also has a virtual size that can be adjusted in the Open Desire program. The virtual size relates to its dimensions in terms of the units of the variables that are to be graphed. The fundamental requirement clearly is to ensure that the range of values taken by these variables is accommodated in the display window. For example, in the case of a time plot, the left boundary of the window corresponds to the initial value of *t* for the simulation run in question (namely t_0) and the right boundary corresponds to $t = (t_0 + \text{TMAX})$. Thus the range of values for the variable *t* is mapped onto the physical distance between the left and right boundaries of the window.

For a time plot, the vertical dimension of the display window is symmetrically divided into positive and negative regions that extend between–**scale** and **+scale** where **scale** is a system parameter whose value is set in the Open Desire program. Suppose that *x* is a variable in the display list for **dispt**and that by some prior insight, it is known that*x* will have a maximum (absolute) value that approaches, but will not likely exceed, 5. Then to properly display *x*a reasonable choice for **scale** is 5.

With scale set at 5, the situation can become awkward when other variables that need to be plotted have maximum (absolute) values that are either far less than, or far greater than, 5. In the first case the displayed graph will appear unreasonably 'small' in the display window whereas in the second case, parts of the displayed graph will extend beyond the window and will therefore not be observable. To deal with such circumstances it is necessary to introduce artificial variables that are 'scaled' versions of the variables that are of interest and then placing these variables in the **dispt** list in place of their counterparts. The purpose of the scaling is to create surrogate variables that have similar maximum (absolute) values and thus have a meaningful presentation in the display window.

To illustrate, consider the following:

dispt x,y

where$x$ is the variable introduced earlier and **scale** has the assigned value of 5. Suppose that it is known that the maximum absolute value for$y$ is approximately 0.2 (in the problem units being used). Define $Y = 25*y$ (hence the maximum value of $Y$is approximately 5) and replace the original **dispt** statement with dispt x,Y.

Although the discussion above has been carried out in the context of the display of time plots, the issues relate equally to the display of cross-plots. It is therefore clearly apparent that an appropriate choice for the value of the parameter **scale** together with the introduction of scaled variables where necessary are essential facets of the management of the graphical display features of Open Desire.

d) *Setting the state of the display window.* In some limited circumstances a **dispt** or **dispxy** statement can spawn its own display window. However, this is not always the case and it is therefore prudent to always include the appropriate 'constructor' statements **display A** (for **dispxy**) or **display B** (for **dispt**) in the EXPERIMENT segmentprior to the initiation of the first simulation run when there is a **dispxy** or **dispt** statement in the DYNAMIC segment. These constructor statements serve to create the display windows into which the graphs resulting from the **dispxy** or **dispt** statements will be placed. The assignment of an appropriate value for **scale** should take place prior to **display A / display B**.

In the simplest case, the data from a single simulation run are placed in the display window as the run executes. It frequently occurs, however, that the code in the EXPERIMENT section initiates a sequence of simulation runs (e.g., corresponding to a prescribed collection of values for some designated parameter). Two alternatives are now available. In the first, the graphical data are presented in a display window for each individual run and cleared between runs or, alternately, the data from all runs are presented in the same display window (a very useful option for comparing the effect of the parameter changes). The first of these two alternatives is set with the statement**display 1**, and the second (superimposed graphs) is set with the statement **display 2**. A third alternative, **display 0**, disables the action of the **dispt** and **dispxy** statements entirely (i.e., the graphical presentation is suppressed). The appropriate **display "k"** statement is typically placed in the EXPERIMENT or OUTPUT segment.

e) *File directed output.* Often it may be of interest to simply collect the data generated by the **type** statement in a file and carry out operations on the data outside the Open Desire environment. This can be easily achieved and two steps are involved. The first is the specification of a file that will hold the data and the other is a minor modification in the syntax of the **type** statement. A typical file specification has the form:

connect 'filename.ext' as output 1

An associated modified **type** statement might then have the form:

type ##1,x,y,z

As is apparent from the discussion above, there is a variety of facets todisplay control, all specified by the assignment of various 'arguments' tothe 'display' command. For ease of reference these are summarised in Table A3.1

TABLE A3.1. Summary of display control options.

Format	Purpose
display 0	Disables whatever display window is currently active
display 1	Activates the display window processing
display 2	Allows the superimposing of curves from multiple runs (otherwise curves from previous run are erased)
display A	Produces the *xy*-axes for a cross-plot display (dispxy --)
display B	Produces the *xt*-axes for a time trajectory display (dispt --)
display F	Erases any display window that may be open
display Q	Data dots in display are 'small'
display R	Data dots in display are 'large'
display W *ix,iy*	The display window origin is set to *(ix,iy)*
display C*n*	Sets colour for the coordinate axes
display N*n*	Sets the colour for the first curve (other colours follow in a prescribed sequence)

A3.7 The Example Revisited

We now revisit our earlier example and introduce the requirement for graphical output in order to illustrate the various conventions in Open Desire that relate to the display of time trajectories, as discussed in Section A3.6. Our specific requirement is to display the time trajectories for $x1$ and $y1$ for each of the ten different release angles considered previously. Furthermore these graphs are to be presented in the same display window in order best to reveal how the behaviour changes with changing values of θ_0. Following this composite presentation, it is required to have one further display of $x1$ and $y1$ for the case of the 'best' release angle that is uncovered (each plotted against t).

The modified Open Desire program is shown in Figure A3.4 and some notes relating to it are given below.

- A value of 501 is selected for the system parameter **NN**. In view of the value assigned to **TMAX** (namely, 2), this yields a value of 0.004 = 2/500 for the communication interval **COMINT**. Thus **COMINT** is twice as large as the integration step size **DT**
- The value of $x1$ (ball's horizontal displacement) relative to the location of the hole is of special interest. A way of presenting this relationship is to provide a horizontal line in the display window that corresponds to the hole location. This has been achieved by introducing an additional state variable $x3$ whose derivative is 0 and whose initial value is H. Thus $x3(t) = H$ and the graph of $x3$ (plotted against time) provides the required horizontal line.
- Although not apparent from information in the problem statement, a value of 4 for the system parameter **scale** and scale factors of 2 and 1, respectively, for $x1$ and $y1$ yield graphs that 'fit' nicely into the display window. The scale factors are handled by displaying the new variables $X1 = x1/2$ and $Y1 = y1$, rather than $x1$ and $y1$. (A third artificial variable $X3 = x3/2$ is similarly introduced to accommodate $x3$.)
- The **OUT** statement provides a means for eliminating unnecessary computation associated with variables that are to be displayed. Assignments of values following the OUT statement occur only at the communication times (which are separated by **COMINT** time

units). As observed above there are half as many such points as there are time steps hence the computational saving.[5]

- The **display B** and the **display 2** statements (fourth line of the EXPERIMENT segment) and the **dispt** statement in the DYNAMIC segment together provide the conditions to generate the required collection of superimposed time trajectories.

```
----->>Bouncing Ball (one bounce; with trajectory plots)
---CONSTANTS
g=9.8 | ---acceleration due to gravity (meters/sec^2)
W=0.15 | m=0.25 | ---wind force (Newtons) and ball mass (kg)
H=5 | ---llocation of hole (meters)
inc=5 | ---increment in initial release angle (degrees)
V0=7.0 | ---initial release velocity (meters/sec)
x1=0 | ---initial horizontal position of ball (meters)
y1=1.0 | ---initial vertical position of ball (meters)
------------------------------------------------------
---EXPERIMENT
thetaD=15 | ---initial release angle (degrees)
x2=V0*cos(thetaD*(PI/180)) | ---initial horizontal velocity (meters/sec)
y2=V0*sin(thetaD*(PI/180)) | ---initial vertical velocity (meters/sec)
MinDist=H
irule=3 | ---fixed step size RK4
t=0 | DT=0.002 | TMAX=2 | NN=501
display W300,80 | display C17 | display N11 | display R
scale=4 | display B | display 2
repeat
  drun
  dist=H-x1
  if abs(dist)<MinDist then bestTD=thetaD | MinDist=abs(dist)
    side=sgn(dist) | timp=t
    else proceed
  reset
  thetaD=thetaD+inc
  x2=V0*cos(thetaD*(PI/180))
  y2=V0*sin(thetaD*(PI/180))
  until thetaD>60
```

FIGURE A3.4. Open Desire simulation model: simplified bouncing ball, enhanced with trajectory displays.

[5] In fact the saving is considerably more because unnecessary computations that would otherwise arise in the numerical solution process are likewise avoided.

```
------------------------------------------------------
---OUTPUT
write 'Best release angle is',bestTD,'degrees'
write 'Minimum distance is: ',MinDist,'meters, with impact to '
if side>0 then write 'left of hole at time t=',timp
  else write 'right of hole at time t=',timp
  proceed
write '>>>>Pause now to reflect on displayed family of curves'
write '>>>>type "go" to continue' | STOP
x2=V0*cos(bestTD*(PI/180))
y2=V0*sin(bestTD*(PI/180))
display 1
drun
------------------------------------------------------
DYNAMIC
d/dt x1=x2
d/dt x2=-W/m
d/dt y1=y2
d/dt y2=-g
term -y1
OUT | X1=x1/2 | Y1=y1 | Z1=H/2
dispt X1,Y1,Z1
```

FIGURE A3.4. Open Desire simulation model: simplified bouncing ball, enhanced with trajectory displays (continued).

- Prior to the generation of the required display of $x1(t)$ and $y1(t)$ corresponding to the 'best' value of θ_0, the program enters a 'Pause' state produced by the **STOP** statement[6] (seventh line of OUTPUT segment). The purpose of the Pause here is to provide time to observe and reflect on the family of curves that is presented in the display window. As the message written on the screen indicates, execution is resumed by typing: "go" (followed by Enter). The **display 1** statement clears the display window and the**drun** statement results in one further solution of the model equations.
- The screen output generated by the simulation program of Figure A3.7 is provided in Figure A3.5. The parabolic-shaped curves

[6] Pausing program execution with the STOP statement is permitted anywhere in the CONSTANTS, EXPERIMENT and OUTPUT segments.

correspond to vertical displacement y_l, and the 'almost' linear curves correspond to the scaled version of horizontal displacement, that is, $x_l/2$. The horizontal line ($H/2$) is a reference for $x_l/2$ and represents the location of the hole. If an $x_l/2$ trajectory ends before reaching $H/2$ then the implication is that the ball strikes the ice surface to the left of (prior to) the hole. If, on the other hand, it ends after crossing $H/2$, the implication is that the ball has passed beyond the hole when it strikes the surface.

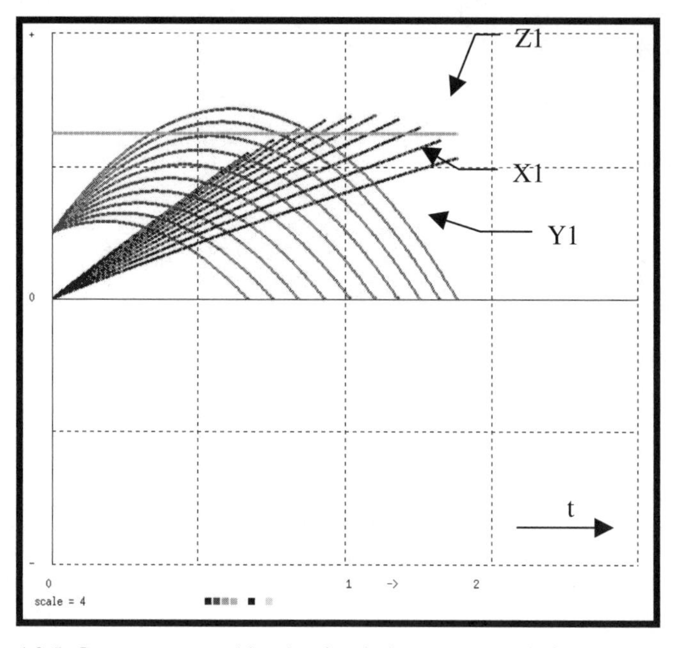

FIGURE A3.5. Output generated by the simulation program of Figure A3.7.

A3.8 The Example Extended

The bouncing ball has not yet 'bounced' because in the two cases considered thus far the ball's trajectory was terminated upon first contact with the ice surface. In order to proceed beyond the collision point, a secondary model has to be developed, namely, a characterisation of what happens when the bounce occurs. In the discussion of the problem in Section 2.2.5 the bounce characterisation is given by

$$\theta(T_C^+) = -\theta(T_C)$$

$$|V(T_C^+)| = \alpha\,|V(T_C)|\ .$$

Here $V(T_C)$ is the ball's velocity vector at the time of collision, T_C, and $\theta(T_C)$ is the velocity vector's orientation with respect to the horizontal. Time T_C^+ is the moment in time that is incrementally beyond T_C. We assume that there is a loss in the ball's kinetic energy as a consequence of the collision and this is reflected in the constant α whose value is in the interval (0, 1). In terms of the state variables of the model, this bounce characterisation becomes:

$$\begin{aligned} x_1(T_C^+) &= x_1(T_C) \\ x_2(T_C^+) &= \alpha\, x_2(T_C) \\ y_1(T_C^+) &= 0 \\ y_2(T_C^+) &= -\,\alpha\, y_2(T_C)\ . \end{aligned} \tag{A3.2}$$

The bounce phenomenon, in fact, is an example one of the challenges that could arise for any numerical procedure that is generating the solution of the differential equations of a CTDS simulation model. The specific issue in question here is an instantaneous change in one or more of the state variables, i.e. a discontinuity (see Section 8.4.3 for a discussion of some important facets of this problem). In the case of the bouncing ball, such an 'event' takes place within the context of the state variables representing the horizontal and vertical velocities of the ball ($x_2(t)$ and $y_2(t)$, respectively) at the moment of contact. The operation of a classical numerical procedure can become severely compromised when attempting to deal with such a circumstance.

Although the procedure may succeed in 'passing over' the discontinuity, considerable error may be introduced. When the numerical process has no intrinsic features to deal with such a discontinuity (the normal case) the best way to proceed is to try to locate the moment of discontinuity accurately and initiate a new solution segment (suitably altered) from that point onwards.

In the case of the bouncing ball, the moment in question is the moment when the ball collides with the ice surface. If T_c represents one such moment (there is one for each bounce) then the initial conditions for the new solution segment are given by Equation (A3.2). However, there still remains a significant hurdle, namely, the determination of the moment of contact T_c. The value of T_c is not explicitly known but instead is implicitly defined by the condition $y_1(T_c) = 0$. This gives rise to the requirement for a suitable numerical process that will solve this equation. Although this certainly is not a particularly difficult undertaking, it is beyond the scope of our current considerations. We instead accept a (rather crude) approximation to T_C that corresponds to the value of time when the occurrence of a negative value for the vertical displacement, $y_1(t)$ is detected. The **term** statement in Open Desire provides an especially convenient means for realising this objective.

With this background, we are now able to consider a more ambitious version of the bouncing ball problem. In this case, our objective is to determine which of a series of release angles results in the third bounce of the ball being closest to the hole. In addition to the explicit identification of the best option, a cross-plot of y_1 versus x_1 is also required for the best option. An Open Desire simulation program that meets these requirements is shown in Figure A3.6. Some notes on features of this program are given below.

- In the context of the current project, a 'solution trajectory' is really a sequence of three subtrajectories. The initial conditions for each new composite trajectory are defined at $t = 0$ with $x_1 = 0$, $y_1 = 1$ and x_2 and y_2 specified in terms of a particular release angle. The initial conditions for the subtrajectories are given by Equation (A3.2) above. It is convenient to encapsulate each such composite trajectory within a PROCEDURE construct. The PROCEDURE called Three has one (output) parameter which is the distance from the hole to the point where ball's third collision with the ice occurs.
- The PROCEDURE outlined above is embedded in a loop in which the six trial values of release angle are each assigned in turn. The best result thus far obtained is monitored at the end of each composite trajectory (i.e., third bounce).
- Within the OUTPUT segment, a final invocation of Three occurs after the **display 2** statement sets up the display window for the required cross-plot.

```
----->>Bouncing Ball (three bounce; with cross-plot of trajectory data)
---CONSTANTS
g=9.8 | ---acceleration due to gravity (meters/sec^2)
W=0.15 | m=0.25 | ---wind force (Newtons) and ball mass (kg)
H=5 | ---location of hole (meters)
inc=2 | ---increment in initial release angle (degrees)
alpha=0.8 | ---energy absorption factor at impact
eps=0.000001 | ---a small number
V0=5.0 | ---initial release velocity (meters/sec)
x1=0 | ---initial horizontal position of ball (meters)
y1=1.0 | ---initial vertical position of ball (meters)
-------------------------------------------------
PROCEDURE Three(;distance)
 drun
 knt=1
 repeat
  y1=eps | x2=alpha*x2 | y2=-alpha*y2
  drun
  knt=knt+1
  until knt=3
 distance=H-x1
 end
-----------------------------------------------------
---EXPERIMENT
thetaD=15 | ---initial release angle (degrees)
x2=V0*cos(thetaD*(PI/180)) | ---initial horizontal velocity (meters/sec)
y2=V0*sin(thetaD*(PI/180)) | ---initial vertical velocity (meters/sec)
MinDist=H
irule=3 | ---fixed step size RK4
t=0 | DT=0.002 | TMAX=6 | NN=501
display W300,80 | display C17 | display N11 | display R
scale=2 | display A | display 0
for i=1 to 6 step 1
 call Three(;dist)
 if abs(dist)<MinDist then bestTD=thetaD | MinDist=abs(dist)
  side=sgn(dist) | t3=t
  else proceed
 t=0 | x1=0 | y1=1.0
 thetaD=thetaD+inc
 x2=V0*cos(thetaD*(PI/180))
 y2=V0*sin(thetaD*(PI/180))
 next
```

FIGURE A3.6. Open Desire simulation model: simplified bouncing ball with three bounces.

```
-----------------------------------------------------
---OUTPUT
write 'Best release angle is: ',bestTD,'degrees'
write 'Minimum distance is: ',MinDist,'meters, with third impact to'
if side>0 then write 'left of hole at time t=',t3
  else write 'right of hole at time t=',t3
  proceed
display 2
t=0 | x1=0 | y1=1.0
x2=V0*cos(bestTD*(PI/180))
y2=V0*sin(bestTD*(PI/180))
call Three(;dist)
-----------------------------------------------------
DYNAMIC
d/dt x1=x2
d/dt x2=-W/m
d/dt y1=y2
d/dt y2=-g
term -y1
OUT | X1=(x1-H)/2.5 | Y1=y1 | Z1=(x1-H)/2.5
dispxy X1,Y1,Z1
```

FIGURE A3.6. Open Desire Simulation Model: simplified bouncing ball with three bounces (continued).

The second phase of the screen output generated by the simulation program of Figure A3.6 is shown in Figure A3.7; the first phase is textual output that is essentially identical to that shown in Figure A3.3. The 'bouncing behaviour' of the ball is apparent. The straight line trajectory (Z1=(x1-H)/2.5)serves to identify the location of the hole, which corresponds to the point where this trajectory crosses the horizontal axis of display window.

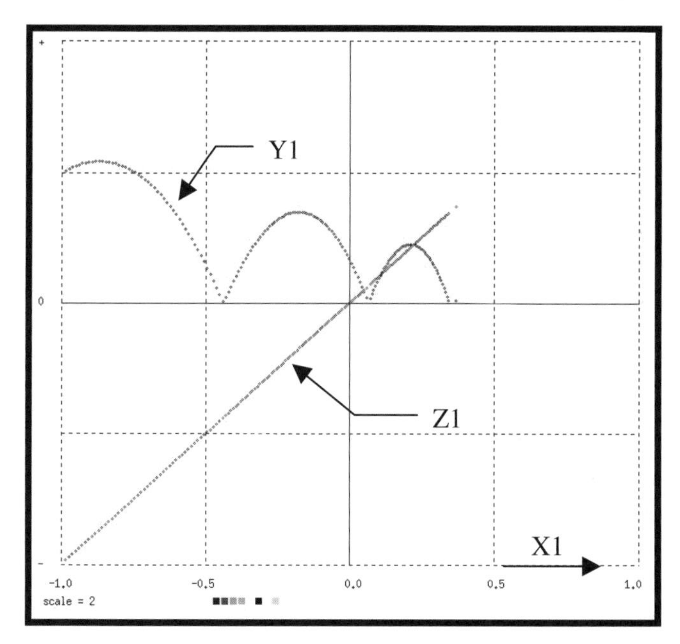

FIGURE A3.7. Output generated by the simulation program of Figure A3.6.

A3.9 Collecting and Displaying Nontrajectory Data

Thus far we have considered the presentation of only time trajectory data, that is, the use of the **type**, **dispt,** and **dispxy** statements as discussed in Section A3.6 above. These statements can, in fact, provide additional functionality inasmuch as they can be used in a broader context. Notice that one of the features of the time trajectory data that we have thus far considered is that its collection is automated by internal mechanisms in Open Desire. In the broader context that we now examine, special programming steps have to be taken explicitly to collect the data that are of interest.

An illustration of the general nature of the data presentation issue being discussed here can be found in the bouncing ball example. Consider the case where the relationship between the release angle thetaD and the miss distance dist = $(H - x1)$ is of special interest. More specifically, suppose the requirement is to obtain a plot of dist versus thetaD. Two steps are involved, namely, the collection of the data and then their subsequent presentation. A prerequisite for the first step is the declaration of two arrays to hold the data values, for example, the statement: **dimension Tvector[50], Dvector[50]** where we assume that the first array is intended to hold values of thetaD and the second is intended to hold corresponding values of dist.

The display of corresponding pairs of (thetaD, dist) values is achieved using a secondary section of the DYNAMIC segment. Such a section is explicitly identified with a user-provided **label** and a reference to the section is provided in the OUTPUT segment following a **drun** invocation. In general, the labelled section may contain an entire new simulation model but our requirements here are far more limited in scope. In fact, the requirement is simply to extract data from the arrays Tvector and Dvector and deliver the data pairs to the Open Desire cross-plot tool. Tentative program code for this task is shown below:

```
label CRSPLOT
get X=Tvector
get Y=Dvector
dispxy X,Y
```

An Open Desire simulation program which illustrates the features discussed above is given in Figure A3.8. This program is a variation on the program of Figure A3.7. One particular extension is that the number of test values for release angle (thetaD) has been increased from 13 to 50 in order to provide a more meaningful display. Some notes on features of this program are given below.

- The last five lines of the OUTPUT segment are especially relevant. Note first that new values for both **scale** and **NN** are assigned in anticipation of the modified data that are to be displayed.
- The data are to be displayed in the standard 'cross-plot' window and, as always, this requires an appropriate scaling of the variable on the horizontal axis (in this case the data stored in Tvector; i.e., values of thetaD ranging from TDmin = 16 to TDmax = 65). The values assigned to the parameters a and b are subsequently used in the scaling operation. Note that the approach demonstrated here has general applicability.
- The last line of the OUTPUT segment (i.e., **drun CRSPLOT**) serves simply to execute the program lines of the DYNAMIC section identified by **label CRSPLOT**. This produces the required display of the (thetaD,dist) pairs.

```
---->>Bouncing Ball (one bounce; with theta vs. dist plot)
---CONSTANTS
g=9.8 | ---acceleration due to gravity (meters/sec^2)
W=0.15 | m=0.25 | ---wind force (Newtons) and ball mass (kg)
H=5 | ---location of hole (meters)
TDmin=16 |TDmax=65 | ---range of release angles (degrees)
inc=1 | ---increment in initial release angle (degrees)
V0=7.0 | ---initial release velocity (meters/sec)
x1=0 | ---initial horizontal position of ball (meters)
y1=1.0 | ---initial vertical position of ball (meters)
-----------------------------------------------------
---EXPERIMENT
thetaD=TDmin | ---initial release angle (degrees)
x2=V0*cos(thetaD*(PI/180)) | ---initial horizontal velocity (meters/sec)
y2=V0*sin(thetaD*(PI/180)) | ---initial vertical velocity (meters/sec)
MinDist=H
irule=3 | ---fixed step size RK4
t=0 | DT=0.002 | TMAX=2 | NN=501
display W300,80 | display C17 | display N11 | display R
scale=4 | display B | display 0
dimension Tvector[50],Dvector[50]
k=0
repeat
  drun
  dist=H-x1
  k=k+1
  Tvector[k]=thetaD
  Dvector[k]=dist
  if abs(dist)<MinDist then bestTD=thetaD | MinDist=abs(dist)
    side=sgn(dist) | timp=t
    else proceed
  reset
  thetaD=thetaD+inc
  x2=V0*cos(thetaD*(PI/180))
  y2=V0*sin(thetaD*(PI/180))
  until thetaD>TDmax
```

FIGURE A3.8. Open Desire simulation model: simplified bouncing ball with plot of miss distance (dist) versus release angle (thetaD).

```
-----------------------------------------------------
---OUTPUT
write 'Best release angle is',bestTD,'degrees'
write 'Minimum distance is: ',MinDist,'meters, with impact to '
if side>0 then write 'left of hole at time t=',timp
  else write 'right of hole at time t=',timp
  proceed
x2=V0*cos(bestTD*(PI/180))
y2=V0*sin(bestTD*(PI/180))
display 1
drun
write 'type "go" to continue' | STOP
scale=1 | NN=50
plus=TDmax+TDmin | minus=TDmax-TDmin
a=plus/2 | b=minus/(2*scale)
drun CRSPLOT
-----------------------------------------------------
DYNAMIC
d/dt x1=x2
d/dt x2=-W/m
d/dt y1=y2
d/dt y2=-g
term -y1
OUT | X1=x1/2 | Y1=y1 | Z1=H/2
dispt X1,Y1,Z1
-----------------------------------------------------
  label CRSPLOT
get thetaD=Tvector
get disp=Dvector
TD=(thetaD-a)/b
dispxy TD,disp
```

FIGURE A3.8. Open Desire simulation model: simplified bouncing ball with plot of miss distance(dist) versus release angle (thetaD) (continued).

Figures A3.9 and Figure A3.10 show the two separate screen outputs resulting from the execution of the simulation program of Figure A3.8. The first of these figures shows the time trajectories of the horizontal and the vertical displacements of the ball up until the first encounter with the ice surface, for the special case of the 'best' initial release angle from among the 50 that are tried. For each of these 50 trials, initial-release-angle/miss-distance pairs are stored. These data, suitably scaled to fit into the display window, are presented as a cross-plot in Figure A3.10.

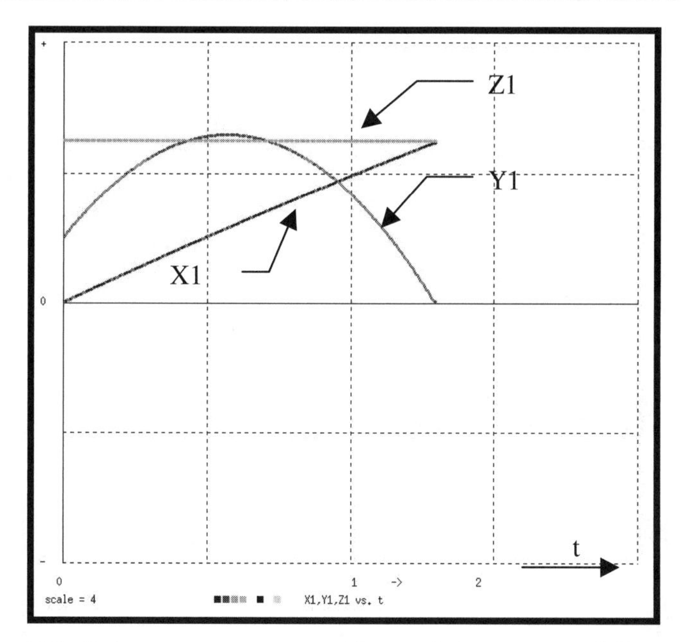

FIGURE A3.9. Trajectory output generated by simulation program of Figure A3.8.

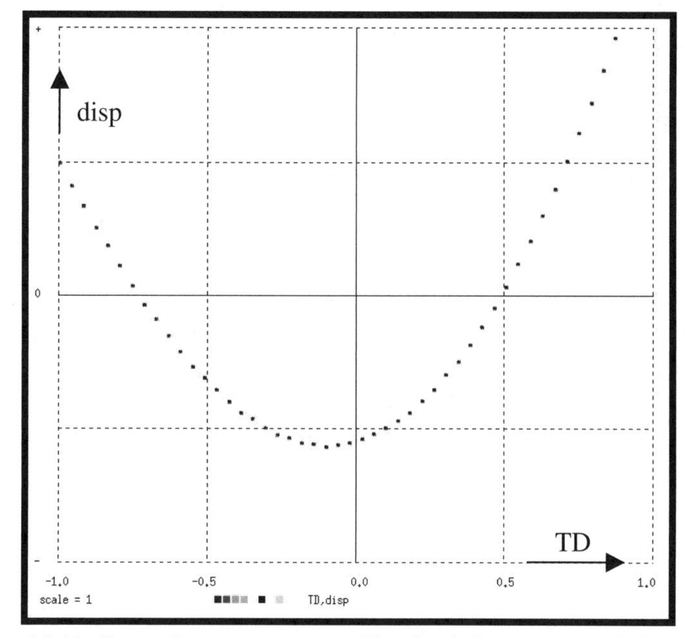

FIGURE A3.10. Cross-plot output generated by simulation program of Figure A3.8.

A3.10 Editing and Execution

In the discussion that follows, we assume that Open Desire is operating within the CYGWIN environment (its host environment) which is available from the textbook Web site. In particular we assume that the CYGWIN environment has been activated according to the instructions included at the Web site. CYGWIN is a UNIX/Linux environment that runs under the Windows operating system and emulates the various UNIX/Linux utilities including the X Window environment. Open Desire has been developed to run under X Window.

FIGURE A3.11. X Terminal with shell interpreter.

Installation will configure CYGWIN to start an X Terminal that runs a UNIX shell command interpreter. Open Desire is launched from the interpreter with the command **desire**, as shown in Figure A3.11. The X Terminal after Open Desire has started is shown in Figure A3.12.

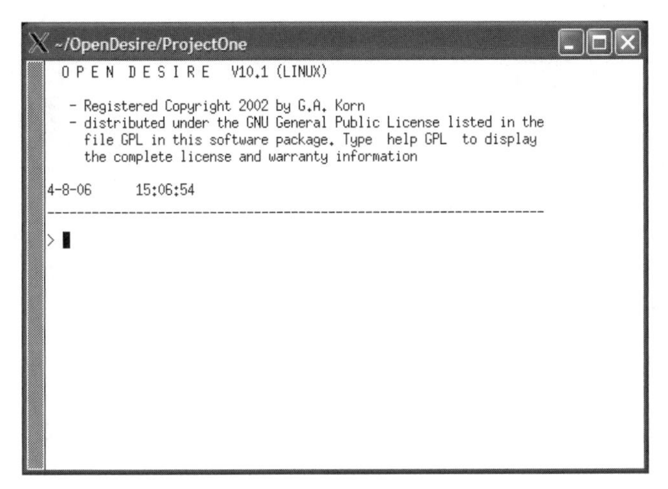

FIGURE A3.12. X Terminal running Open Desire.

Central to any UNIX/Linux environment is the notion of the 'current directory' which corresponds to a particular Windows file folder (i.e., UNIX/Linux directory) in which the editing process looks for referenced files. The default current directory created by the installation of the Open Desire module as provided at the Web site is called Open Desire. However, we assume in the following discussion that the user has changed the current directory (using the cd command) in the CYGWIN environment to ProjectOne prior to running Open Desire. We assume also that Open Desire has been initiated (typically via the command **desire** issued in the X Terminal window and that the Open Desire prompt symbol ('>') appears as in Figure A3.7. Note that the current working directory for Open Desire is the working directory just prior to starting Open Desire.

We consider first the case where a new Open Desire simulation program is to be created using the emacs editor (another X Window application). The editor is invoked from within Open Desire with the command **ed** and results in the display of a text file called SYSPIC.lst. The resulting editor window is shown in Figure A3.13.

The file SYSPIC.lst is the standard working file in the current working directory for the editor and is established by Open Desire. If there is no program file in the Open Desire workspace at the time of invocation of the editor (as in our present scenario), then SYSPIC.lst is empty. Generally this file is simply a copy of the file in the Open Desire workspace and the name of that file is displayed on the line labelled PIC (program identification code). In our current example the Open Desire workspace was empty hence the generic file name 'no_name.lst' appears in Figure A3.13.

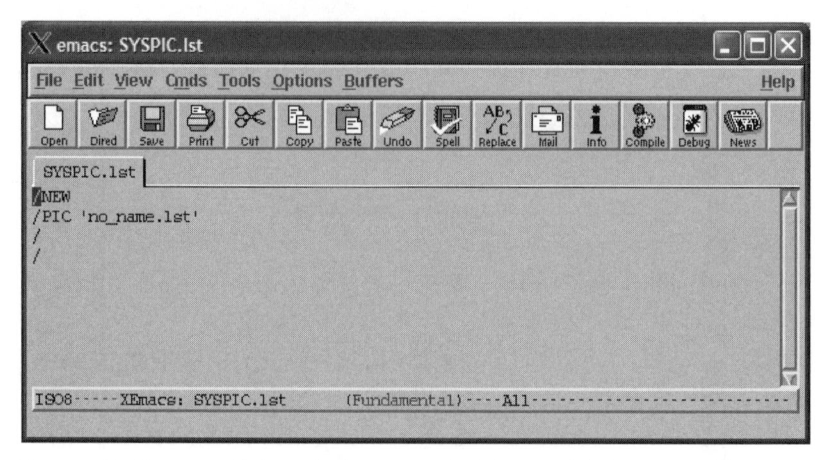

FIGURE A3.13. emacs Editor window opened with a new file (no_name.lst).

Coding of the program in the file SYSPIC.lst begins after the first '/' (the second '/' serves as a program terminator). When coding is completed, the editor's **save** button results in SYSPIC.lst being saved in the current directory (namely, ProjectOne). Closing (or minimising) emacs results in the return to the '>' prompt of Open Desire.

An explicit step is required now to copy the file SYSPIC.lst into Open Desire's workspace in order to permit various operations to be carried out. This is achieved with the command **load**. Three particular options that become available following the **load** command are:

list: Displays a listing of the program file in the Open Desire workspace.

keep 'filename': Stores the program file in the Open Desire workspace in the current directory (namely, ProjectOne) with the file name 'filename.lst'.

run: Executes the program file.

The emacs editor may, of course, be invoked at any time from the Open Desire prompt ('>') using the **ed** command to carry out further program development or program changes. It must be emphasised that after each exit from the editor (i.e., following the **save** operation), the **load** command is essential to ensure that the modified program (as contained in the editor's working file, SYSPIC.lst) becomes the active program within the Open Desire workspace. Note also that if the emacs editor were simply minimised, then restoring it gives immediate access to the earlier version of SYSPIC.lst that remained resident in the emacs workspace.

When the sequence of program development changes and executions has ended, the 'final' program (in the Open Desire workspace) is typically stored in the current directory under some meaningful filename, such as ICdesign. This can be achieved with the command **keep 'ICdesign'**. The result is program storage in the file ICdesign.lst in the current directory ProjectOne. Exit from the Open Desire environment is via the command **bye**.

As an alternate case, suppose that the intent is to continue work on an existing Open Desire simulation program (e.g., stored in ICdesign.lst within the directory ProjectOne as discussed above). The easiest approach is to first establish ProjectOne as the current directory in the X Terminal window. Following invocation of Open Desire (via the host environment command **desire**), the simulation program of interest must be 'loaded'. This is achieved with the command **reload 'ICdesign'** (or simply, **rld 'ICdesign'**). This places the program of interest in Open Desire's workspace and the various operations outlined above can then be carried out (i.e., editing/execution). Note that invocation of the editor (via the **ed** command) copies the program from the Open Desire workspace into the file SYSPIC.lst in the current directory and, as previously noted, this file becomes the editor's working file.

An overview of the various commands and operations discussed above is provided in Figure A3.14. Note, however, that neither our discussion in this Annex nor this overview should be regarded as a fully comprehensive presentation of the editing/execution aspects of Open Desire. Our primary purpose is simply to provide some initial guidance.

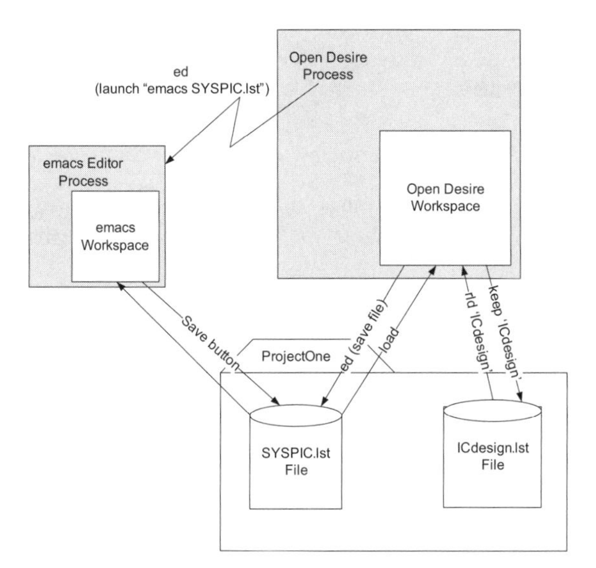

FIGURE A3.14. A view of the Open Desire editing and execution process.

A3.11 Concluding Remarks

It is important to again stress that this primer has provided an overview of only a subset of the features available in Open Desire. The focus has been on capabilities that are essential for, and typically provided by, any simulation tool with an orientation to the continuous time dynamic system context. Many important and powerful features of Open Desire have been omitted from our discussion, for example, features that accommodate difference equations, Fast Fourier transforms, complex number operations, neural networks, fuzzy logic, and vector-matrix operations. The interested reader is urged to explore these topics in the references listed below and, as well, in publications provided at: www.members.aol.com/gatmkorn.

Although not emphasised in this primer, it nevertheless should be appreciated that Open Desire has been developed with a view towards achieving interactive simulation. This, in particular, implies software design that strives for computational efficiency, that is, speed. One noteworthy feature in this respect is the manner in which an Open Desire simulation program is managed. Inasmuch as the principal computational burden occurs in the solution of the differential equations, the program code in the DYNAMIC segment of the program executes in compiled form. All other sections of the program are interpreted. Note also that all computations are carried out in 64-bit floating point double-precision arithmetic.

A3.12 References

A3.1 Korn, G.A., (1989), *Interactive Dynamic System Simulation*, McGraw-Hill, New York.

A3.2 A3.2 Korn, G.A., (1998), *Interactive Dynamic-System Simulation under Microsoft Windows 95 and NT*, Gordon and Breach, London.

INDEX

P

Q

R

S

Printed in the United States of America